New National Framework

MATHEMATICS

7+

M. J. Tipler   K. M. Vickers

J. Douglas

OXFORD
UNIVERSITY PRESS

Great Clarendon Street, Oxford, OX2 6DP, United Kingdom

Oxford University Press is a department of the University of Oxford.
It furthers the University's objective of excellence in research, scholarship,
and education by publishing worldwide. Oxford is a registered trade mark of
Oxford University Press in the UK and in certain other countries

British Library Cataloguing in Publication Data
Data available

978-0-7487-6752-6

21

Printed and bound by CPI Group (UK) Ltd, Croydon, CR0 4YY

Acknowledgements

Illustrations: Angela Lumley, Oxford Designers and Illustrators, Harry Venning
Page make-up: Mathematical Composition Setters Ltd

The authors and publishers would like to thank Jocelyn Douglas for her contribution
to the development of this book.

The publishers thank the following for permission to reproduce copyright material.

Albrecht Durer/Bridgeman Art Library: 64; Casio: 6, 55, 109; Corel (NT): 7, 35, 155,
287, 345; Corel 671 (NT): 165; Corel 776 (NT): 189; Corel 578 (NT): 455; Digital Vision
5 (NT): 74; Image 100 (NT): 81; Imagin (NT): 274; Photodisc 31 (NT): 196; Redferns
Music Picture Library: 428; Stephen Frink/Digital Vision LU (NT): 36.

Although we have made every effort to trace and contact all copyright holders
before publication this has not been possible in all cases. If notified, the publisher
will rectify any errors or omissions at the earliest opportunity.

Links to third party websites are provided by Oxford in good faith and for
information only. Oxford disclaims any responsibility for the materials contained in
any third party website referenced in this work.

Contents

Introduction   v

Number Support   1
Place value – whole numbers, Reading and writing whole numbers and decimals, Multiplying and dividing by 10, 100 and 1000, Putting numbers in order, Rounding to the nearest 10, 100 or 1000, Negative numbers, Multiples, factors, primes, squares, triangular numbers, Divisibility, Mental calculation, Estimating, Written calculation, Fractions, Decimals, Percentages, Ratio and proportion, Using a calculator, Practice questions

1 Place Value, Ordering and Rounding   15
In days gone by   15
Decimal place value   16
Adding and subtracting multiples of 0.1, 0.01 and 0.001   18
Multiplying and dividing by 10, 100, and 1000   19
× and ÷ by multiples of 10, 100 and 1000   21
Putting decimals in order   23
Rounding to the nearest 10, 100, 1000, 10000   26
Rounding to the nearest whole number or to one or two decimal places   28
Summary of key points   31
Test yourself   32

2 Special Numbers   34
All in a spin   34
Adding and subtracting integers   35
Multiplying and dividing integers   40
Divisibility   43
Prime numbers   46
Factors   48
Highest common factor and lowest common multiple   50
Squares, square roots, cubes, cube roots   52
Calculator squares and square roots   55
Summary of key points   58
Test yourself   59

3 Mental Calculation – Whole Numbers and Decimals   62
Much ado about something   62
Adding and subtracting whole numbers   63
Adding and subtracting decimals   67
Multiplying and dividing whole numbers   69
Multiplying and dividing decimals   72
Solving problems mentally   74
Order of operations   76
Making estimates   80
Estimating answers to calculations   82
Summary of key points   84
Test yourself   85

4 Written and Calculator Calculations   87
Kaprekar's number   87
Adding and subtracting   88
Multiplication   91

Division   96
Dividing by a decimal   99
Mixed calculations   100
Checking if answers are sensible   101
Checking answers using inverse operations   103
Checking answers using an equivalent calculation   105
Using a calculator   106
Brackets on a calculator   108
Summary of key points   110
Test yourself   112

5 Fractions   114
Bits and pieces   114
Fractions of shapes   115
One number as a fraction of another   116
Fractions and decimals   121
Comparing fractions   124
Adding and subtracting fractions   126
'Fraction of' and multiplying an integer by a fraction   128
Dividing an integer by a fraction   131
Summary of key points   133
Test yourself   134

6 Percentages, Fractions and Decimals   136
A bit off   136
Percentages to fractions and decimals   137
Fractions and decimals to percentages   138
Percentage of – mentally   142
Percentage of – using a calculator   144
Percentage increase and decrease   146
Summary of key points   149
Test yourself   150

7 Ratio and Proportion   152
All mixed up   152
Proportion   153
Writing ratios   155
Equivalent ratios   156
Ratio and proportion   159
Solving ratio and proportion problems   161
Dividing in a given ratio   164
Summary of key points   166
Test yourself   167

Algebra Support   169
Unknowns, Expressions, Formulae, Counting on and back, Practice questions

8 Expressions, Formulae and Equations   174
Get the message   174
Understanding algebra   175
Collecting like terms   177
Simplifying expressions by cancelling   180
Brackets   182
Substituting   185
Formulae   188
Writing expressions and formulae   190
Writing equations   195
Solving equations using inverse operations   197
Solving equations by transforming both sides   204

Summary of key points 207
Test yourself 209

9 Sequences and Functions 212
Wheel of fortune 212
Sequences 213
Counting on and counting back 214
Showing sequences with geometric patterns 216
Writing sequences from rules 218
Writing sequences given the nth term 221
Sequences in practical situations 224
Finding the rule for the nth term 228
Functions 229
Finding the function given the input and the output 234
Finding the input given the output 237
Summary of key points 241
Test yourself 242

10 Graphs 245
Working out the plot 245
Coordinate pairs 246
Drawing straight-line graphs 247
Lines parallel to the x- and y-axes 253
Reading real-life graphs 254
Plotting real-life graphs 258
Interpreting and sketching real-life graphs 262
Summary of key points 267
Test yourself 268

Shape, Space and Measures Support 270
Angles, Lines, Triangles, Quadrilaterals, Polygons, 3-D shapes, Nets, Coordinates, Symmetry, Transformations, Measures, Perimeter and area, Practice questions

11 Lines and Angles 284
Straight or curved 284
Naming lines and angles 285
Parallel and perpendicular lines 287
Constructing lines and bisectors 288
Angles 291
Angles made with parallel lines 294
Angles in triangles 297
Summary of key points 302
Test yourself 304

12 Shape. Construction 305
Three of a kind 305
Naming triangles 306
Describing and sketching 2-D shapes 307
Properties of triangles, quadrilaterals and polygons 308
Tessellations 316
Constructing triangles and quadrilaterals 318
Describing and sketching 3-D shapes 320
Nets 322
Summary of key points 326
Test yourself 327

13 Coordinates and Transformations 329
That sinking feeling 329
Coordinates 330
Reflection 332
Rotation 336
Translation 339
Combinations of transformations 340
Symmetry 342
Enlargement 345
Summary of key points 350
Test yourself 351

14 Measures 355
Topsy turvy 355
Metric measurements 356
More metric units 358
Solving measures problems 359
Metric and imperial equivalents 363
Reading scales 365
Units, measuring instruments and accuracy 368
Estimating measurements 371
Perimeter and area of a triangle, parallelogram and trapezium 375
Perimeter and area of shapes made from rectangles 380
Surface area 383
Volume 387
Summary of key points 390
Test yourself 392

Handling Data Support 395
Collecting and displaying data, Databases, Sorting diagrams, Mode and range, Probability, Practice questions

15 Planning and Collecting Data 402
Viewing figures 402
Surveys – planning and collecting data 403
Discrete and continuous data 406
Grouping continuous data 407
Summary of key points 409
Test yourself 410

16 Mode, Range, Mean, Median. Displaying Data 411
Averaging it out 411
Mode and range 412
Mean 413
Median 418
Mean, median, mode and range 420
Bar charts and line graphs 421
Frequency diagrams 424
Drawing pie charts 426
Summary of key points 429
Test yourself 431

17 Interpreting Graphs. Comparing Data 433
A date with data 433
Interpreting graphs 434
Comparing data 440
Surveys 443
Summary of key points 445
Test yourself 446

18 Probability 448
Can I play? 448
How likely. Probability scale 449
Outcomes 452
Calculating probability 454
Probability from experiments 458
Comparing calculated probability with experimental probability 460
Summary of key points 463
Test yourself 464

Test Yourself Answers 466

Index 473

Introduction

We hope that you enjoy using this book. There are some characters you will see in the chapters that are designed to help you work through the materials.

These are

 This is used when you are working with information.

 This is used where there are hints and tips for particular exercises.

 This is used where there are cross references.

 This is used where it is useful for you to remember something.

 These are blue in the section on number.

 These are green in the section on algebra.

 These are red in the section on shape, space and measures.

 These are yellow in the section on handling data.

Number Support

Place value — whole numbers

This is a place value chart.

This chart shows **place value**.

In 5 632 479   the value of the 5 is 5 millions or 5 000 000
the value of the 6 is 6 hundreds of thousands or 600 000
the value of the 3 is 3 tens of thousands or 30 000
the value of the 2 is 2 thousands or 2000
the value of the 4 is 4 hundreds or 400
the value of the 7 is 7 tens or 70
the value of the 9 is 9 units or 9.

Practice Questions 1, 2, 19, 28

Reading and writing whole numbers and decimals

Large numbers are **read** in groups of three.

MILLIONS			THOUSANDS					
Hundreds	Tens	Units	Hundreds	Tens	Units	Hundreds	Tens	Units
		9	6	7	4	1	0	8

The number 9 674 108 is read as 'nine million, six hundred and seventy-four thousand, one hundred and eight'.

17·384 is said as 'seventeen point three eight four'.
Sixteen thousand and two and thirty-five hundredths is written as 16 002·35.
$14\frac{37}{100}$ is said as 'fourteen and thirty-seven hundredths'. As a decimal we write this as 14·37.

Practice Questions 5, 7, 12, 40

Multiplying and dividing by 10, 100 and 1000

We use **place value** to **multiply and divide by 10, 100 and 1000**.
When we **multiply by 10, 100 and 1000** the digits move to the **left** by one place for 10, two places for 100 and three places for 1000.

Examples

$$46 \times 10 = 460$$

$$60 \times 1000 = 60\,000$$

When we **divide by 10, 100, 1000**, the digits move to the **right** by one place for 10, two places for 100 and three places for 1000.

Examples

$$860 \div 10 = 86$$

$$56\,000 \div 100 = 560$$

Practice Question 17

Putting numbers in order

To **put numbers in order** we compare digits with the same place value. We start at the left.

7546 > 7364   Starting at the left, the thousands digit in both numbers is 7.
                  The hundreds digits are 3 and 5.   5 > 3 so 7546 > 7364

8·73 > 8·69   Starting at the left, the units digit in both numbers is 8.
                  The tenths digits are 7 and 6.   7 > 6 so 8·73 > 8·69.

Practice Questions 11, 25, 38, 69

Rounding to the nearest 10, 100 or 1000

27 cm is closer to 30 cm than to 20 cm.
27 cm to the nearest 10 cm is 30 cm.

518 mm is closer to 500 mm than to 600 mm.
518 mm to the nearest 100 mm is 500 mm.

£6870 is closer to £7000 than to £6000.
£6870 to the nearest £1000 is £7000.

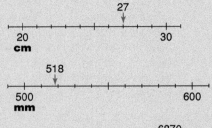

Numbers exactly halfway between are rounded up.
465 to the nearest 10 is 470.   250 to the nearest hundred is 300.   7500 to the nearest 1000 is 8000.

Sometimes in practical problems we must decide whether to **round up or round down**.

Example   8-seater minibuses are hired to take 107 people to the theatre.
           $\frac{107}{8}$ = 13·375. So **14** are needed. 13 would not be enough because some people would be left over.

Practice Questions 15, 46

Negative numbers

A temperature of ⁻5 °C means 5 °C below zero.

We can show positive and negative numbers on a number line.

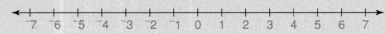

The numbers ... ⁻4, ⁻3, ⁻2, ⁻1, 0, 1, 2, 3, 4, ... are called **integers**.

Example   The temperature is ⁻7 °C.
           If the temperature rises 5 °C it will then be ⁻2 °C.

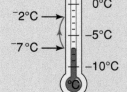

We can put **positive and negative numbers in order** by comparing them.

Example   These temperatures are in order from coldest to warmest.
           ⁻10 °C   ⁻2 °C   4 °C

To order positive and negative numbers, we can put them on a number line.

Example ⁻1, ⁻4, 2, 0, ⁻3 are shown on a number line.

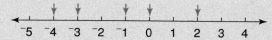

In increasing order, these integers are ⁻4, ⁻3, ⁻1, 0, 2.

Practice Questions 10, 14, 20, 23, 29, 35

Multiples, factors, primes, squares, triangular numbers

The **multiples** of a number are found by multiplying the number by 1, 2, 3, 4, 5, ...

Example The multiples of 6 are 6, 12, 18, 24, 30, 36, ...

The **factors** of a number are all of the numbers that will divide into that number leaving no remainder. We usually list them as pairs.

Example The factor pairs of 24 are 1 and 24, 2 and 12, 3 and 8, 4 and 6.

A **prime number** has exactly two factors, itself and 1. 1 is not a prime number.

> 1 is not a prime because it has only got one factor.

The prime numbers less than 30 are 2, 3, 5, 7, 11, 13, 17, 19, 23 and 29.

A whole number, when multiplied by itself, gives a **square number**.

Example $4 \times 4 = 16$. 16 is a square number.

Square numbers can be represented as squares.
The first 12 square numbers are 1, 4, 9, 16, 25, 36, 49, 64, 81, 100, 121 and 144.

The **triangular numbers** can be represented as triangles.

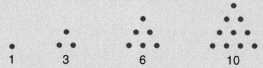

The first six triangular numbers are 1, 3, 6, 10, 15, 21.

Practice Questions 6, 13, 45, 58, 62, 66

Divisibility

A number is **divisible by 2** if it is an even number.
A number is **divisible by 3** if the sum of its digits is divisible by 3.
Example 198 is divisible by 3 since $1 + 9 + 8 = 18$ and 18 is divisible by 3.

A number is **divisible by 4** if the last two digits are divisible by 4.
Example 248 is divisible by 4 since 48 is.

A number is **divisible by 5** if its last digit is 0 or 5.
A number is **divisible by 6** if it is divisible by both 2 and 3.
A number is **divisible by 8** if half of it is divisible by 4.
Example Half of $432 = 216$. 216 is divisible by 4 so 432 is divisible by 8.

A number is **divisible by 9** if the sum of its digits is divisible by 9.
Example 279 is divisible by 9 since $2 + 7 + 9 = 18$ and 18 is divisible by 9.

A number is **divisible by 10** if the last digit is 0.

Practice Questions 65, 74

Number

Mental calculation

Always try to work out a calculation **in your head** before using a written method.
You should know   addition and subtraction facts to 20
multiplication and division facts up to 10×10
addition and subtraction are inverse operations
multiplication and division are inverse operations.

When two numbers add to 10 they are **complements in 10**.
When two numbers add to 100 they are **complements in 100**.
When two numbers add to 1000 they are **complements in 1000**.
When two numbers add to 1 they are **complements in 1**.

Examples   7 and 3 are complements in 10. $7 + 3 = 10$
53 and 47 are complements in 100. $53 + 47 = 100$
250 and 750 are complements in 1000. $250 + 750 = 1000.$
0·68 and 0·32 are complements in 1. $0·68 + 0·32 = 1$

For mental calculation we often **partition** a number.

$$68 = 60 + 8$$

tens   units

$$352 = 300 + 50 + 2$$

hundreds   tens   units

Practice Questions 3, 4, 8, 16, 18, 27, 43, 44, 54, 59, 64, 67, 70, 73

Estimating

An **estimate** is an approximate answer.
Always **estimate** the answer to a calculation first.

Example   8·16 × 3·98 is approximately $8 \times 4 = 32$.

Sometimes there are different ways to find an estimate.

Example   348 + 197 is approximately $300 + 200 = 500$ **or** $350 + 200 = 550$.

Practice Questions 22, 31

Written calculation

Examples

```
  4843        ⁴5·⁶8¹
 +2791        -3·95
 ─────        ─────
  7634         1·73
   11
```

Line up the digits underneath one another carefully.

Example   683 × 27
683 × 27 is approximately equal to $700 \times 30 = 21\,000$.

	600	80	3
20	12 000	1600	60
7	4200	560	21

or

```
      683
   ×   27
   ─────
   13660    ← 683 × 20
    4781    ← 683 × 7
   ─────
   18441
     11
```

Answer $1200 + 1600 + 60 + 4200 + 560 + 21 = \mathbf{18441}$

4

Example   3270 ÷ 7

3270 ÷ 7 is approximately 3500 ÷ 7 = 500.

```
       467
   7 ) 3270
      -2800      7 × 400
       470
      - 420      7 × 60
        50
      -  49      7 × 7
         1
```

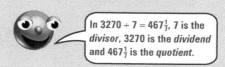

In 3270 ÷ 7 = 467$\frac{1}{7}$, 7 is the *divisor*, 3270 is the *dividend* and 467$\frac{1}{7}$ is the *quotient*.

Answer   **467 R 1** or **467$\frac{1}{7}$**

Practice Questions 21, 32, 36, 37, 39, 41, 47, 50, 56, 68, 72, 75, 76

Fractions

$\frac{7}{8}$ means 7 out of 8 or 7 ÷ 8.

$\frac{7}{8}$ ← numerator
    ← denominator

$\frac{2}{3}, \frac{4}{6}, \frac{6}{9}, \frac{8}{12}, \frac{10}{15}, \ldots$ are **equivalent fractions**.

We find **equivalent fractions** by multiplying or dividing both the numerator and denominator by the same number.

Example

$\frac{4}{5} = \frac{8}{10}$ $\frac{4}{5} = \frac{20}{25}$ $\frac{24}{30} = \frac{4}{5}$

$\frac{4}{5}, \frac{8}{10}, \frac{20}{25}, \frac{24}{30}$ are equivalent fractions.

A fraction is in its **lowest terms** if the numerator and denominator have no common factors except 1.

Lowest terms is the same as simplest form.

We simplify a fraction to its lowest terms by **cancelling**.

Example   In its lowest terms, $\frac{24}{30} = \frac{4}{5}$.

$\frac{24^{12}}{30_{15}} = \frac{12}{15}$ Both 24 and 30 have been divided by 2.

$\frac{12^4}{15_5} = \frac{4}{5}$ Both 12 and 15 have been divided by 3.

We could have cancelled $\frac{24}{30}$ to $\frac{4}{5}$ in one step by dividing numerator and denominator by 6.

We can find a **fraction of a quantity**.

Examples $\frac{1}{10}$ of 40 = 40 ÷ 10
                    = **4**

$\frac{2}{5}$ of 35 = **14** $\frac{1}{5}$ of 35 = 35 ÷ 5
                            = 7
                    $\frac{2}{5}$ of 35 = 2 × 7
                            = 14

We find $\frac{1}{10}$ by dividing by 10.

Sometimes a division is written as a fraction.

Example $\frac{53}{4}$ means 53 ÷ 4.

$\frac{7}{8}$ is called a **proper fraction**.
The numerator is smaller than the denominator.

$\frac{23}{4}$ is called an **improper fraction**.
The numerator is larger than the denominator.

$3\frac{2}{5}$ is called a **mixed number**.
A mixed number has a whole number and a fraction.

Practice Questions 9, 24, 26, 30, 34, 42, 48, 49, 51, 57, 63, 71

Decimals

In 0·25, the digit 2 means 2 tenths
       the digit 5 means 5 hundredths.

We can write **fractions as decimals**.
You should know these.

$\frac{1}{2} = 0{\cdot}5$ $\frac{1}{4} = 0{\cdot}25$ $\frac{3}{4} = 0{\cdot}75$ $\frac{1}{5} = 0{\cdot}2$

$\frac{1}{10} = 0{\cdot}1$ $\frac{1}{100} = 0{\cdot}01$

$\frac{2}{10} = 0{\cdot}2$ $\frac{2}{100} = 0{\cdot}02$

$\frac{3}{10} = 0{\cdot}3$ and so on. $\frac{68}{100} = 0{\cdot}68$ and so on.

Practice Questions **60, 61**

Percentages

7% is read as 'seven per cent' and means 7 out of 100.
Rachel got 69 out of 100 in her maths test. This is 69%.

You should know these.
100% = 1   50% = $\frac{1}{2}$ = 0·5   25% = $\frac{1}{4}$ = 0·25   75% = $\frac{3}{4}$ = 0·75   10% = $\frac{1}{10}$ = 0·1

Practice Questions **52, 53**

Ratio and proportion

There is one red square for every two blue squares.

Jack gets **3** apples for every **2** apples that Zoe gets.
If Jack gets 12 apples (**3** × 4) then Zoe gets 8 apples (**2** × 4).

Practice Questions **33, 55**

Using a calculator

We use a calculator to find the answers to difficult
calculations.
**Always try to use a mental or written way
first**.

Practice Questions **77–80**

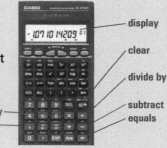

display
clear
divide by
multiply by
add
subtract
equals

Practice Questions

1 What is the value of the 8 in these?
   a 34 859   **b** 584 392   **c** 123 468   **d** 4 348 213   **e** 38 050 712
   f 1 800 904   **g** 7 342 814   **h** 8 234 620   **i** 85 263 047

2 Which digit in each of the numbers in question **1** has place value of tens of thousands?

3 Use complements to find the missing numbers.
   a 37 + ☐ = 100   **b** 100 − ☐ = 56   **c** 4·1 + ☐ = 10   **d** 0·6 + ☐ = 1   **e** 1 − ☐ = 0·84

4 Find the answers mentally.

Mentally means 'in your head'.

a 30 + 10 + 10 **b** 50 + 20 − 10 **c** 80 − 20 − 10 **d** 80 + 40 − 40
e 40 + 50 + 60 **f** 80 + 30 + 20 **g** 100 − 60 + 40 **h** 90 − 30 + 50
i 60 + 50 − 30 **j** 160 − 50 **k** 130 + 70 + 40

5 Write these in figures.

a eight point six
b seventeen point four one
c three hundred and seventy point zero two
d five thousand and two point nine
e four hundred and twenty thousand and sixty-four point zero zero seven
f two hundred and eleven thousand point six eight
g three million, six hundred and fourteen thousand point five two

6 Use a copy of this.
Shade
a the prime numbers
b the numbers that have 3 as a factor
c the factors of 56
d the multiples of 7.
What shape does the shading make?

65	100	76	82	92	44
87	29	15	8	42	99
35	6	28	60	11	56
300	21	17	7	30	33
77	4	1	63	72	14
110	55	74	40	200	76

7 Some Olympic times are given in red. Write them in words.

a Yordanka Donkova — Hurdles — **12·38 sec**
b Marie-Jose Perec — 400 m — **48·25 sec**
c Shiva Keshavan — Luge — **52·315 sec**

8 Find the answers to these mentally.

a 60 × 2 **b** 40 × 3 **c** 50 × 9 **d** 300 × 5 **e** 600 × 4
f 40 × 50 **g** 80 × 90 **h** 320 ÷ 8 **i** 280 ÷ 7 **j** 180 ÷ 3
k 240 ÷ 6 **l** 540 ÷ 9 **m** 630 ÷ 7 **n** 560 ÷ 8

9 What fraction of each shape is shaded?

a   **b**   **c**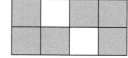

10

°C
⁻90 ⁻80 ⁻70 ⁻60 ⁻50 ⁻40 ⁻30 ⁻20 ⁻10 0 10 20 30 40

Use a copy of this number line and mark these temperatures.
A ⁻89 °C. The coldest temperature ever measured (at a place in Antarctica).
B ⁻27 °C. The lowest temperature ever measured in Britain (on 11 February 1985 at Braemar).
C 32 °C. 13 days in a row were hotter than this in the heatwave in 1976.

11 Which of < or > goes in the box?

a 16·75 ☐ 16·57 **b** 7·526 ☐ 7·52 **c** 8·07 ☐ 8·071
d 0·643 ☐ 0·634 **e** 0·012 ☐ 0·02

12 Write these as decimal numbers.
    a three hundred and five and four tenths
    b twenty-five and six tenths
    c seventeen and three tenths
    d two thousand, five hundred and six and five hundredths
    e sixty-five and seven hundredths
    f seven thousand and forty-one and eight thousandths

13 Write down the factor pairs of
    a 8   **b** 20   **c** 45   **d** 60.

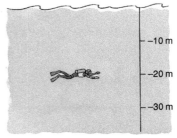

14 A scuba diver is 20 m below sea level.
Her position is given as $^-20$ m.
Give her new position as a negative number if she
    a goes down a further 10 m
    b comes up 10 m.

15 This list gives the number of people at an art gallery. Round them
    a to the nearest 10   **b** to the nearest 100   **c** to the nearest 1000.
    Monday 2489   Tuesday 1124   Wednesday 7500   Thursday 1485   Friday 2850

16 a $2 \cdot 3 + 6 \cdot 9 = 9 \cdot 2$
    Write down three more facts that are true using the numbers $2 \cdot 3$, $6 \cdot 9$ and $9 \cdot 2$.
   b $4 \cdot 2 \times 3 \cdot 5 = 14 \cdot 7$
    Write down three more facts that are true using the numbers $4 \cdot 2$, $3 \cdot 5$ and $14 \cdot 7$.

17 Calculate
    a 52×10 **b** $750 \div 10$ **c** 2×100 **d** $46\,000 \div 100$ **e** $8 \cdot 2 \times 100$
    f 7×1000 **g** $5600 \div 1000$ **h** $5 \cdot 2 \times 1000$ **i** $72 \div 10$ **j** $830 \div 100$
    k $5 \cdot 2 \times 10$ **l** $6 \cdot 43 \times 100$ **m** $4100 \div 1000$ **n** $860 \div 1000$.

$\boxed{\text{T}}$ **18** Use a copy of this.
The number in any square is found by adding the numbers in the two circles on either side of the square.
Find the missing numbers mentally.
    a   **b**

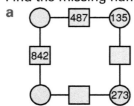 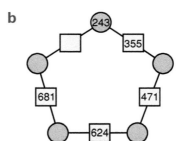

19 What is
    a 100 more than 5824   **b** 10 less than 7269   **c** 1000 less than 60 427
    d 1000 more than 29 842   **e** 100 less than 6080   **f** 10 less than 14 007?

20 Which of > or < goes between these?
    a $2 \quad ^-2$ **b** $^-1 \quad 1$ **c** $^-8 \quad 0$ **d** $0 \quad ^-5$
    e $^-9 \quad ^-4$ **f** $^-6 \quad ^-7$ **g** $^-14 \quad ^-7$ **h** $^-5 \quad ^-15$

21 a $\begin{array}{r} 5 \cdot 2 \\ \times\ 3 \\ \hline \end{array}$ **b** $\begin{array}{r} 2 \cdot 8 \\ \times\ 6 \\ \hline \end{array}$ **c** $\begin{array}{r} 3 \cdot 7 \\ \times\ 8 \\ \hline \end{array}$ **d** $\begin{array}{r} 4 \cdot 6 \\ \times\ 5 \\ \hline \end{array}$ **e** $\begin{array}{r} 9 \cdot 8 \\ \times\ 4 \\ \hline \end{array}$

22 There are 884 men and 703 women at a conference.
Rory estimated that there were about 1400 people altogether.
Explain why Rory is wrong.

23

	6 am	midday	6 pm	midnight
York	⁻7 °C	⁻3 °C	⁻3 °C	⁻4 °C
Manchester	⁻5 °C	0 °C	4 °C	⁻3 °C
Gloucester	2 °C	4 °C	0 °C	2 °C
Bristol	5 °C	6 °C	1 °C	1 °C
Glasgow	⁻4 °C	⁻1 °C	⁻2 °C	⁻1 °C

Put the temperatures at these times in order from coldest to warmest.
a 6 a.m.　　**b** midday　　**c** 6 p.m.　　**d** midnight

24 Peter took $3\frac{1}{4}$ oranges to share with his football team.
How many quarters will these oranges make?

25 This table gives the weights of five of Rochelle's kittens.

Cheeky	Wolf	Tricks	Panda	Lucy
0·56 kg	0·58 kg	0·61 kg	0·5 kg	0·6 kg

a Which kitten weighs the most?
b Which kitten weighs the least?
c Put the weights of the kittens in order from lightest to heaviest.
d Rochelle has a sixth kitten, Missy. Missy's weight is exactly halfway between Panda's and Lucy's weights.
How much does Missy weigh?

26 Find
a $\frac{1}{4}$ of 24　　**b** $\frac{1}{8}$ of 64　　**c** $\frac{2}{3}$ of 36　　**d** $\frac{3}{4}$ of 36　　**e** $\frac{7}{8}$ of 96.

27 Find the answers to these mentally.
a What is the sum of 83 and 58?
b Find the difference between 93 and 76.
c Increase 286 by 119.
d Decrease 372 by 239.
e Find the number halfway between 45 670 and 45 680.

28 What must be added to or subtracted from the first number to get the second number?
The answer to **a** is **add 40**.
a 528, 568　　　　**b** 8325, 8125　　　**c** 7432, 7932　　　**d** 59 624, 51 624
e 609 231, 409 231　**f** 724 164, 720 164　**g** 80 326, 78 326　**h** 87 541, 92 541
i 27·68, 37·68　　**j** 53·71, 52·71　　**k** 59·86, 89·86

29 The temperature was ⁻15 °C. How much has it risen if it is now
a 0 °C　**b** ⁻8 °C　**c** 10 °C　**d** ⁻5 °C　**e** ⁻3 °C　**f** 27 °C?

30 Use a copy of this.
Mark these fractions on the line.
a $\frac{1}{5}, \frac{3}{5}$

b $\frac{1}{6}, \frac{5}{6}, \frac{1}{3}, 1\frac{2}{3}$

31 Estimate the answers to these.
a 916 + 898　　**b** 872 − 342　　**c** 58 × 22　　**d** 924 ÷ 31
e 4·68 + 3·02　**f** 17·47 − 5·62　**g** 4·71 × 21

32 Calculate

    a 176 + 94   **b** 283 – 57   **c** 817 + 594   **d** 588 – 197   **e** 982 + 97

    f 1797 + 87   **g** 402 – 53   **h** 8325 – 1214   **i** 5107 – 56   **j** 8280 + 469

    k 6217 – 364   **l** 5003 – 586.

33 For every $1\frac{1}{2}$ kg of strawberries, Mrs Thom makes four jars of jam.
How many jars of jam does $4\frac{1}{2}$ kg of strawberries make?

34 Find the answers to these mentally.

    a Only about $\frac{1}{6}$ of suitable organ donors actually donate their organs. Out of 48 suitable donors, about how many will donate their organs?

    b About $\frac{3}{4}$ of Amy's drama group went for a pizza at the end of term. If there were 20 people in the drama group, about how many went for a pizza?

35 Put the integers $^-5$, $^-9$, $^-3$, 1, 2, $^-6$ on a number line.
Which number is halfway between these?

    a $^-5$, $^-9$ **b** $^-3$, 1   **c** 2, $^-6$ **d** $^-6$, $^-3$ **e** $^-7$, $^-4$

36 Find the answers to these.

    a 321×9 **b** 864×7 **c** 392×8 **d** 586×9 **e** 52×12

    f 82×14 **g** 27×24 **h** 293×72 **i** 486×12

37 **a** There are 100 students at a tennis coaching school.
They are put into teams of 6.
How many full teams can be made?

    b All of the 100 students are taken by 12-seater minibuses to a tournament.
How many minibuses are needed?

38 Put these lists in order from smallest to largest.

    a 0·82, 0·28, 0·2, 0·8   **b** 0·7 m, 0·69 m, 0·72 m, 0·68 m

    c 13·4 ℓ, 14·05 ℓ, 14·5 ℓ, 13·38 ℓ

39 **a** 312 ÷ 4   **b** 295 ÷ 5   **c** 234 ÷ 6   **d** 486 ÷ 9

    e 1715 ÷ 7   **f** 1707 ÷ 3   **g** 2008 ÷ 8   **h** 835 ÷ 4

    i 371 ÷ 6   **j** 5099 ÷ 9   **k** 3725 ÷ 7

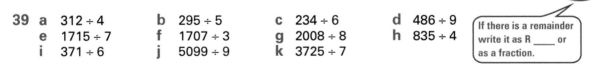

If there is a remainder write it as R ____ or as a fraction.

40 Write these in words and then as a decimal number.

    a $4\frac{7}{10}$ **b** $12\frac{59}{100}$ **c** $114\frac{77}{100}$

    d $42\frac{367}{1000}$ **e** $617\frac{21}{1000}$

41 Eight friends went on holiday. They paid a total of £3384 for their travel.
How much did each pay?

42 Find the missing numbers.

    a $\frac{3}{5} = \frac{12}{\Box}$ **b** $\frac{2}{3} = \frac{\Box}{12}$ **c** $\frac{24}{30} = \frac{\Box}{5}$ **d** $\frac{36}{45} = \frac{12}{\Box}$ **e** $1 = \frac{\Box}{18}$

    f $\frac{45}{50} = \frac{9}{\Box}$ **g** $\frac{2}{3} = \frac{\Box}{300}$ **h** $\frac{8}{7} = \frac{64}{\Box}$ **i** $\frac{72}{54} = \frac{\Box}{6}$

43 Find the answers to these mentally.
At a bird sanctuary the keeper weighs injured birds before and after they feed to see how much they ate. Find the amount that these birds ate at a feeding.
   a **Seamus** before 590 g after 647 g
   b **Pecker** before 389 g after 417 g
   c **Tramp** before 1864 g after 1958 g
   d **Rat** before 1479 g after 1601 g

44 The tallest Ferris wheel in the world has 60 arms. Each arm seats eight people.
How many people does the Ferris wheel seat in total?

45 **a**

1	6	9	16	24	25	36	49	55	64	100	111	121	144

   Write down the square numbers from the box.
   b Write down the first five triangular numbers.

46 Jan's bank balance, to the nearest £10, is £870.
Which of these amounts might Jan have in her bank?
   £871   £875   £865   £868   £862

47 Eight people paid a total of £128 for a minibus tour.
   a How much change did they get from £150?
   b What was the cost for each person?

48

$$\frac{15}{24} \quad \frac{35}{50} \quad \frac{24}{36} \quad \overset{A}{\frac{2}{4}} \quad \frac{12}{45} \qquad \frac{24}{64} \quad \frac{12}{27} \quad \overset{A}{\frac{2}{4}} \quad \frac{12}{45} \quad \frac{60}{100} \qquad \overset{A}{\frac{2}{4}} \quad \frac{12}{45} \quad \frac{12}{27}$$

$$\frac{24}{36} \quad \frac{12}{27} \quad \frac{5}{9} \quad \frac{12}{48} \qquad \frac{3}{4} \quad \overset{A}{\frac{2}{4}} \quad \frac{28}{28} \quad \frac{4}{5} \quad \frac{12}{27} \quad \frac{4}{5}$$

Use a copy of this box.
Write the letter beside each of these, above the equivalent fraction in the box.
   A $\frac{1}{2} = \frac{2}{4}$ **S** $\frac{3}{5}$ **L** $\frac{2}{3}$ **D** $\frac{12}{15}$ **N** 1   **T** $\frac{25}{100}$ **E** $\frac{4}{9}$
   P $\frac{5}{8}$ **O** $\frac{7}{10}$ **R** $\frac{4}{15}$ **B** $\frac{6}{16}$ **F** $\frac{45}{81}$ **H** $\frac{48}{64}$

49 Use cancelling to write these fractions in their lowest terms.
   a $\frac{4}{8}$ **b** $\frac{7}{21}$ **c** $\frac{6}{8}$ **d** $\frac{16}{20}$ **e** $\frac{10}{15}$
   f $\frac{80}{100}$ **g** $\frac{18}{24}$ **h** $\frac{20}{24}$ **i** $\frac{5}{45}$ **j** $\frac{15}{40}$
   k $\frac{42}{56}$ **l** $\frac{24}{40}$ **m** $\frac{27}{45}$ **n** $\frac{42}{63}$

50 The Quarayak glacier in Greenland moves about 24 metres per day.
How far does the glacier move in
   a 4 days   **b** 3 weeks   **c** 12 hours   **d** a year (not a leap year)?

51 Write these as fractions in their lowest terms.
   a Tim had £18. He spent £6 at the fair.
      What fraction did he spend at the fair?
   b There are 28 pupils in Lorna's class.
      21 of them gave something for the fair raffle.
      What fraction did not give something?

52 Write these as decimals.
  a 50%   **b** 25%   **c** 75%   **d** 10%   **e** 57%   **f** 89%

53 Write these as fractions.
  a 50%   **b** 25%   **c** 75%   **d** 10%   **e** 83%   **f** 60%

54 A car park charges 80p for the first hour and 50p for each half-hour, or part half-hour after that. Rob parked there from 11:15 a.m. until 2:55 p.m. How much did he pay?

55 In Lisa's class there are two boys for every three girls.
  a If there are four boys, how many girls are there in Lisa's class?
  b If there are ten boys, how many girls are there?

56 **a** $8\cdot6 + 3\cdot2$ **b** $9\cdot7 + 0\cdot42$ **c** $17\cdot5 + 3\cdot64$ **d** $8\cdot52 + 9\cdot63$
   e $4\cdot9 - 3\cdot7$ **f** $4\cdot2 - 3\cdot8$ **g** $5\cdot47 - 2\cdot86$ **h** $12\cdot7 - 1\cdot38$

T

57

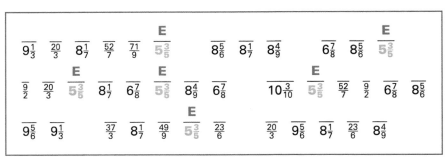

$$\overline{9\tfrac{1}{3}}\quad \overline{\tfrac{20}{3}}\quad \overline{8\tfrac{1}{7}}\quad \overline{\tfrac{52}{7}}\quad \overline{\tfrac{71}{9}}\quad \overset{E}{\overline{5\tfrac{3}{5}}}\qquad \overline{8\tfrac{5}{6}}\quad \overline{8\tfrac{1}{7}}\quad \overline{8\tfrac{4}{9}}\qquad \overline{6\tfrac{7}{8}}\quad \overline{8\tfrac{5}{6}}\quad \overset{E}{\overline{5\tfrac{3}{5}}}$$

$$\overline{\tfrac{9}{2}}\quad \overline{\tfrac{20}{3}}\quad \overset{E}{\overline{5\tfrac{3}{5}}}\quad \overline{8\tfrac{1}{7}}\quad \overline{6\tfrac{7}{8}}\quad \overset{E}{\overline{5\tfrac{3}{5}}}\quad \overline{8\tfrac{4}{9}}\quad \overline{6\tfrac{7}{8}}\qquad \overline{10\tfrac{3}{10}}\quad \overset{E}{\overline{5\tfrac{3}{5}}}\quad \overline{\tfrac{52}{7}}\quad \overline{\tfrac{9}{2}}\quad \overline{6\tfrac{7}{8}}\quad \overline{8\tfrac{5}{6}}$$

$$\overline{9\tfrac{5}{6}}\quad \overline{9\tfrac{1}{3}}\qquad \overline{\tfrac{37}{3}}\quad \overline{8\tfrac{1}{7}}\quad \overline{\tfrac{49}{9}}\quad \overset{E}{\overline{5\tfrac{3}{5}}}\quad \overline{\tfrac{23}{6}}\qquad \overline{\tfrac{20}{3}}\quad \overline{9\tfrac{5}{6}}\quad \overline{8\tfrac{1}{7}}\quad \overline{\tfrac{23}{6}}\quad \overline{8\tfrac{4}{9}}$$

Use a copy of this box. Write the letter beside each fraction above its answer in the box. Write these as mixed numbers.

  E $\tfrac{28}{5} = 5\tfrac{3}{5}$ **A** $\tfrac{57}{7}$ **T** $\tfrac{55}{8}$ **H** $\tfrac{53}{6}$ **S** $\tfrac{76}{9}$ **L** $\tfrac{103}{10}$ **O** $\tfrac{59}{6}$ **F** $\tfrac{84}{9}$

Write these as improper fractions.

  G $4\tfrac{1}{2}$ **R** $6\tfrac{2}{3}$ **D** $3\tfrac{5}{6}$ **N** $7\tfrac{3}{7}$ **V** $5\tfrac{4}{9}$ **C** $7\tfrac{8}{9}$ **P** $12\tfrac{1}{3}$

58 Jane and Lata are nurses.
Jane works three days, then has a day off.
Lata works two days, then has a day off.
They both have a day off on July 4th.
When do they next have a day off together?

59 Find the answers to these mentally.
  a $48 \div 6$ **b** $72 \div 8$ **c** $90 \div 6$ **d** half of 98   **e** one tenth of 560
  f $470 \div 2$ **g** $3900 \div 4$ **h** $300 \div 12$ **i** $378 \div 21$

> Hint for i: divide by 3 then by 7.

60 Write these as decimals.
  a $\tfrac{1}{2}$ **b** $\tfrac{1}{4}$ **c** $\tfrac{3}{4}$ **d** $\tfrac{1}{10}$ **e** $\tfrac{1}{100}$ **f** $2\tfrac{1}{2}$ **g** $4\tfrac{3}{4}$
  h $\tfrac{3}{10}$ **i** $\tfrac{42}{100}$ **j** $\tfrac{39}{100}$ **k** $\tfrac{6}{100}$ **l** $\tfrac{4}{100}$

61 Write these as fractions in their simplest form.
  a $0\cdot3$ **b** $0\cdot7$ **c** $0\cdot9$ **d** $0\cdot8$ **e** $0\cdot5$ **f** $0\cdot6$
  g $0\cdot41$ **h** $0\cdot77$ **i** $0\cdot83$ **j** $0\cdot25$ **k** $0\cdot14$ **l** $0\cdot18$
  m $0\cdot64$ **n** $0\cdot74$ **o** $0\cdot45$ **p** $0\cdot72$ **q** $0\cdot85$ **r** $0\cdot94$

62 a The numbers 246, 264 and 426 all have the digits 2, 4 and 6.
What other numbers can be made with these three digits?
b What is the largest multiple of 8 that you can make from the digits 2, 4 and 6?
c What is the largest multiple of 7 that you can make from these digits?

63 Write the equivalent fraction that has a denominator of 24.
a $\frac{1}{2}$ **b** $\frac{2}{3}$ **c** $\frac{3}{4}$ **d** $\frac{5}{6}$ **e** $\frac{3}{8}$

64 Find the answers mentally.
a $3 \times 4 \times 5$ **b** $5 \times 6 \times 4$ **c** 35×2 **d** 48×2
e 42×5 **f** 51×3 **g** 64×6 **h** 53×4
i 62×7 **j** 46×20 **k** 57×60 **l** 120×30
m 13×11 **n** 21×12 **o** 24×25

65 Choose a number from 168, 174, 177, 200, 388, 418 and 423 for each gap.
Use each number only once.
a ____ is divisible by 2. **b** ____ is divisible by 3.
c ____ is divisible by 4. **d** ____ is divisible by 5.
e ____ is divisible by 6. **f** ____ is divisible by 8.
g ____ is divisible by 9.

66 There are just four numbers, less than 100, with exactly three factors.
One of these is 9. What are the other three?
What do you notice about these four numbers?

Think about prime numbers.

67 Pete has three 35p stamps and five 25p stamps.
Find all the different amounts he could stick on a parcel.

68 Find two consecutive numbers with a product of 156.

69 5 7 4 3 6 9
a What is the smallest even number that can be made using all these digits?
b What is the largest odd number that can be made?

70 Use all the digits 2, 3, 4, 7, 8 and 9 to make two 3-digit numbers so that
a the difference between these numbers is as small as possible
b the sum of these numbers is 1032.
Is there more than one answer to each of these?

71 a I am equivalent to $\frac{2}{3}$.
My denominator is 10 more than my numerator.
What fraction am I?
b I am equivalent to $\frac{80}{100}$.
My denominator is a prime number.
What fraction am I?

72 Put the digits 3, 4, 5, 6 and 7 in the boxes to make each true.
Use each digit only once.
a □□□ × □□ = 25 228 **b** □□□ × □□ = 24 310

73 Use only the digits 3, 5, 7 and 9 to make each sum correct.
You may use each digit as often as you like.
a □□ + □□ = 114 **b** □□ + □□ = 90 **c** □□ + □□ = 172
d □□ + □□ = 136 **e** □□ + □□ = 130

74 The sum of four even numbers is divisible by 4.
When is this true? When is it false?

75   ★ 2 ★ 2   Replace the ★ with the same digit to make this subtraction correct.
     − 2 ★ 2 ★
      ‾‾‾‾‾‾
       3 ★ 3 ★

76   1111
     3333
     5555   Replace ten of the digits with 0 so that the sum comes to 1111.
     7777
   +9999
     ‾‾‾‾

Use your calculator to answer questions 77–80.

77   a   Fourteen people ate a meal at this cafe.
         How much did it cost altogether?
    b   A group all had the special.
         It cost £92·65 altogether.
         How many were in the group?
    c   One morning 27 people ate breakfast at the cafe.
         How much did this cost altogether?

CAFE

Meals £9·70

Breakfast £6·70

Special £5·45

78   Each ◯ is a different missing digit.

    ◯◯ × 5 ◯ = 4056

    What is the calculation?

79   a   Find different ways of completing   ▢▢ × ▢ = 252.
    b   Each ◯ represents one of the digits 0, 2, 3, 4, 5, 6.
         Each digit is used only once.
         Replace each ◯ to make this true.

     ◯◯ × ◯ = ◯◯◯

80   Two whole numbers when divided give the answer 2·727 272 72 ...
    What are the two numbers?

1 Place Value, Ordering and Rounding

You need to know

✓ place value — whole numbers   page 1
✓ reading and writing whole numbers and decimals   page 1
✓ multiplying and dividing by 10,100 and 1000   page 1
✓ putting numbers in order   page 2
✓ rounding to the nearest 10, 100 or 1000   page 2
✓ rounding up or down   page 2

Key vocabulary

ascending, billion, compare, decimal place, descending, greater than(>), hundredth, less than(<), nearest, one decimal place, order, place holder, place value, round, tenth, thousandth, two decimal places, value

In days gone by

Many ancient civilisations had number systems that used symbols.

The Egyptian number system had symbols for 1, 10, 100, 1000, 10 000, 100 000, 1 000 000.

The symbol for 1 was a stroke |

         10 was a heel bone ∩

         100 was a coiled rope ☉

         1000 was a lotus flower ⚱

         10 000 was a bent finger ⌠

         100 000 was a tadpole ✎

         1 000 000 was an excited man ✹

The order in which the symbols were written was not important.

For instance, 23 could be written as ∩ ∩ ||| or | ∩ ∩ || or || ∩ ∩ | etc.

9 could be written as ||||| over ||| or |||| over ||||| or ||||||||| etc.

Write these numbers using the Egyptian system.

      8   16   33   58   386   4900   56 423

Write other numbers using the Egyptian system.

Decimal place value

T

The **place** of a digit tells you its value.

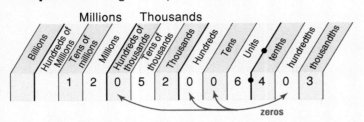

Millions   Thousands

Billions	Hundreds of Millions	Tens of millions	Millions	Hundreds of thousands	Tens of thousands	Thousands	Hundreds	Tens	Units	tenths	hundredths	thousandths
	1	2	0	5	2	0	0	6	4	0	3	

zeros

This is a *place value* chart.

A billion is a thousand million.
This chart shows the number 12 052 006·403.
We read this as twelve million, fifty-two thousand and six point four zero three.
The zeros are **place holders**.

We use zeros to tell us there are no hundreds of thousands, no hundreds, no tens and no hundredths.

Worked Example
How many times larger is the first 6 than the second 6 in 526 047·63?

Answer
The first 6 means 6 thousands.
The second 6 means 6 tenths.

$10 \times 10 \times 10 \times 10$ moves a digit 4 places left.

Each place on the place value chart is 10 times larger than the place to its right.
6 thousands is $10 \times 10 \times 10 \times 10$ times larger than 6 tenths.
The first 6 is ten thousand times larger than the second 6.

Exercise 1

1   What is the value of the digit 4 in these?
   a 7·462   b 410·057   c 930·704   d 4672·30   e 7105·04

2   What is the value of the digit 7 in the numbers in **question 1**?

3   a   Make the largest number you can with Mark's cards.
       There must be at least one number after the
       decimal point and the last digit must not be zero.
   b   Repeat a but make the third largest number.
   c   Make the smallest number you can with Mark's
       cards. There must be at least one number before
       the decimal point and the first digit must not be
       zero.
   d   Repeat c but make the third smallest number.
   e   How many different decimal numbers, each with
       two decimal places, can you make using all of
       Mark's cards? Write them down.

Make the largest number you can with your cards.

4   Repeat **question 3a–d** for these cards.

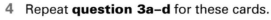

4 7 0 1 3 6 8 .

*5   Repeat **question 3a–d** for these cards.

3 8 2 2 1 6 0 0 9 .

6 Nathan cut three pieces of string. Each was between 1·12 m and 1·13 m in length.
Write down how long each piece might be?

7 How many times larger is the first 8 than the second 8 in these?
  a 818·3   **b** 188·05   **c** 18·83   **d** 0·88   **e** 0·838
  f 885·6   **g** 846·82   **h** 983·08   **i** 58·804

8 Write in figures the number that is
  a 8 more than two and a half thousand   **b** 5 more than four and a half million
  c 6 more than one and a quarter thousand   ***d** 6 less than four hundred thousand
  ***e** 8 less than eleven and a half million   ***f** 4 less than two and a quarter billion.

Review 1   What is the value of the digit 8 in these?
a 6·84   **b** 0·08   **c** 3·048   **d** 8·02   **e** 5800·9
f 4·832   **g** 83 241·7   **h** 194·83   **i** 18 057·6

Review 2   `4` `9` `0` `3` `6` `4` `·`

a Make the second largest number you can with these cards.
  There must be at least one number after the decimal point and the last digit must not be
  zero.
b Make the third smallest number you can with these cards.
  There must be at least one number before the decimal point and the first digit must not be
  zero.

Review 3   How many times larger is the first 4 than the second 4 in these?
a 454·63   **b** 164·42   **c** 0·44   **d** 453·42   **e** 504·354

Review 4   Write in figures the number that is
a 7 more than two and a quarter thousand   ***b** 9 more than one and half billion
***c** 3 less than eight hundred thousand   ***d** 1 less than four and a quarter million.

***Review 5**   Repeat Review **2** with these cards.   `8` `3` `6` `4` `3` `1` `1` `7` `0` `·`

? **Puzzle**

1 I am a decimal number. Just two of my digits are the same.
  I have two digits before and two digits after the decimal point.
  My tens digit is smaller than my unit digit.
  My unit digit is the same as my tenths digit.
  My tens digit is one less than my hundredths digit.
  My digits add up to 11.
  What number am I?

***2** A number has 5 digits.
  The number is between 20 and 30.
  The difference between the tenths digit and the unit digit is the same as the
  hundredths digit.
  The tens digit is the same as the thousandths digit.
  The tenths digit is three times greater than the tens digit.
  The sum of all the digits is 16.
  There are seven numbers this could be.
  What are they?

Adding and subtracting multiples of 0·1, 0·01 and 0·001

To **add or subtract 0·1, 0·2, 0·3**, ... we add or subtract 1, 2, 3, ... to or from the tenths.

Example 8·632 + 0·1 = 8·732

To **add or subtract 0·01, 0·02, 0·03**, ... we add or subtract 1, 2, 3, ... to or from the hundredths.

Example 4·306 – 0·05 = 4·256

To **add or subtract 0·001, 0·002, 0·003**, ... we add or subtract 1, 2, 3, ... to or from the thousandths.

Examples 2·119 + 0·001 = 2·120   38·042 – 0·005 = 38·037

> We usually write 2·120 as 2·12.

Exercise 2   **This exercise is to be done mentally.**

1 What length is 0·1 m less than   **a** 27·68 m   **b** 4·12 m   **c** 7 m?

2 What length is 0·01 m greater than   **a** 5·72 m   **b** 8·69 m   **c** 3·09 m?

3 What length is 0·001 km greater than   **a** 4·723 km   **b** 8·406 km   **c** 0·589 km?

4 **a** 4·67 + 0·01   **b** 8·325 + 0·1   **c** 27·29 – 0·1   **d** 8·23 – 0·01   **e** 8·09 + 0·01
   f 5·40 – 0·01   **g** 8·632 + 0·001   **h** 5·752 – 0·001   **i** 3·68 + 0·001   **j** 4·79 + 0·01
   k 4·340 – 0·001   **l** 8·999 + 0·001   **m** 5·0 – 0·1   **n** 16·0 – 0·01   **o** 4 – 0·001

5 **a** 8·42 + 0·4   **b** 3·847 + 0·05   **c** 4·73 – 0·6   **d** 5·301 + 0·07   **e** 7·653 + 0·004
   f 8·729 – 0·007   **g** 0·052 – 0·003   **h** 8·413 + 0·008   **i** 5·230 – 0·002   **j** 7·299 + 0·005
   k 4·000 – 0·006   **l** 15·737 + 0·06   **m** 53·64 – 0·9   **n** 14·017 – 0·02   **o** 6·0 – 0·04
   p 8 – 0·7   **q** 15·9 – 0·007   **r** 15 – 0·009   **s** 12 – 0·008

6 **a** What number is 0·01 less than 26·5?
   b Find the sum of 18·692 and 0·04.
   c Find the difference between 13·056 km and 0·1 km.
   d The difference between two lengths is 0·1 m. One length is 8·64 m.
        What might the other be?
   e What number is 0·008 less than 5·7?

7 What needs to be added or subtracted to change
   e 4·624 to 4·644   **b** 8·759 to 8·959   **c** 9·632 to 9·232   **d** 18·54 to 18·04
   e 7·862 to 7·869   **f** 9·099 to 9·091   **g** 8·78 to 8·8   **h** 4 to 3·999
   i 6·598 to 6·668   ***j** 4·002 to 3·994   ***k** 8·106 to 8·097?

8 **a** Graham wanted a length of wood 1·83 m long. He measured
        a piece he had in the garage. It was 10 cm too short.
        How long was the piece in the garage?
   ***b** Another piece in the garage was 8 mm too short.
        How long was this piece?

*9 Barry's Street is being repaved. At present it has 1·38 km of paving.
    This paving is to be extended by 40 m.
    How long in total will the new paving be?

Review 1

$\overline{4 \cdot 996}$ $\overline{4 \cdot 93}$ $\overline{4 \cdot 107}$ $\overline{5 \cdot 81}$ $\overline{4 \cdot 698}$ $\overline{7 \cdot 422}$ $\overline{2 \cdot 404}$ $\overset{E}{\overline{5 \cdot 84}}$ $\overline{8 \cdot 307}$ $\overline{4 \cdot 923}$ $\overline{4 \cdot 698}$

$\overline{4 \cdot 684}$ $\overline{7 \cdot 134}$ $\overline{7 \cdot 422}$ $\overline{4 \cdot 996}$ $\overline{4 \cdot 694}$ $\overline{4 \cdot 923}$ $\overline{4 \cdot 698}$ $\overline{4 \cdot 299}$

$\overline{5 \cdot 81}$ $\overline{0 \cdot 042}$ $\overline{4 \cdot 684}$ $\overset{E}{\overline{5 \cdot 84}}$ $\overline{4 \cdot 694}$ $\overline{5 \cdot 81}$ $\overline{8 \cdot 307}$ $\overset{E}{\overline{5 \cdot 84}}$

$\overline{4 \cdot 684}$ $\overline{4 \cdot 996}$ $\overline{4 \cdot 923}$ $\overline{4 \cdot 694}$ $\overline{4 \cdot 684}$ $\overline{3 \cdot 663}$

Use a copy of this box.
Write the letter that is beside each question above its answer.

E $5 \cdot 83 + 0 \cdot 01 = 5 \cdot 84$ **O** $7 \cdot 234 - 0 \cdot 1$ **T** $4 \cdot 683 + 0 \cdot 001$ **I** $4 \cdot 924 - 0 \cdot 001$
N $4 \cdot 697 + 0 \cdot 001$ **F** $0 \cdot 043 - 0 \cdot 001$ **Y** $3 \cdot 683 - 0 \cdot 02$ **A** $5 \cdot 804 + 0 \cdot 006$
S $7 \cdot 382 + 0 \cdot 04$ **G** $8 \cdot 310 - 0 \cdot 003$ **K** $4 \cdot 3 - 0 \cdot 001$ **R** $4 \cdot 689 + 0 \cdot 005$
B $2 \cdot 399 + 0 \cdot 005$ **M** $4 \cdot 099 + 0 \cdot 008$ **U** $5 \cdot 0 - 0 \cdot 07$ **H** $5 - 0 \cdot 004$

Review 2

a What mass is 0·06 kg less than 28·352 kg?
b Find the sum of 8·647 km and 0·07 km.
c Find the difference between 18·072 ℓ and 0·004ℓ.

Review 3 What must be added to or subtracted from the first number to get the second number?

a 18·65, 18·55   **b** 204·831, 204·871   **c** 3·999, 4   **d** 5·911, 5·903

Multiplying and dividing by 10, 100 and 1000

Investigation

10, 100 and 1000

You will need a calculator, a spreadsheet package or a place value board and a copy of the table.

1 Fill in the table.
You could use a calculator, spreadsheet package or place value board to help you.

Number	×10	×100	×1000	÷10	÷100	÷1000
8·56						
7·03						
5·06						
2·954						
0·03						
0·042						

2 Explain what happens to the digits when you multiply a number by 10, 100 or 1000. Does the number get bigger or smaller?

3 Explain what happens to the digits when you divide a number by 10, 100 or 1000. Does the number get bigger or smaller?

continued ...

If you used a spreadsheet to fill in the table, change the numbers in column A and see if you get the same answers to questions 2 and 3.

	A	B	C	D	E	F	G
1	Number	×10	×100	×1000	÷10	÷100	÷1000
2	8.56	=A2*10	=A2*100	=A2*1000	=A2/10	=A2/100	=A2/1000
3	7.03						
4	5.06						

Use the fill down feature on your spreadsheet.

To multiply by 10, move each digit one place to the left.
To multiply by 100, move each digit two places to the left.
To multiply by 1000, move each digit three places to the left.

Example   46·37 × 10 = **463·7**

To divide by 10, move each digit one place to the right.
To divide by 100, move each digit two places to the right.
To divide by 1000, move each digit three places to the right.

Example   17·3 ÷ 100 = **0·173**

Worked Example
a 203·6 × 100   b 2·74 × 1000   c 46·2 ÷ 100   d 3·04 ÷ 1000

Answer
a **20 360**   Move each digit two places to the left.
b **2740**   Move each digit three places to the left.
c **0·462**   Move each digit two places to the right.
d **0·00304**   Move each digit three places to the right.

Exercise 3

1 Find the answer to these.
   a 59·2 × 10   b 59·2 × 100   c 59·2 ÷ 10   d 59·2 ÷ 1000   e 59·2 × 1000
   f 59·2 ÷ 100   g 1·23 × 100   h 10·8 ÷ 10   i 351 ÷ 1000   j 724·3 × 100
   k 0·021 ÷ 100   l 23·04 × 10   m 48·24 × 100   n 2 ÷ 100   o 67·9 × 1000
   p 841·05 ÷ 100   q 67·03 ÷ 100   r 8 ÷ 1000   s 0·06 × 100

2 What number goes in the box?
   a ☐ ÷ 10 = 0·52   b ☐ × 10 = 604   c ☐ × 1000 = 60 132   d ☐ ÷ 100 = 0·074
   e ☐ ÷ 1000 = 0·89   f ☐ ÷ 1000 = 0·0024   g ☐ × 1000 = 0·21

3 Nick bought plastic to cover his books. He needed 41·5 cm for each book.
   How much plastic did he need for 10 books?

4 a Dianne measured the thickness of 1000 sheets of paper as 264 mm.
      How thick is each of these sheets?
   b She measured 500 layers of slate as 502 mm. How thick is each layer?

*5 Choose three different numbers between 8550 and 8600.
   Divide each number by 1000.

T | **Review**

		A						**A**							
$\overline{1\cdot2}$	$\overline{0\cdot12}$	$\overline{620}$	$\overline{14}$	$\overline{12}$	$\overline{1\cdot2}$		$\overline{620}$	$\overline{14}$	$\overline{6200}$		$\overline{6\cdot2}$	$\overline{0\cdot12}$	$\overline{6200}$		

												A		
$\overline{0\cdot062}$	$\overline{62}$	$\overline{140}$	$\overline{1\cdot4}$		$\overline{0\cdot0012}$	$\overline{0\cdot62}$	$\overline{1\cdot2}$	$\overline{0\cdot12}$		$\overline{6\cdot2}$	$\overline{0\cdot12}$	$\overline{620}$	$\overline{6\cdot2}$	

	A							
$\overline{0\cdot14}$	$\overline{620}$	$\overline{62}$		$\overline{120}$	$\overline{140}$	$\overline{0\cdot62}$	$\overline{62}$	$\overline{12}$

Use a copy of this box.
Find the answer to these in the box. Write the letter above the answer.

A $62 \times 10 = 620$	**E** 62×100	**O** $0\cdot62 \div 10$	**T** $6200 \div 1000$	**N** $0\cdot062 \times 1000$
I $62 \div 100$	**L** $1\cdot4 \times 100$	**Y** $14 \div 10$	**C** $140 \div 1000$	**R** $0\cdot014 \times 1000$
F $1\cdot2 \div 1000$	**H** $12 \div 100$	**S** $0\cdot012 \times 100$	**K** $0\cdot012 \times 1000$	**B** $1\cdot2 \times 100$

× and ÷ by multiples of 10, 100 and 1000

Discussion

Sarah was told the average weight of a sixteen-year-old student was 60 kg.
She wanted to know how much a bus load of 40 of these students would weigh.
How could Sarah use her knowledge of multiplying and dividing by 10, 100, 1000, ... to
help her find the answer? **Discuss**.

We can use multiplication and division by 10, 100, 1000, ... to help us **multiply and divide by
multiples of 10, 100, 1000, ...**

Worked Example
Find **a** 40×70 **b** $3\cdot2 \times 40$ **c** $60 \div 200$ **d** $1\cdot8 \div 300$

Answer

a $40 \times 70 = 4 \times 10 \times 7 \times 10$
$\qquad = 4 \times 7 \times 10 \times 10$
$\qquad = 28 \times 100$
$\qquad = \mathbf{2800}$

b $3\cdot2 \times 40 = 3\cdot2 \times 4 \times 10$
$\qquad = 12\cdot8 \times 10$
$\qquad = \mathbf{128}$

or 4 | 3   0·2 |
           | 12   0·8 | answer 12·8
$12\cdot8 \times 10 = \mathbf{128}$

c $60 \div 200 = 0\cdot3$

because $60 \div 2 = 30$
$\qquad\qquad 30 \div 100 = \mathbf{0\cdot3}$

d $1\cdot8 \div 300 = 1\cdot8 \div 3 \div 100$
$\qquad\qquad = 0\cdot006$

because $1\cdot8 \div 3 = 0\cdot6$
$\qquad\qquad 0\cdot6 \div 100 = \mathbf{0\cdot006}$

or $1\cdot8 \div 3 = 0\cdot6$
$\qquad 1\cdot8 \div 30 = 0.06$
$\qquad 1\cdot8 \div 300 = \mathbf{0.006}$

| Exercise 4 | | **This exercise is to be done mentally. You may use jottings.** |

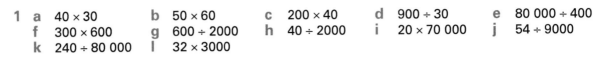

1 a 40×30 b 50×60 c 200×40 d $900 \div 30$ e $80\,000 \div 400$
   f 300×600 g $600 \div 2000$ h $40 \div 2000$ i $20 \times 70\,000$ j $54 \div 9000$
   k $240 \div 80\,000$ l 32×3000

2 a 4.1×20 b 5.3×30 c 1.1×400 d $8.4 \div 20$ e $0.42 \div 300$
   f $16.4 \div 40$ g $0.56 \div 80$ h 0.052×2000 i $7.2 \div 900$ j $4.8 \div 6000$
  *k 8.1×6000 *l $18.6 \div 600$ *m 6.7×8000 *n $9.24 \div 40$

3 Jamie taught himself to type using a program on a computer. He can
   now type 20 words in one minute.
   How many words can Jamie type in
   a 30 minutes   b 70 minutes   c 50 minutes?

4 Angela trained on a 400 m track. Each week she ran a total of 12000 m.
   How many laps of the track did Angela run each week?

5 a Melanie reads a scale on a balance in science as 8.4 g. Another student reads it
      correctly as 20 times larger than this. What is the correct reading?
   b Melanie divides *her* reading by 40. What answer did she get?
   c What is the correct reading divided by 40?

*6 What two numbers, to the nearest 100, could you multiply together to get an answer of
    80 000?

*7 Make up some divisions that have an answer of
    a 20   b 0.2   c 0.02.

> You could make up a word
> problem for each.

Review 1
a 80×20 b 30×40 c 400×50 d 700×300 e $600 \div 30$
f $8000 \div 20$ g $400 \div 2000$ h $12.4 \div 40$ i 2.7×300 j $3.6 \div 90$
*k 6.2×400 *l $25.5 \div 500$ *m 5.7×300

Review 2   At a supermarket, 80 boxes were unloaded from a lorry. Each of these contained
40 tins of soup.
How many tins of soup were there altogether?

Review 3   A warehouse had a total of 30 000 packets of tea.
These were stored in boxes of 50.
How many boxes of tea were in the warehouse?

Review 4   Make up some multiplications that have an answer of 16 000.

Review 5   Make up some divisions that have an answer of 0.08.

Putting decimals in order

T

Remember
To put decimals in order we **compare** digits with the same place value.
Always work from left to right to find the first digits that are different.

Worked Example
Which of < or > goes between these   a 0·02567 ___ 0·02589   b 8·1763 ___ 8·1709?

Answer
a Starting at the left, the digits are the same until we get to the digits 6 and 8.   0·025**6**7
   6 is less than 8, so **0·02567** < **0·02589**.   0·025**8**9
b Starting at the left, the digits are the same until we get to the digits 6 and 0.   8·17**6**3
   6 is bigger than 0, so **8·1763** > **8·1709**.   8·17**0**9

Worked Example
This list gives the masses of the five largest fish caught in a contest. Put the masses in descending order.

3·1895 kg   3·1878 kg   3·1825 kg   3·0985 kg   3·1899 kg

 Descending order means from largest to smallest.

Answer
Compare the units, then tenths, then hundredths, then thousandths, then ten thousandths.
The numbers in descending order are

3·1899 kg, 3·1895 kg, 3·1878 kg, 3·1825 kg, 3·0985 kg.

Exercise 5

1 Which of < or > goes between these?
   a 0·803 m   0·83 m
   b 4·27 ℓ   4·297 ℓ
   c 3·8462 mm   3·845 mm
   d 7·0309 km   7·031 km
   e 5·2064 kg   5·2604 kg
   f 0·05491 km   0·055 km
   g 0·4723 ℓ   0·47229 ℓ
   h 4·60304 km   4·6034 km

2 a This list gives the masses, in kilograms, of the mice in a 'cutest
      mouse' contest. Put the masses in ascending order.
         0·081 kg   0·128 kg   0·11 kg   0·082 kg   0·08 kg
   b This list gives the distances, in metres, that six boys threw a welly
      in a welly-throwing contest. Put the distances in ascending order.
         20·65 m   21·057 m   24·792 m   21·07 m   20·563 m

Ascending means from smallest to largest.

 *To compare measurements the units must be the same.*

3 Which of < or > goes in the box?
   a 7·2 m ☐ 7204 mm
   b 4·68 km ☐ 497 m
   c 7·63 cm ☐ 768 mm
   d 9·04 ℓ ☐ 909 mℓ
   e 5·029 m ☐ 52·79 cm
   f 5869 g ☐ 0·5896 kg

4 A, B, C, D are four numbers on a number line.
   Which of these would be at B?
   a 0·72   0·702   0·722   0·7
   b 8·06   8·053   8·099   8·038

⟵ A   B   C D ⟶

23

5 Estimate the decimal number that goes in each box.

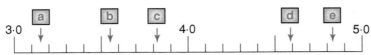

6 2·7 < 2·8 **BUT** ⁻2·7 > ⁻2·8

Are these true or false?
a   ⁻2·7 < 2·6
b   ⁻6·1 > ⁻6·4
c   ⁻6·8 < ⁻6·83
d   ⁻15·23 > ⁻15·2
e   ⁻0·02 > ⁻0·024

7 Which number is halfway between these?

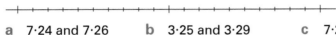

a   7·24 and 7·26   b   3·25 and 3·29   c   7·2 and 7·3
d   0·08 and 0·09   e   0·02 and 0·03   f   4·01 and 4·02
g   3·25 and 3·28   h   5·72 and 5·75   i   ⁻1·2 and ⁻1·4
j   ⁻0·3 and ⁻0·4   k   ⁻0·03 and ⁻0·04

Use a number line to help.

8 0·45 lies halfway between two other numbers.
What could they be?
Find at least five different answers.

9 81·43 ⩽ z ⩽ 81·53
Give 10 values that z could have.

10 Ben did a survey on the heights of students at a dance school. This is his frequency table.

Height (m)	Tally	Frequency
1.50 < x ⩽ 1.55		
1.55 < x ⩽ 1.60		
1.60 < x ⩽ 1.65		
1.65 < x ⩽ 1.70		
1.70 < x ⩽ 1.75		

a   Explain what 1·50 < x ⩽ 1·55 means.
b   Joanne is 1·55 m. In which colour box would Ben put the tally mark for Joanne?
c   April is 1·70 m. In which colour box would Ben put the tally mark for April?
d   In which colour box would the tally mark for each of these go?
    1·63 m   1·54 m   1·72 m   1·69 m

11 a Fran has these cards. Write down all the numbers with three
decimal places that she could make using all the cards. Do
not have zero as the first or last digit.
Put the numbers in order from largest to smallest.

b Repeat **a** for these cards.

T   ***12** Use a graphical calculator to generate ten random decimal numbers with three decimal
places. Put the numbers in descending order.

Review 1   Are these true or false?
a   5·864 < 5·86   b   3·4709 < 3·478   c   6·3024 > 6·2024
d   0·0647 > 0·06407   e   0·96004 > 0·9603   f   ⁻3·4 < ⁻3·5
g   ⁻0·27 > ⁻0·26

Review 2   Which of < or > goes between these?
a   3·7291   3·7289   b   0·02437   0·02452   c   3·1895   3·1825

Review 3   _____ < _____ < _____
Put these numbers in the gaps.
a   7·234, 7·259, 7·212   b   6·832, 6·7941, 6·8032
c   0·3096, 0·309 61, 0·308 94   d   0·0045, 0·00057, 0·00507

Review 4   Put these in descending order.
a   6·324 kg, 6320 g, 6·3024 kg, 6·3421 kg, 6243·1 g
b   0·864 km, 0·8927 km, 0·8641 km, 806·4 km
c   14·2794 ℓ, 14·2789 ℓ, 14300 mℓ, 14279 mℓ, 14·2783 ℓ

Review 5   Which number is halfway between these?
a   0·7 and 0·8
b   0·06 and 0·09
c   ⁻0·5 and ⁻0·6
d   ⁻2·3 and ⁻1·8

Ladders — a game for a group

You will need a ladder like this.

To play
● Choose a leader.
● The leader chooses ten decimal numbers between 0 and 10 and calls them out one at a time.
● As each number is called the other players put it on their ladder.
● The numbers must be put on the ladder in order.
● The winner is the player with the most numbers on his or her ladder.

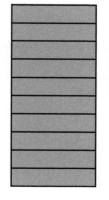

The leader could use the Ran # key on the calculator to help.
Shift Ran# ⊗ 1 0 = =

Example   Paul was leader. The first five numbers he called were 5·84, 7·3, 5·486, 8·371 and 8·7.
Jasmine had put them on her ladder like this.
The next number Paul called was 8·583.
Jasmine could not put this on her ladder.

| 8.7 |
| 8.371 |
| |
| 7.3 |
| |
| 5.84 |
| 5.486 |

25

Rounding to the nearest 10, 100, 1000, 10 000, ...

Ben's father bought a new car for £5863.
He wanted to insure it.
He was asked to round the price he paid to the nearest £100.
£5863 is between £5800 and £5900.

£5863 to the nearest £100 is £5900.

4572 is closer to 5000 than to 4000.
4572 rounded to the nearest 1000 is 5000.

Examples   648 to the nearest hundred is 600.
8642 to the nearest thousand is 9000.
68 241 to the nearest ten thousand is 70 000.
4 086 423 to the nearest million is 4 million.

When a number is exactly halfway between two numbers we round up.
45 to the nearest 10 is 50.
8500 to the nearest thousand is 9000.
25 000 to the nearest ten thousand is 30 000.
4 500 000 to the nearest million is 5 million.

Worked Example
To the nearest 10 000, the number of cars in a town is 140 000.
What is the largest and smallest number of cars that could be in the town?

Answer
The largest number of cars is **144 999**.
One more car would be 145 000, which would be rounded up to 150 000.
The smallest number of cars is **135 000**. One less than this would be 134 999, which would be
rounded down to 130 000.

Exercise 6

1 This table gives the length of the ten longest
   rivers in the world.
   a Round each length to the nearest 100 km.
   b Put the rivers in order from longest to
      shortest.
   c Which five rivers have the same length to
      the nearest 1000 km?

	Km		Km
Amur	4416	Ob-Irtysh	5410
Amazon	6570	Paraná	4500
Huang He	4840	Yangtze	6380
Mississippi	6020	Yenisei	5310
Nile	6670	Zaire(Congo)	4630

2 This table shows the diameters of the planets in kilometres.

Mercury	Venus	Earth	Mars	Jupiter	Saturn	Uranus	Neptune	Pluto
4878	12 104	12 756	6794	142 800	120 000	52 000	48 400	2300

   a Round each diameter to the nearest thousand kilometres.
   b Put the planets in order from smallest to largest.

3 Round
   a 324 to the nearest 10   b 5862 to the nearest 100
   c 5682 to the nearest 1000   d 1825 to the nearest 10
   e 28 562 to the nearest 10 000   f 856 427 to the nearest 100 000
   g 43 250 to the nearest 100   h 89 500 to the nearest 1000
   i 725 000 to the nearest 100 000   j 68 423 to the nearest 100
   k 58 740 to the nearest 1000   l 69 782 to the nearest 1000
   m 138 987 to the nearest 100 000   n 5 864 321 to the nearest million
   o 5 500 000 to the nearest million.

4 72 680 people went to a concert.
   a The local newspaper reported that about 73 000 people went to the concert.
      Was this given to the nearest 100, nearest 1000 or nearest 10 000?
   b Another newspaper reported that about 70 000 people went to the concert.
      Was this given to the nearest 1000 or nearest 10 000?

5 £187 642 was raised to give to charity.
   a The charity reported that nearly £188 000 had been raised.
      Was this amount given to the nearest 100, nearest 1000 or nearest 10 000?
   b Another newspaper reported that the charity had raised about £190 000.
      Was this given to the nearest 1000 or nearest 10 000?

6 This chart gives the five most widely spoken languages in the world.
  What is the approximate number of people who speak each language?

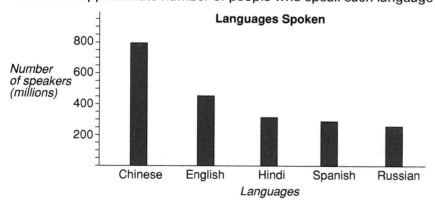

7 A newsletter said that, to the nearest hundred, 700 tickets to a fund-raising evening had been sold.
   a What is the greatest number of tickets that could have been sold?
   b What is the smallest number of tickets that could have been sold?

8 The population of Hereford to the nearest thousand is 48 000.
   a What is the smallest number of people who might live there?
   b What is the greatest number of people?
   c What is the greatest and smallest number of people who might
      live in a town with a population of 17 600 to the nearest hundred?

Find the population of your nearest city. Round to the nearest thousand.

9 How long is this string to the nearest   **a** 10 cm   **b** cm   **c** 100 cm   **d** mm?

123   124   125   126   127 cm

10 This table gives the number of fans at some football games.

Manchester United	61 864
Arsenal	33 575
Liverpool	39 504
Southampton	27 398

   a Round the Arsenal number to the nearest 10.
   b Round the Southampton number to the nearest 10 000.
   c Round the Liverpool number to the nearest thousand.
   d Add the numbers of fans at all four matches then round to the nearest 100.
   e Round each of the four numbers to the nearest 100, then add these rounded numbers.
  *__f__ Explain why the answers to **d** and **e** are different.

Review 1 Round
a 386 to the nearest 10
b 4821 to the nearest 100
c 4724 to the nearest 100
d 8894 to the nearest 1000
e 35 062 to the nearest 10 000
f 4685 to the nearest 10
g 47 850 to the nearest 100
h 19 500 to the nearest 10 000
i 856 842 to the nearest 100 000
j 6 327 541 to the nearest million

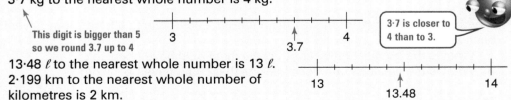

Review 2 35 682 people went to a football game.
A newspaper said 35 700 people were at the game.
Was this number given to nearest hundred or nearest thousand?

Review 3 The size of a crowd at a cricket match is given as 24 700 to the nearest hundred.
a What is the smallest number the crowd could be?
b What is the largest number the crowd could be?

Rounding to decimal places

To **round** to the **nearest whole number**, we look at the digit in the first **decimal place**.
If this digit is 5 or more, we round the whole number up. Otherwise, it is left unchanged.

Examples   3·7 kg to the nearest whole number is 4 kg.

This digit is bigger than 5
so we round 3.7 up to 4

3   3.7   4

3·7 is closer to
4 than to 3.

13·48 ℓ to the nearest whole number is 13 ℓ.
2·199 km to the nearest whole number of kilometres is 2 km.

13   13.48   14

To **round** to **one decimal place**, we look at the digit in the second decimal place.
If this digit is 5 or more, we round the number in the first decimal place up. Otherwise it stays the same.

Examples   4·682 to one decimal place is 4·7

8 ⩾ 5 so we round the first decimal place up.

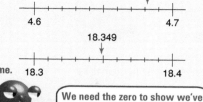

4.682

4.6   4.7

18·349 to one decimal place is 18·3.

18.349

4 < 5 so the first decimal place stays the same.   18.3   18.4

4·97 to one decimal place is 5·0.

> We need the zero to show we've rounded to one decimal place and not to the nearest whole number.

To **round** to two decimal places, we look at the digit in the third decimal place.

Examples   4·7312 to two decimal places is 4·73.

1 < 5 so the second decimal place stays the same.

17·159 to two decimal places is 17·16.

9 ⩾ 5 so the second decimal place is rounded up to 6.

0·735 to 2 d.p. is 0·74

5 ⩾ 5 so the second decimal place is rounded up to 4.

Decimal place is often written as **d.p.**

Example   8·69 to 1 d.p. is 8·7.

Exercise 7   **Except for questions 9, 10 and 11.**

1 Round these to the nearest whole number.
   a 14·71   b 8·34   c 17·20   d 19·684   e 62·867
   f 13·50   g 26·285   h 633·851   i 526·4 g   j 82·913 kg
   k 904·56 m   l 350·67 m   m 1086·7 mm   n 1492·0359 ℓ   o 0·7187 m
   p 0·48 km   q 0·3862 ℓ   r 0·7443 kg   s 0·81134 km   *t 9·6527 ℓ
  *u 0·08451 km   *v 0·99862 m   *w 99·99143 kg

2 Round the numbers in question **1** to one decimal place.

> Look at the second decimal place.

3 Some readings were taken from scientific instruments. Round them to 2 d.p.
   a 2·8237   b 6·827   c 0·7348   d 1·292   c 1·835
   f 2·896   g 12·994   h 4·0038   i 1·00376   j 22·88143
   k 0·8598   l 0·8957   m 0·98534   n 325·0925   *o 18·99929

4 This is the answer to a calculation.
   Rob rounded this to the nearest whole number.
   He said 'The 8 is 5 or more so 3·4 becomes 3·5.
   Now 3·5 to the nearest whole number is 4.'
   Explain why Rob is wrong.

3.48

5 What goes in the gaps?
   a 4·98 rounded to the nearest whole number is _____ and to 1 d.p. is _____ .
   b 0·96 rounded to the nearest whole number is _____ and to 1 d.p. is _____ .

6 This table shows the points given to the five dogs in the final of a 'most loveable pooch' competition.

Name	Misty	Dougal	Katie	Hound Dog	Muss
Points	93·42	86·87	91·95	88·60	92·00

Round each to
a the nearest whole number   b one decimal place.

7 These show the answer Kiran got for some calculations.
Round these to two decimal places.
a 424.389   b 17.9307   c 0.7645   d 35.098

8 Gwyneth bought the following for her party.
Give each amount to the nearest tenth of a kilogram.
a 2840 g of cocktail sausages   b 1356 g of sweets   c 1720 g of sausage rolls

9 Round the answers to these to 2 d.p.
a 56 ÷ 17   b 824 ÷ 27   c 9174 ÷ 824   d 96·8 ÷ 3·24

*10 A boot-lace worm can grow up to 55 m in length. The average garden worm is about 9 cm long. How many garden worms, lying end to end, would it take to equal the length of the boot-lace worm? Round the answer sensibly.

*11 Round the answers to these to 2 d.p.
a Martha's gas bill was £31·58 for a 42 day billing period. How much was this per day?
b Flannel, the elephant weighed 7·6 tonnes. His keeper, Momo, weighed 76·4 kg. How many times heavier is Flannel than Momo?
c Six test tubes in an experiment have masses of 8·3 g, 10·4 g, 7·6 g, 12·9 g, 9·5 g and 11·7 g.
Find the mean mass.

> There is more about mean on page 413.

Review 1   Round these to the nearest whole number.
a 15·83   b 7·04   c 19·55   d 36·147   e 43·996

Review 2   Round the numbers in **Review 1** to one decimal place.

Review 3   This table shows the distances five people threw in a 'welly throwing' competition.

Name	B.Smith	R.Johnson	M.Carter	S.Patel	M.O'Reagan
Distance (m)	17·23	20·49	16·58	19·98	18·00

Round each distance to
a the nearest whole number   b 1 d.p.

Review 4   Karen rounded these calculator displays to two decimal places. What answers should she have got?
a 6.2544   b 45.26612
c 0.3678   d 0.0205
e 10.0392   f 32.9955

*Review 5   The average width of a human is about 33 cm. In a horror movie, a giant snake could open its jaws 2·73 m wide. How many humans, standing side by side, could it hold in its jaws? Round the answer sensibly.

Summary of key points

 This chart shows **place value**.

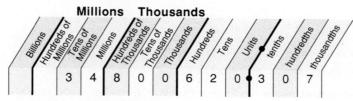

Millions   Thousands

Billions	Hundreds of Millions	Tens of Millions	Millions	Hundreds of Thousands	Tens of Thousands	Thousands	Hundreds	Tens	Units	tenths	hundredths	thousandths
	3	4	8	0	0	6	2	0	3	0	7	

A billion is a thousand million.

The zeros in 34 800 620·307 are **place holders**.

B To **add or subtract**   0·1, 0·2, ..., add or subtract 1, 2, ... to or from the tenths

0·01, 0·02, ..., add or subtract 1, 2, ... to or from the hundredths

0·001, 0·002, ..., add or subtract 1, 2, ... to or from the thousandths

Examples   8·473 + 0·01 = 8·483   9·64 − 0·001 = 9·639   28·79 + 0·6 = 29·39

C To **multiply by 10, 100, 1000**, ... move each digit one place to the left for each zero in 10, 100, 1000, ...

Examples   46 × 10 = 460   0·8 × 100 = 80   0·520 × 1000 = 520

To **divide by 10, 100, 1000**, ... move each digit one place to the right for each zero in 10, 100, 1000, ...

Examples   530 ÷ 10 = 53   4·8 ÷ 100 = 0·048   834 ÷ 1000 = 0·834

D We can use multiplication and division by 10, 100, 1000, ... to **multiply and divide by multiples of 10, 100, 1000**, ...

Examples   80 × 600 = 8 × 10 × 6 × 100   4·8 ÷ 300 = 0·016
             = 8 × 6 × 10 × 100   because   4·8 ÷ 3 = 1·6
             = 48 000   1·6 ÷ 100 = 0·016

E To put **decimals in order**, compare digits with the same place value. Work from left to right to find the first digits that are different.

Example   6·8724 < 6·8731   Starting at the left, the digits are the same until we get to the digits 2 and 3.   2 is less than 3 so 6·8724 < 6·8731.

F **Rounding to the nearest 10, 100, 1000,**

4 872 487   to the nearest ten thousand is 4 870 000

to the nearest hundred thousand is 4 900 000

to the nearest million is 5 million.

When a number is halfway between two numbers we round up.

Example   250 000 to the nearest hundred thousand is 300 000.

G To **round to the nearest whole number** we look at the tenths digit.

If the tenths digit is 5 or more, we round the whole number up. Otherwise, it stays the same.

Example 16·54

The tenths digit is 5 or more so 16 becomes 17.

16·54 to the nearest whole number is 17.

To **round to one decimal place** we look at the digit in the second decimal place. If this digit is 5 or more, we round the digit in the first decimal place up. Otherwise, it stays the same.

Examples 4·58 to 1 d.p. is 4·6
4·82 to 1 d.p. is 4·8
5·97 to 1 d.p. is 6·0

To **round to two decimal places** we look at the digit in the third decimal place.

Test yourself

1 What is the value of the 8 in these? **A**
a 43 825 b 508·72 c 396·84 d 13·08 e 2·918

2 Make the third largest number you can with these cards. You must have at least one number after the decimal point. The last digit must not be zero. **A**

3 Write, in figures, the number that is **A**
a 4 more than three and a half million
b 7 less than two and a quarter million.

4 a 5·374 + 0·001 b 8·986 − 0·001 c 15·239 + 0·001 d 15·590 − 0·001 **B**
e 4·238 + 0·004 f 17·283 − 0·005

5 What must be added to 9·638 to get 9·642? **B**

6 Which of < or > goes between these? **E**
a 5·6832 5·6829 b 4·3872 4·3879 c ⁻1·81 ⁻1·79

7 Put these in ascending order. **E**
a 4·2741, 4·7241, 4·72, 4·0741 b 0·8692, 0·6892, 0·06, 0·0078, 0·798

8 Here are some first run times for the luge event at the 1998 Winter Olympics.
Mueller 49·954, Holm 50·224, Hackl 49·619, Prock 49·861,
Zoeggeler 49·714, Kleinheinz 50·016, Huber 50·100
Who had the   **a** fastest time   **b** second fastest time   **c** slowest time?

9 Which number is halfway between
a   4·6 and 4·7   **b** 0·07 and 0·08   **c** ⁻0·8 and ⁻0·9?

10 Calculate
a 48 × 1000   **b** 96 ÷ 10   **c** 54 ÷ 100   **d** 83 ÷ 1000   **e** 5·7 × 10
f 0·89 × 100   **g** 0·52 × 1000   **h** 6·37 ÷ 1000   **i** 52 ÷ 1000   **j** 300 × 50
k 1200 ÷ 40   **l** 1·2 × 600   **m** 0·72 ÷ 90   **n** 33·6 ÷ 6000.

11 Ten friends won £36·50 in a raffle. It was to be divided equally.
How much did each get?

12 Enid did a survey on the weights of cats. This shows part of her frequency table.

Weight (kg)	Tally	Frequency
$1.0 < x \leqslant 1.5$		
$1.5 < x \leqslant 2.0$		
$2.0 < x \leqslant 2.5$		

a   Explain what $1 \cdot 0 < x \leqslant 1 \cdot 5$ means.
b   Enid's kitten weighs 1·5 kg. In which colour box should Enid put the tally mark?

13 Round these to   **i** the nearest thousand   **ii** the nearest hundred thousand
iii the nearest million.
a 58 623 471   **b** 4 932 314   **c** 7 003 741   **d** 885 500

14 April chose a 4-digit number.
She rounded it to the nearest 1000 and got 8000.
a   What is the smallest number April could have chosen?
b   What is the biggest number April could have chosen?

15 Approximate these to the given number of decimal places.
a 54·386 (2 d.p.)   **b** 72·354 (2 d.p.)   **c** 4·031 (2 d.p.)   **d** 16·257 (1 d.p.)
e 4·605 (1 d.p.)   **f** 186·052 (1 d.p.)   **g** 5·097 (2 d.p.)   **h** 16·423 (0 d.p.)
i 105·004 (2 d.p.)   **j** 9·998 (2 d.p.)

16 Peter tried to beat the Guinness Book of Records for talking non-stop.
He talked for 116 hours. How many days is this? Round your answer sensibly.

Did he beat
the record?

2 Special Numbers

All in a spin

		14	13
5	4	3	12
6	1	2	11
7	8	9	10

- Barry began to make a spiral of numbers from 1 to 100.
  Use a copy of this. Finish the spiral of numbers.

- Colour all the multiples of 2.
  What do you notice?

- Use another copy of the spiral.
  Colour all the square numbers.
  What do you notice?

- Use more copies of the spiral.
  Find other number patterns.

Adding and subtracting integers

Discussion

- $4 + 1 = 5$
  $4 + 0 = 4$
  $4 + {}^-1 = 3$
  $4 + {}^-2 = 2$
  $4 + {}^-3 = 1$
  $4 + {}^-4 = 0$
  $4 + {}^-5 = {}^-1$

  ${}^-3 + 5 = 2$
  ${}^-3 + 4 = 1$
  ${}^-3 + 3 = 0$
  ${}^-3 + 2 = {}^-1$
  ${}^-3 + 1 = {}^-2$
  ${}^-3 + 0 = {}^-3$
  ${}^-3 + {}^-1 = {}^-4$

  ${}^-3 + {}^-1$ is said as negative 3 plus negative 1.

  $2 - 2 = 0$
  $2 - 1 = 1$
  $2 - 0 = 2$
  $2 - {}^-1 = 3$
  $2 - {}^-2 = 4$
  $2 - {}^-3 = 5$
  $2 - {}^-4 = 6$

  ${}^-3 - 2 = {}^-5$
  ${}^-3 - 1 = {}^-4$
  ${}^-3 - 0 = {}^-3$
  ${}^-3 - {}^-1 = {}^-2$
  ${}^-3 - {}^-2 = {}^-1$
  ${}^-3 - {}^-3 = 0$
  ${}^-3 - {}^-4 = 1$

  ${}^-3 - {}^-2$ is said as negative 3 minus negative 2.

  What are the next three lines of each of these number patterns? **Discuss.**
  Use your patterns to find the answers to these.
  $4 + {}^-9$ ${}^-3 + {}^-5$ $2 - {}^-8$ ${}^-3 - {}^-8$

- How could you use the number line to help you find the answers to these? **Discuss.**
  $4 + {}^-3$ ${}^-3 + 4$ ${}^-3 + {}^-2$ $3 - 4$ ${}^-2 - 1$ $2 - {}^-3$ ${}^-3 - {}^-4$

  ${}^-7\ {}^-6\ {}^-5\ {}^-4\ {}^-3\ {}^-2\ {}^-1\ 0\ 1\ 2\ 3\ 4\ 5\ 6\ 7$

  Beth wrote this.

  Adding and subtracting are inverse operations.
  To add a positive number move →. To subtract a positive number move ←.
  To add a negative number move ←. To subtract a negative number move →.
  What does Beth mean? **Discuss.**

- Sonia owed her friend £473. She earned £87 for busking one weekend.
  To work out how much she still owed she keyed this on her calculator.

   to get $-386.$

  What does Sonia's answer mean?
  Did she key correctly? **Discuss.**

Exercise 1

You could check your answers using the calculator.

Place	Height above sea level (m)
Ayers Rock	900
Caspian Sea	−30
London	30
Nairobi	1800
Rotterdam	−5
The Dead Sea	−400

1 The height above sea level of some places is shown.
   a What is the difference in height between London and the Caspian Sea?
   b What is the difference in height between The Dead Sea and Ayers Rock?
   c How much higher is Nairobi than Rotterdam?

2 Priya's mother played 9 holes of golf.
   The table shows how far above or below par she was for each hole.

Hole number	1	2	3	4	5	6	7	8	9
Above/below par	+1	${}^-2$	+3	${}^-1$	${}^-2$	${}^-1$	+2	${}^-2$	+1

Work out how far above or below par Pryia's mother was at the end of the 9 holes.
You could use this number line to help.

PAR
${}^-5\quad {}^-4\quad {}^-3\quad {}^-2\quad {}^-1\qquad +1\quad +2\quad +3\quad +4\quad +5$

3 The temperature is below zero. It rises 3 degrees then falls 5 degrees. What could the temperature be now?

4 Write down the next 3 lines of these patterns.

a $2 + 1 = 3$
$2 + 0 = 2$
$2 + {}^-1 = 1$
$2 + {}^-2 = 0$
$2 + {}^-3 = {}^-1$
$2 + {}^-4 = {}^-2$

b $5 + {}^-1 = 4$
$5 + {}^-2 = 3$
$5 + {}^-3 = 2$
$5 + {}^-4 = 1$
$5 + {}^-5 = 0$
$5 + {}^-6 = {}^-1$

c $2 - 1 = 1$
$2 - 0 = 2$
$2 - {}^-1 = 3$
$2 - {}^-2 = 4$
$2 - {}^-3 = 5$

d $3 - 1 = 2$
$3 - 0 = 3$
$3 - {}^-1 = 4$
$3 - {}^-2 = 5$
$3 - {}^-3 = 6$

e ${}^-2 - 1 = {}^-3$
${}^-2 - 0 = {}^-2$
${}^-2 - {}^-1 = {}^-1$
${}^-2 - {}^-2 = 0$
${}^-2 - {}^-3 = 1$

5 Use the number patterns in question 4 to find the answers to these.

a $3 + {}^-8$
b $3 + {}^-9$
c $2 + {}^-7$
d $2 + {}^-9$
e $5 + {}^-9$
f $2 - {}^-7$
g $2 - {}^-9$
h $3 - {}^-5$
i $3 - {}^-7$
j ${}^-2 - {}^-5$
k ${}^-2 - {}^-8$
l ${}^-2 - 2$
m ${}^-2 - 4$

You could use a number line to help.

6 a ${}^-3 + 4$
b ${}^-1 + 4$
c ${}^-2 + 8$
d $2 + {}^-5$
e ${}^-1 + {}^-4$
f $7 - 8$
g $2 - 4$
h ${}^-3 - 1$
i ${}^-1 - {}^-2$
j ${}^-8 - 9$
k ${}^-16 - {}^-12$
l ${}^-18 + 7$
m ${}^-37 + 26$
n ${}^-17 - {}^-36$
o ${}^-142 + 142$
p ${}^-83 + 90$
q ${}^-47 + 60$
r ${}^-24 - {}^-40$
s ${}^-98 - {}^-100$
t ${}^-11 + 14 - {}^-7$
u $18 + {}^-7 - {}^-10$
v ${}^-24 + {}^-16 - 10$
w $30 - {}^-18 - 50$
x ${}^-11 + {}^-16 - {}^-11$

7 a ${}^-1{\cdot}2 + 3{\cdot}4$
b $3 - 3{\cdot}6$
c ${}^-5{\cdot}2 + 3{\cdot}2$
d $4{\cdot}5 + {}^-2{\cdot}5$
e ${}^-3{\cdot}4 + {}^-2{\cdot}1$
f $3{\cdot}6 - {}^-2{\cdot}4$
g ${}^-2{\cdot}9 - {}^-1{\cdot}3$
h ${}^-6{\cdot}4 - {}^-2{\cdot}9$
i ${}^-8{\cdot}4 - 3{\cdot}7$
j ${}^-6{\cdot}3 - {}^-8{\cdot}7$
k $5{\cdot}2 - 4{\cdot}7 - 5{\cdot}2$
l $8{\cdot}6 + {}^-5{\cdot}1 - 3{\cdot}7$
m ${}^-3{\cdot}7 + {}^-3{\cdot}9 + 1{\cdot}4$
n ${}^-3{\cdot}1 - 2{\cdot}4 + 1{\cdot}6$

8 Use your calculator to find the answers to these.

a $128 - 864$
b ${}^-237 + 149$
c $422 - {}^-587$
d $279 + {}^-384$
e $641 - {}^-237$
f ${}^-386 + {}^-249$
g $241 - 796$
h ${}^-384 + {}^-279$
i ${}^-814 - {}^-1024$
j $683 - {}^-719 - 412$

Remember ⁻7 is keyed as (−) 7

9 A diver was at ⁻64 feet. A shark was 17 feet below him.
The diver rose 15 feet and the shark rose 9 feet.
How far apart are the diver and the shark now?

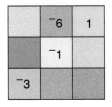

T

10 Use a copy of this magic square.

Remember Each row, each column and each diagonal must add to the same number.

a What does the diagonal ⁻3, ⁻1, 1 add up to?
b Complete the magic square.

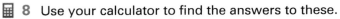

11

a   Choose a number card to give the answer ⁻19. $^-4 - \square = {}^-19$

b   Choose a number card to give the answer 7. $^-4 - \square = 7$

c   Choose two number cards to give the answer ⁻4. $\square - \square = {}^-4$

d   Choose another two number cards to give the answer ⁻4.

e   Choose three different number cards so this
    answer is as large as possible. $\square + \square - \square = \underline{\quad}$

12 Use a copy of this.
    The numbers in the squares add to the number in the
    circle between them.
    a   Find the numbers in the circles.
    b   What is the total of the numbers in the circles?
    c   What is the total of the numbers in the squares?
    d   Subtract the total in the squares from the total in the
        circles.

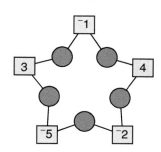

13 Use a copy of this.

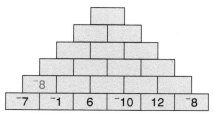   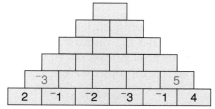

On this green pyramid, each pair
of numbers is added to get the
number above.

On this blue pyramid, the number
on the left is subtracted from the
number on the right to get the
number above.

a   What number is at the top of the green pyramid?
b   What number is at the top of the blue pyramid?

14 Use a copy of this.

a	⁻2	3	⁻1	4	⁻3	5	⁻6	1	⁻7	⁻4
b	3	⁻1	4	⁻2	5	⁻3	2	⁻4	⁻2	⁻5
a–b	⁻5	4								

The numbers in the second row are subtracted from the numbers in the first row, to get
the third row.
a   Copy and complete this list of numbers for the third row.
    ⁻5, 4, ___, ___, ___, ___, ___, ___, ___, ___
b   Find the sum of the numbers in the first row.
c   Find the sum of the numbers in th second row.
d   Explain how you can find the sum of the numbers in the third row without adding them.

15 What numbers are Tania and Gareth thinking of?

a

b

16 a Two negative numbers are added.
The answer is ⁻18.
What might the numbers be?

b The difference between
two numbers is ⁻15.
What might the numbers be?

17 Find the difference between these.
  a Absolute zero (⁻273°C) and the freezing point of hydrogen (⁻194°C).
  b The temperature at which dry ice evaporates (⁻73°C) and human body temperature (37°C).
  c The height of the highest mountain in Africa, Kilimanjaro 5642 m, and the depth of the deepest trench in the Atlantic Ocean, 9220 m below sea level.
  ＊**d** The founding of the city of Rome (753 BC) and the year Great Britain introduced the Gregorian calendar (AD 1752). Remember there is no year 0.

The Gregorian calendar is the one we use today.

T

＊**18** Use two copies of this pyramid.
  a On this pyramid, each pair of numbers is added to get the number above.
    Fill it in so the top number is ⁻86.
  b On this pyramid the number on the right is subtracted from the number on the left to get the number above.
    Fill it in so the top number is ⁻86.

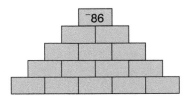

Review 1
a ⁻24 + 31   **b** ⁻14 + 11   **c** 4·6 + ⁻5·2   **d** 8 − ⁻4·7   **e** ⁻37 + 24
f ⁻84 − ⁻60   **g** ⁻167 + 167   **h** ⁻100 − ⁻84   **i** 4 + 6 − 10 − 8   **j** ⁻12 + ⁻4 − 1
k ⁻17 − ⁻9 − 23   **l** 3 + 5 − 16 − ⁻12   **m** ⁻30 + ⁻60 − ⁻15   **n** ⁻11 + 14 − ⁻12 + 17

Review 2 Use your calculator to find the answers to these.
a ⁻387 + ⁻472   **b** 842 + ⁻981   **c** 347 − 572   **d** ⁻517 − 412   **e** ⁻317 − ⁻281
f ⁻527 − ⁻862   **g** ⁻394 + ⁻472   **h** ⁻419 − ⁻153   **i** ⁻417 + ⁻211 − ⁻186

T

Review 3

a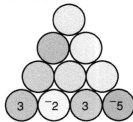

b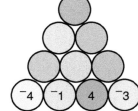

c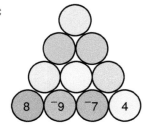

Use a copy of this.
Add two numbers to get the number above. The second row of **a** will be .
What number goes in the top circle?

T **Review 4** Use another copy of the circles in **Review 3**.

Subtract the number on the right from the number on the left to get the number above.

The second row of **a** will be

What number goes in the top circle?

T **Review 5** Use two copies of this.

a Add two numbers to get the number above. Fill in the circles so that the top number is ⁻111.

b Subtract the number on the right from the number on the left to get the number above. Fill in the circles so that the top number is ⁻111.

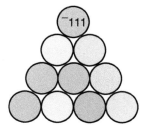

Integers – a game for a group or class

You will need to draw a diagram like this for each turn.

To play Choose a leader.
The leader calls out four integers between ⁻10 and 10.
As each one is called, put it in one of your blue boxes.
Whoever gets the biggest total is the leader for the next turn.

Practical

Make up a game that uses negative numbers.
Write clear instructions for your game.
Play your game to make sure it works.

Puzzle

1 Each line of circles adds to the number given in red.
Replace the letters with numbers to give the totals shown.

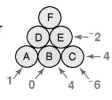

2 This is a magic square. Each row, column and diagonal adds to the same total. What might the missing numbers be?

		⁻2
	⁻5	
		⁻6

Multiplying and dividing integers

Discussion

Look for patterns in the rows and columns.

- Use a copy of this table.
  Discuss how to fill it in.
  Shade the positive answers one colour, the negative answers another colour and zeros another colour.

×	⁻3	⁻2	⁻1	0	1	2	3
3	⁻9	⁻6	⁻3	0	3	6	9
2				0	2	4	6
1				0	1	2	3
0				0	0	0	0
⁻1							⁻3
⁻2							⁻6
⁻3							⁻9

  Use the table to find the answers to these.
  $3 \times {}^-2$, $2 \times {}^-1$, ${}^-3 \times 2$, ${}^-2 \times 3$

  What do you think the answers to these would be?
  Discuss.
  $4 \times {}^-3$, $5 \times {}^-2$, ${}^-2 \times 6$, ${}^-3 \times 5$, ${}^-6 \times 4$

  Use the table to find the answers to these.
  ${}^-3 \times {}^-2$, ${}^-2 \times {}^-2$, ${}^-3 \times {}^-1$, ${}^-2 \times {}^-1$

  What do you think the answers to these would be? **Discuss**.
  ${}^-3 \times {}^-4$, ${}^-5 \times {}^-4$, ${}^-3 \times {}^-6$

- Katie wrote this.

  Multiplying a positive number and a negative number gives a negative number.

  Examples ${}^-2 \times 3 = {}^-6$ $2 \times {}^-5 = {}^-10$

  Multiplying two negative numbers gives a positive number.

  Examples ${}^-3 \times {}^-2 = 6$ ${}^-2 \times {}^-1 = 2$

  Is Katie correct? **Discuss**.

Multiplying and dividing are inverse operations.

Dividing by ⁻2 is the inverse of multiplying by ⁻2.

Inverse operation means one undoes the other.

$4 \times {}^-2 = {}^-8$ so $\dfrac{{}^-8}{{}^-2} = 4$ ${}^-6 \times {}^-2 = 12$ so $\dfrac{12}{{}^-2} = {}^-6$

Worked Example
How many ⁻3s are in ⁻6?

Answer

This can be written as ${}^-6 \div {}^-3$ or $\dfrac{{}^-6}{{}^-3}$. $\dfrac{{}^-6}{{}^-3} = 2$ since ${}^-6 = {}^-3 \times 2$.

The answer is **2**.

Exercise 2

1 The 3 times table is on the left.
Copy and complete these multiplication tables.

3 × 1 = 3	3 × ⁻1 = ⁻3	⁻3 × 1 = ⁻3	⁻3 × ⁻1 =
3 × 2 = 6	3 × ⁻2 = ⁻6	⁻3 × 2 = ⁻6	⁻3 × ⁻2 =
3 × 3 = 9	3 × ⁻3 = ⁻9	⁻3 × 3 =	⁻3 × ⁻3 =
3 × 4 =	3 × ⁻4 =	⁻3 × 4 =	⁻3 × ⁻4 =
3 × 5 =	3 × ⁻5 =	⁻3 × 5 =	⁻3 × ⁻5 =
3 × 6 =	3 × ⁻6 =	⁻3 × 6 =	⁻3 × ⁻6 =
3 × 7 =	3 × ⁻7 =	⁻3 × 7 =	⁻3 × ⁻7 =
3 × 8 =	3 × ⁻8 =	⁻3 × 8 =	⁻3 × ⁻8 =
3 × 9 =	3 × ⁻9 =	⁻3 × 9 =	⁻3 × ⁻9 =
3 × 10 =	3 × ⁻10 =	⁻3 × 10 =	⁻3 × ⁻10 =

2 Write down the next three lines of these.

a
$2 \times 4 = 8$
$1 \times 4 = 4$
$0 \times 4 = 0$
$^-1 \times 4 = {}^-4$
$^-2 \times 4 = {}^-8$

b
$2 \times {}^-4 = {}^-8$
$1 \times {}^-4 = {}^-4$
$0 \times {}^-4 = 0$
$^-1 \times {}^-4 = 4$
$^-2 \times {}^-4 = 8$

c
$^-2 \times 4 = {}^-8$
$^-1 \times 4 = {}^-4$
$0 \times 4 = 0$
$1 \times 4 = 4$
$2 \times 4 = 8$

d
$^-2 \times {}^-4 = 8$
$^-1 \times {}^-4 = 4$
$0 \times {}^-4 = 0$
$1 \times {}^-4 = {}^-4$
$2 \times {}^-4 = {}^-8$

3 **a** $3 \times {}^-5$ **b** $4 \times {}^-6$ **c** $5 \times {}^-3$ **d** $^-2 \times 8$ **e** $^-4 \times 7$
f $^-3 \times 9$ **g** $^-2 \times 9$ **h** $^-3 \times {}^-4$ **i** $^-4 \times {}^-5$ **j** $^-4 \times {}^-8$
k $^-10 \times 3$ **l** $10 \times {}^-3$ **m** $^-10 \times {}^-3$

T 4 Use a copy of this.

a **b** **c** **d**

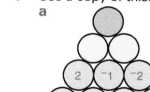 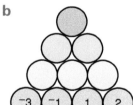 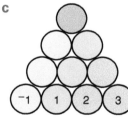 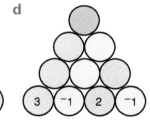

Each pair of numbers is multiplied to get the number above.
What numbers go in the red circles?

T 5 Use a copy of these multiplication squares.
Fill them in.

a

×	2	⁻5	⁻3	4
⁻1	2	5		
3	6			
⁻2				
5				

b

×	⁻1	2	3	⁻4
3				
⁻1				
⁻4				
⁻6				

c

×	⁻6	2	⁻1	8
⁻4				
7				
⁻6				
⁻1				

d

×	⁻3	7	⁻2	⁻8
6				
⁻1				
⁻7				
2				

6 $3 \times {}^-2 = {}^-6$ $^-2 \times 3 = {}^-6$ $^-6 \div 3 = {}^-2$ $^-6 \div {}^-2 = 3$
These four facts all use the numbers 3, ⁻2 and ⁻6.

Write three more facts using the numbers given in these.
a $4 \times {}^-1 = {}^-4$ **b** $^-9 \times 2 = {}^-18$ **c** $^-7 \times {}^-3 = 21$

7 Use multiplication tables to help answer these.

a $^-9 \div 3$ b $6 \div {}^-2$ c $^-4 \div 2$ d $6 \div {}^-3$ e $^-9 \div {}^-3$

f $^-8 \div {}^-2$ g $^-6 \div {}^-3$ h $8 \div {}^-4$ i $^-8 \div 2$ j $^-6 \div {}^-2$

k $^-8 \div 4$ l $^-2 \div {}^-1$ m $^-4 \div 1$ n $4 \div {}^-1$

8 a $24 \div {}^-3$ b $32 \div {}^-8$ c $^-16 \div 2$ d $^-18 \div 3$ e $^-24 \div {}^-6$

f $^-50 \div {}^-5$ g $^-27 \div 3$ h $28 \div {}^-7$ i $\frac{^-24}{6}$ j $\frac{^-56}{8}$

k $\frac{36}{9}$ l $\frac{^-35}{7}$ m $\frac{^-40}{^-5}$

T

9 Use a copy of this.

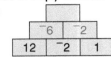 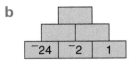 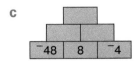 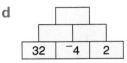

a   b   c   d

In each pair of numbers, the one on the left is divided by the one on the right to get the number above.

What numbers go in the top squares?

10 a How many $^-2$s are in $^-6$?   b How many $^-3$s are in $^-12$?

c How many $^-5$s are in $^-30$?   d How many $^-4$s are in $^-48$?

e How many $^-1$s are in $^-9$?

T

11 Use a copy of this.

How might this multiplication square be completed?

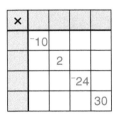

12 Make up some multiplications and some divisions with Brad's answer.

Brad

T

Review 1 Use a copy of this.

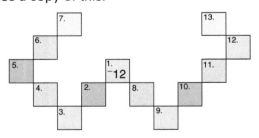

Use the clues to find the numbers in the yellow squares.

Clues

Square 2. The answer to **1** divided by 2.   Square 8. The answer to **1** multiplied by $^-2$.

Square 3. The answer to **2** divided by $^-2$.   Square 9. The answer to **8** divided by $^-6$.

Square 4. The answer to **3** multiplied by $^-1$.   Square 10. The answer to **9** divided by 4.

Square 5. The answer to **4** multiplied by $^-5$.   Square 11. The answer to **10** multiplied by 3.

Square 6. The answer to **5** divided by 3.   Square 12. The answer to **11** multiplied by $^-4$.

Square 7. The answer to **6** multiplied by $^-4$.   Square 13. The answer to **12** divided by $^-3$.

Review 2

a	$^-6 \times ^-4$	b	$6 \times ^-7$	c	$^-5 \times 8$	d	$^-3 \times ^-7$
e	$9 \times ^-3$	f	$^-8 \times 8$	g	$7 \times ^-1$	h	$^-4 \times ^-1$
i	$^-24 \div 4$	j	$^-30 \div ^-5$	k	$42 \div ^-7$	l	$\frac{25}{5}$
m	$\frac{^-36}{4}$	n	$\frac{^-18}{^-9}$				

Review 3

a How many $^-4$s are in $^-28$?

b $^-8 \times 3 = ^-24$. Use the numbers $^-8$, 3 and $^-24$ to write down three more facts.

Divisibility

T

Remember
The rules for divisibility are on page 3.

345 pupils at Buttersfield School were to travel by 15-seater mini-buses to a sports day.
Mr Cameron wanted to know if 345 is divisible by 15.

For **divisibility by larger numbers** we can check for divisibility by factors.
Mr Cameron checked if 345 is divisible by 5 and then 3.
345 is divisible by 5 because the last digit is 5.
It is also divisible by 3 because $3 + 4 + 5 = 12$ and 12 is divisible by 3.
345 is divisible by 15.

Example   To check if 1362 is divisible by 18, check for divisibility by 2 and 9.
1362 is divisible by 2 but not by 9.   1362 is not divisible by 18.

$15 = 3 \times 5$
$18 = 2 \times 9$

Discussion

- Damion said 410 is divisible by 20 because 410 is
  divisible by both 2 and 10.
  Zenta said 410 is not divisible by 20 because 410 is
  divisible by 5 but not by 4.
  Damion used $20 = 2 \times 10$.
  Zenta used $20 = 4 \times 5$.
  Zenta is correct and Damion is wrong.
  Why do Zenta's numbers work but Damion's don't?
  Discuss.

Hint:
Think about
common factors.

- To check if 258 is divisible by 12, Josh checked with 3 and 4.
  Diana checked with 2 and 6.
  Whose numbers are the right ones to use? Why? **Discuss**.

Exercise 3

1

| 1278 | 3600 | 4536 | 6804 | 12 645 | 24 056 |

Which numbers in the box are divisible by
a 2   b 3   c 4   d 5   e 8   f 9   g 10?

2 a Caroline wanted to know if the number of stamps she has is divisible by 24.
Which numbers would you use to check for divisibility by 24?
Explain.
A 3 and 8   B 4 and 6
b Angela wanted to know if her page would divide into 40 mm wide columns.
Which numbers would you use to check for divisibility by 40?
Explain.
A 4 and 10   B 5 and 8

3 Which of the numbers in question **1** are divisible by
a 6   b 15   c 12   d 18   e 20   f 24   g 40?

Use your answers to *question 1* to help.

4 Which of
a	516	2346	3468	4332	10 736	are divisible by 12
b	1782	2680	3510	8465	14 715	are divisible by 15
c	1008	3564	6570	12 510	14 061	are divisible by 18
d	980	1460	2330	5906	19 470	are divisible by 20
e	1659	2352	5646	16 480	23 568	are divisible by 24
f	1360	2548	4615	58 360	815 815	are divisible by 40?

*5

| 10 488 | 14 586 | 38 808 | 40 500 |

This could take time.

Choose a number from this list for each gap. Use each number only once.
a _____ is divisible by 6.   b _____ is divisible by 15.
c _____ is divisible by 18.   d _____ is divisible by 24.

*6 A school of over 300 pupils is able to line up in complete rows of 6, 8 or 9.
What is the smallest possible number of pupils at this school?

Review 1   Which numbers would you use to check divisibility by 18? Explain.
A 3 and 6   B 2 and 9

*Review 2

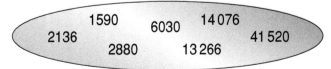

1590	6030	14 076
2136		41 520
2880	13 266	

Which of the numbers in the ring are divisible by
a 6   b 15   c 12   d 18   e 20   f 24   g 30?

Review 3   A drama group, with over 120 members, can split into equal groups of 4, 6 or 9.
What is the smallest possible number of members in the group?

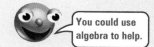

Investigation

Consecutive numbers

22, 23, 24, 25, 26 are five consecutive numbers.

22 + 23 + 24 + 25 + 26 = 120
$$= 5 \times 24$$

The sum of these five numbers is divisible by 5.

Choose other sets of five consecutive numbers.
Show that the sums are divisible by 5.

*Prove that the sum of *any* five consecutive numbers is divisible by 5.

> You could use algebra to help.

*Investigation

Divisibility by 7 and 11

1 To test whether 6146 is divisible by 7, follow these steps.

Step 1 Remove the units digit (6).

Step 2 Subtract twice this digit (2 × 6 = 12).

```
 614
- 12
 602
```

Repeat *Steps 1* and *2*. Remove the units digit.
          Subtract twice this digit (2 × 2 = 4).

```
 60
- 4
 56
```

56 is divisible by 7. So 6146 is divisible by 7.

The numbers in this list are all divisible by 7.
Test to see if this method works for them.

> 0 is divisible by 7.

| 112 | 154 | 182 | 245 | 511 | 994 | 1078 | 1274 | 1757 | 2051 |
| 2464 | 2513 | 2520 | 6223 | 6447 | 11823 | 16506 |

Investigate to find if this method works for other numbers that are divisible by 7.

2 A number is divisible by 11 if the sum of the numbers made by *pairing* from the right is divisible by 11.

Examples   572 paired and added is 72 + 5 = 77 which is divisible by 11.
            284 669 paired and added is 69 + 46 + 28 = 143.
            143 paired and added is 43 + 1 = 44 which is divisible by 11.

There is another way of testing for divisibility by 11.
Investigate to find this way.

> Hint: Look at every second digit.

Prime numbers

T

Remember
A **prime number** has exactly two factors, itself and 1.

Examples   5, 11, 19, 37
1 is *not* a prime number.

The prime numbers less than 30 are 2, 3, 5, 7, 11, 13, 17, 19, 23, 29.

The only rectangle that can be drawn using a prime number of dots is a straight line.

Example   11   • • • • • • • • • • •

35   35 can be divided by 5 and 7 as well as 1 and 35.
35 is **not** a prime number.

To **test if a number is prime**, try to divide it by each of the prime numbers in turn.

Example   To test if 149 is prime, try dividing it by 2, 3, 5, 7 and 11.

$149 ÷ 2$ 74.5
$149 ÷ 3$ 49.66666667
$149 ÷ 5$ 28.8

Discussion

In the example above how do we know that we can stop testing at 11? **Discuss**.

Exercise 4

1 Which of these numbers are prime? You could test by trying to draw a rectangle of dots that is not just a straight line.
   a 17   **b** 27   **c** 29   **d** 31   **e** 39   **f** 43   **g** 51

2 Test whether these numbers are prime.
   a 33   **b** 41   **c** 45   **d** 47   **e** 51   **f** 52   **g** 157
   h 163   **i** 273   **j** 378   **k** 487   **l** 587   **m** 3917

3 A lawn has an area of 41 m². What must its length and breadth be if they are both whole numbers?

4 Explain why 2 is the only even prime number.

5 Two prime numbers are added.
   The answer is 50.
   What could the numbers be?

   Find all sets of answers.

41m²

6 Find pairs of primes with a difference of 4.

Review 1 Test whether these numbers are prime.

a 57   b 87   c 107   d 155   e 269   f 383

Review 2 Two prime numbers are added.
The answer is 36.
What could the numbers be?

Find all possible answers.

Practical

There is a lot of information about prime numbers on the Internet.
Use the Internet to write a project on primes.

You could find out about
● primes and codes
● history of primes
● largest known primes
● formulae for finding primes and the exceptions to these etc.

Investigation

Primes

1 **Investigate** whether these are true or false. Draw a table to help.
    a Every even number greater than 4 is the sum of two prime numbers.
    b Every odd number greater than 7 is the sum of three prime numbers.

These are called Goldbach's conjectures.

2 Primes with a difference of two are called *Twin Primes*.
Use a copy of this table and finish filling it in.

Twin Primes	Sum	Product
3 and 5	8	15
5 and 7	12	35
11 and 13	24	143

Hint   15 is 16 − 1
        35 is 36 − 1
       143 is 144 − 1

What do you notice about the sums? **Investigate**.
What do you notice about the products? **Investigate**.

3 Between every number greater than 1, and its double, there is at least one prime number
Investigate to see if this is true. For example, 6 is double 3. Between 3 and 6 there is the prime number, 5.

4 Every prime number greater than 3 is one more than or one less than a multiple of __.
Investigate to find what goes in the gap.

5 **Emirp** numbers are prime numbers that give a different prime number when the digits are written in reverse order.
37 and 73 are emirp numbers.
Investigate to find all the 2-digit emirp numbers.
* Are there any 3-digit emirp numbers? **Investigate**.

? Puzzle

1 Two different prime numbers, each less than 100, are multiplied.
The answer ends in 2.
What could the numbers be?

Find the five possible sets of answers.

* 2 Two different 2-digit prime numbers are multiplied.
The answer ends in 1.
What could the numbers be?

Find the 50 possible sets of answers.

* 3 Two 2-digit prime numbers are multiplied.
The answer is 4717.
Which two prime numbers were multiplied?

Factors

[T]

Remember
A **prime factor** is a factor that is a prime number.

Example   The factors of 12 are 1, 2, 3, 4, 6 and 12.
Of these, 2 and 3 are prime numbers.
The prime factors of 12 are 2 and 3.

12 can be written as a **product of prime factors**.
$12 = 2 \times 2 \times 3 = 2^2 \times 3$

In 2^2 the little 2 is called an index.

We can use a **table** or **factor tree** to help us write a number as a product of prime factors.

Example

2	180
2	90
3	45
3	15
	5

or

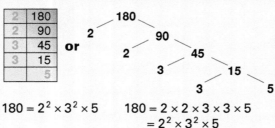

Divide by 2, the smallest prime that divides into 180.
Divide by 2, the smallest prime that divides into 90.
Divide by 3, the smallest prime that divides into 45.
Divide by 3, the smallest prime that divides into 15.

$180 = 2^2 \times 3^2 \times 5$

$180 = 2 \times 2 \times 3 \times 3 \times 5$
$= 2^2 \times 3^2 \times 5$

Exercise 5

1 Use a table to write these numbers as a product of prime factors.
a 88   b 148   c 168   d 300   *e 396   *f 4680

2 What are the missing numbers, A, B and C, on these factor trees?

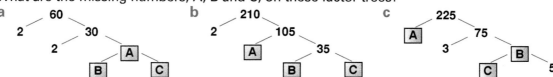

3 Copy these factor trees.
Fill in the numbers in the boxes.

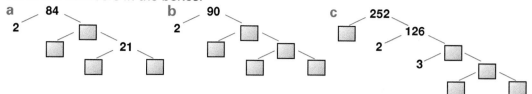

4 Use the factor trees in question **3** to write each of these as a product of prime factors.
a 84 b 90 c 252

5 Use tables or factor trees to help you write these as products of prime factors.
a 110 b 45 c 70 d 100 e 350 f 93 *g 1224 *h 1725 *i 2580

Review 1 Use a table to write these numbers as a product of prime factors.
a 24 b 120 c 198 d 144 *e 1836

Review 2 What are the missing numbers, A, B, C and D, on these factor trees?

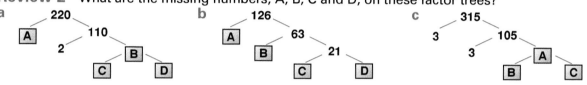

Review 3 Use a table or factor tree to write each of these as a product of prime factors.
a 220 b 126 c 315 *d 702 *e 812 *f 1755

 Puzzle

1 There are just four numbers less than 100, with exactly three factors.
One of these is 9. What are the other three?
What do you notice about these four numbers?

Think about prime numbers.

2 There are eight 2-digit numbers that have three prime factors.
One of these is 30, since 2, 3 and 5 are the prime factors of 30.
What are the other seven numbers?

Investigations

Factors

1 The sum of the digits in 42 is 6 and 6 is a factor of 42.
How many 2-digit numbers can you find like this?
Investigate.

Score	23	Excellent
	17–22	Very good
	10–16	Good
	less than 16	Keep trying

2 Marty said that 84 was the only 2-digit number with exactly 12 factors.
Are there any other 2-digit numbers with exactly 12 factors?
Investigate.

3 Marty discovered that the sum of the digits of 84 is 12 *and* 84 has exactly 12 factors.
Are there any other 2-digit numbers like this?
Investigate.

∗4 Take any 2 by 2 square like this.

$15 \times 24 = 360$
$20 \times 18 = 360$

The products of the numbers on opposite corners are the same.

What if you choose another square?

Prove that it is always true.

Hint Use factors.

Try 3 by 3 squares or even bigger ones.
Try rectangles instead of squares.

Use algebra to help prove or disprove this.

×	1	2	3	4	5	6	7	8	9	10
1	1	2	3	4	5	6	7	8	9	10
2	2	4	6	8	10	12	14	16	18	20
3	3	6	9	12	15	18	21	24	27	30
4	4	8	12	16	20	24	28	32	36	40
5	5	10	15	20	25	30	35	40	45	50
6	6	12	18	24	30	36	42	48	54	60
7	7	14	21	28	35	42	49	56	63	70
8	8	16	24	32	40	48	56	64	72	80
9	9	18	27	36	45	54	63	72	81	90
10	10	20	30	40	50	60	70	80	90	100

Highest common factor and lowest common multiple

The **HCF (highest common factor)** of two numbers is the largest factor common to both.
The factors of 18 are 1, 2, 3, 6, 9, 18.
The factors of 30 are 1, 2, 3, 5, 6, 10, 15, 30.
The HCF of 18 and 30 is 6.

We use HCF when cancelling fractions. See page 116.

The **LCM (lowest common multiple)** of two numbers is the smallest number that is a multiple of both.
The multiples of 9 are 9, 18, 27, 36, 45, 54, 63, ...
The multiples of 15 are 15, 30, 45, ...
The LCM of 9 and 15 is 45.

We can use prime factors to find the HCF and LCM.

Example
$504 = 2 \times 2 \times 2 \times 3 \times 3 \times 7$
$700 = 2 \times 2 \times 5 \times 5 \times 7$
HCF of 504 and 700 $= 2 \times 2 \times 7$
$= \mathbf{28}$.
LCM of 504 and 700 $= 2 \times 2 \times 2 \times 3 \times 3 \times 5 \times 5 \times 7$
$= \mathbf{12600}$

$2 \times 2 \times 7$ is in both 504 and 700.

These are the factors left once the HCF is taken out.

We could write 504 as $2^3 \times 3^2 \times 7$ and 700 as $2^2 \times 5^2 \times 7$.
The HCF of 504 and 700 could be written as $2^2 \times 7$.
The LCM could be written as $2^3 \times 3^2 \times 5^2 \times 7$.

We can use a **diagram** to find the HCF and LCM.

1 Write the numbers that are prime factors of both 504 and 700 here.
2 Write the other prime factors of 504 in the 504 circle.
3 Write the other prime factors of 700 in the 700 circle.

504   700

2 3   2
       2   5
  3   7   5

To find the HCF, multiply the numbers where the circles overlap. HCF $= 2 \times 2 \times 7$.
To find the LCM, multiply all the numbers on the diagram. LCM $= 2 \times 3 \times 3 \times 2 \times 2 \times 7 \times 5 \times 5$.

Exercise 6

1 Copy and complete the diagrams.
  a $84 = 2 \times 2 \times 3 \times 7$
      $120 = 2 \times 2 \times 2 \times 3 \times 5$

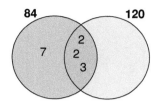

  b $144 = 2 \times 2 \times 2 \times 2 \times 3 \times 3$
      $150 = 2 \times 3 \times 5 \times 5$

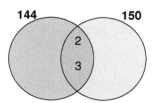

  c $180 = 2 \times 2 \times 3 \times 3 \times 5$
      $770 = 2 \times 5 \times 7 \times 11$

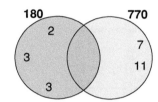

  d $630 = 2 \times 3 \times 3 \times 5 \times 7$
      $720 = 2 \times 2 \times 2 \times 2 \times 3 \times 3 \times 5$

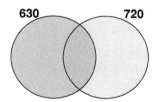

2 Use your diagrams from question **1** to find the HCF of
  a 84 and 120   **b** 144 and 150   **c** 180 and 770   **d** 630 and 720.

3 Use your diagrams from question **1** to find the LCM of
  a 84 and 120   **b** 144 and 150   **c** 180 and 770   **d** 630 and 720.

4 Draw diagrams like the ones in question **1**, for these pairs of numbers. Find the HCF.
  a 60 and 63   **b** 48 and 72   **c** 75 and 90   **d** 126 and 300   **e** 275 and 525.

5 Find the LCM of the pairs of numbers in question **4**.

Write as products of
prime factors first.

6 Cancel these fractions, by finding the HCF.
  a $\frac{64}{120}$ **b** $\frac{75}{200}$ **c** $\frac{130}{455}$ **d** $\frac{180}{405}$ **e** $\frac{120}{264}$

*7 $400 = 2 \times 2 \times 2 \times 2 \times 5 \times 5$ $360 = 2 \times 2 \times 2 \times 3 \times 3 \times 5$
        $= 2^4 \times 5^2$ $= 2^3 \times 3^2 \times 5$
  HCF of 400 and 360 is $2^3 \times 5$. LCM of 400 and 360 is $2^4 \times 3^2 \times 5^2$.
  Write the HCF and LCM of these in the same way.
  a   144 and 216   **b**   180 and 600   **c**   90 and 165   **d**   252 and 294
  e   450 and 1500   **f**   462 and 490   **g**   1800 and 6000   **h**   2450 and 2695
  *i   1305 and 5075   *j   3157 and 2255

Review 1   What is the HCF of the pairs of numbers shown in red?

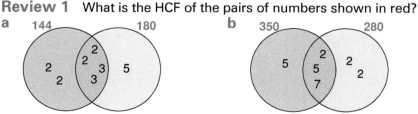

Review 2   Find the LCM of the pairs of numbers in Review **1**.

T | **Review 3** Use a copy of this.

Across
1. LCM of 21 and 28
2. LCM of 84 and 126
4. HCF of 720 and 648
6. HCF of 300 and 990
7. HCF of 168 and 196
10. HCF of 132 and 330
11. LCM of 264 and 72

Down
1. LCM of 120 and 168
2. HCF of 110 and 132
3. HCF of 144 and 120
4. LCM of 150 and 125
5. HCF of 216 and 420
8. HCF of 900 and 792
9. HCF of 756 and 1485

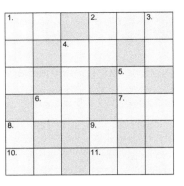

Squares, square roots, cubes, cube roots

Remember

3^2 is read as 'three **squared**' and means 3×3.

$\sqrt{16}$ is read as 'the **square root** of 16'.
To find $\sqrt{16}$ we find what number squared equals 16.
$4 \times 4 = 16$ and also $^-4 \times {}^-4 = 16$
4 and $^-4$ are both square roots of 16.   4 is the positive square root.
$\sqrt{16} = 4$ or $^-4$.

> Squaring and finding the square root are inverse operations.
>
> $3 \rightarrow \boxed{\text{square}} \rightarrow 9$
> $3 \leftarrow \boxed{\text{square root}} \leftarrow 9$

When we are asked for the square root of a number we usually just give the positive square root, but you need to know that there is also a negative square root.

Examples $\sqrt{49} = 7$ or $^-7$ $\sqrt{144} = 12$ or $^-12$

8^3 is read as 'eight **cubed**' and means $8 \times 8 \times 8$.
$1^3 = 1$, $2^3 = 8$, $3^3 = 27$, $4^3 = 64$, $5^3 = 125$, $10^3 = 1000$
1, 8, 27, 64, 125 and 1000 are **cube numbers**.

> Cubing and finding the cube root are inverse operations.
>
> $4 \rightarrow \boxed{\text{cubed}} \rightarrow 64$
> $4 \leftarrow \boxed{\text{cube root}} \leftarrow 64$

$\sqrt[3]{27}$ is read as 'the **cube root** of 27'.

To find $\sqrt[3]{27}$ we ask 'what number cubed equals 27?'

$3 \times 3 \times 3 = 27$ so $\sqrt[3]{27} = 3$.

$\sqrt[3]{1} = 1$ $\sqrt[3]{8} = 2$ $\sqrt[3]{27} = 3$ $\sqrt[3]{64} = 4$ $\sqrt[3]{125} = 5$ $\sqrt[3]{1000} = 10$

In 5^2 and 8^3, the small 2 and 3 are called **indices**. Indices is the plural of **index**.

Exercise 7

1 Write these using indices.
   a 6×6 **b** 4×4 **c** $5 \times 5 \times 5$ **d** 2×2 **e** $8 \times 8 \times 8$ **f** $12 \times 12 \times 12$

2 Find the missing numbers.
   a The square root of 36 is ___.
   b 4 is the square of ___.
   c The square root of ___ is 7.
   d ___ is the square of 9.
   e 12 is the square root of ___.
   f ___ is the square root of 1.
   g The square of 10 is ___.
   h 5 is the square root of ___.
   i ___ is the square of 11.

> Give the positive square root only.

3 Find

a 2^2 b 7^2 c 8^2 d 12^2 e 9^2
f $\sqrt{36}$ g $\sqrt{64}$ h $\sqrt{144}$ i $\sqrt{121}$ j $\sqrt{1}$

4 Find the missing numbers.

a 3 cubed equals ___.   b 64 is the cube of ___.
c ___ is the cube of 3.   d 1000 is the cube of ___.
e 4 cubed equals ___.   f 125 is the cube of ___.
g The cube root of 27 is ___.   h The cube root of 64 is ___.
i ___ is the cube root of 8.   j ___ is the cube root of 27.
k The cube root of 1 is ___.   l The cube root of 64 is ___.
m ___ is the cube root of 1000.

5 a Find the next three lines of this number pattern.

$$1^2 = 1$$
$$2^2 = 1 + 3$$
$$3^2 = 1 + 3 + 5$$

b What goes in the gap?
27^2 is equal to the sum of the first ____ odd numbers.

6 Which is bigger, the sum of the squares of 3 and 5, or the square of the sum of 3 and 5?

7 Which is bigger, the square root of the sum of 9 and 16, or the sum of the square roots of 9 and 16?

8 a $\sqrt[3]{8}$ b 2^3 c $\sqrt[3]{64}$ d 3^3 e $\sqrt[3]{1000}$
f 5^3 g 1^3 h $\sqrt[3]{125}$ i $\sqrt[3]{1}$ j 4^3

9 There is just one 2-digit number that is both a square number and a cube number. Which 2-digit number is this?

10 Belinda wrote this number pattern. The last two lines are smudged. What are they?

$$1^3 = 1$$
$$2^3 = 3 + 5$$
$$3^3 = 7 + 9 + 11$$

Review 1 Write these using indices.
a 1×1 b $7 \times 7 \times 7$ c 9×9 d $11 \times 11 \times 11$ e 6×6

Review 2 Find
a 8^2 b 9^2 c 7^2 d 1^3 e 2^3 f 8^2
g 10^2 h 10^3 i 5^3 j 6^2 k 4^3.

Review 3 Find
a $\sqrt{49}$ b $\sqrt{100}$ c $\sqrt[3]{27}$ d $\sqrt[3]{8}$ e $\sqrt{81}$ f $\sqrt[3]{64}$.

Review 4 Which is smaller, the sum of the squares of 5 and 6, or the square of the sum of 5 and 6?

Puzzle

1 a The square number 100 can be written as the sum of two prime numbers in six different ways. One way is 100 = 59 + 41.
Find the other five ways.

   b Which of the square numbers from 1 to 144 *cannot* be written as the sum of two prime numbers?

2 17, 41, 53 and 97 can be written as the sum of two square numbers.
17 = 1 + 16   41 = 16 + 25   53 = 4 + 49   97 = 16 + 81
Another six 2-digit primes can be written in the same way.
Find these six numbers.

Investigation

* Sum and difference of two squares

1 $65 = 4^2 + 7^2$

   Which 1-digit and 2-digit numbers can be written as the sum of two square numbers? Are there any patterns in the numbers that you can write like this? **Investigate**.

2 $28 = 8^2 - 6^2$

   Which numbers between 1 and 30 can be written as the difference of two square numbers? **Investigate**.

$4 \times 4 = 16,$ $\qquad$ $0.4 \times 0.4 = 0.16$ $\qquad$ $\sqrt{0.16} = 0.4$
$\qquad$ $0.1^2 = 0.01$ and $\sqrt{0.01} = 0.1$
$\qquad$ $0.2^2 = 0.04$ and $\sqrt{0.04} = 0.2$
$\qquad$ $0.3^2 = 0.09$ and $\sqrt{0.09} = 0.3$
$\qquad$ and so on up to $0.9^2 = 0.81$ and $\sqrt{0.81} = 0.9$

We can use place value to find the **squares and square roots of multiples of 10 and 100**.

Worked Example
Find a 40^2 b 400^2

Answer
a $40^2 = (4 \times 10)^2$
$\qquad = 16 \times 100$
$\qquad = \mathbf{1600}$

b $400^2 = (4 \times 100)^2$
$\qquad = (4 \times 10 \times 10)^2$
$\qquad = 16 \times 100 \times 100$
$\qquad = \mathbf{160\ 000}$

Worked Example
Find a $\sqrt{6400}$ b $\sqrt{640\ 000}$

Answer
a $\sqrt{6400} = \sqrt{64 \times 100}$
$\qquad = \sqrt{64} \times \sqrt{100}$
$\qquad = 8 \times 10$
$\qquad = \mathbf{80}$

b $\sqrt{640\ 000} = \sqrt{64 \times 10\ 000}$
$\qquad = \sqrt{64 \times 100 \times 100}$
$\qquad = \sqrt{64} \times \sqrt{100} \times \sqrt{100}$
$\qquad = 8 \times 10 \times 10$
$\qquad = \mathbf{800}$

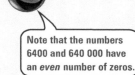

Note that the numbers 6400 and 640 000 have an *even* number of zeros.

Exercise 8

1 Copy and complete these tables.

a

0.1^2	0.2^2	0.3^2	0.4^2	0.5^2	0.6^2	0.7^2	0.8^2	0.9^2
0.01	0.04	0.09						0.81

b

$\sqrt{0.01}$	$\sqrt{0.04}$	$\sqrt{0.09}$	$\sqrt{0.16}$	$\sqrt{0.25}$	$\sqrt{0.36}$	$\sqrt{0.49}$	$\sqrt{0.64}$	$\sqrt{0.81}$
	0.2			0.5				

2 Find **a** 20^2 **b** 50^2 **c** 80^2 **d** 100^2.

Remember the first twelve squares and square roots.

3 Find **a** 300^2 **b** 500^2 **c** 700^2 **d** 900^2.

4 Find **a** $\sqrt{400}$ **b** $\sqrt{900}$ **c** $\sqrt{2500}$ **d** $\sqrt{4900}$.

5 Find **a** $\sqrt{90\,000}$ **b** $\sqrt{360\,000}$ **c** $\sqrt{640\,000}$

Review 1 Find **a** $0{\cdot}2^2$ **b** $0{\cdot}8^2$ **c** $\sqrt{0.09}$ **d** $\sqrt{0{\cdot}49}$.

Review 2 Find **a** 30^2 **b** 90^2 **c** 600^2 **d** 800^2.

Review 3 Find **a** $\sqrt{1600}$ **b** $\sqrt{8100}$ **c** $\sqrt{490\,000}$ **d** $\sqrt{810\,000}$.

Calculator squares and square roots

Squares on a calculator

To find 7^2, **key** 7 x^2 = to get 49.

To find $6{\cdot}3^2$, **key** 6.3 x^2 = to get 39·69.

To find $(^-8{\cdot}4)^2$, **key** ((–) 8.4) x^2 = to get 70·56.

CASIO

Square roots on a calculator

To find $\sqrt{81}$, **key** √ 81 = to get 9.

To find $\sqrt{5{\cdot}29}$, **key** √ 5.29 = to get 2·3.

To find $\sqrt{29}$, **key** √ 29 = to get 5.385164807

$\sqrt{29} = 5{\cdot}4$ (1 d.p.)

Always estimate the answer first.

Example $\sqrt{89}$ is between $\sqrt{81}$ and $\sqrt{100}$ or between 9 and 10

It is closer to 9 than 10.

Exercise 9

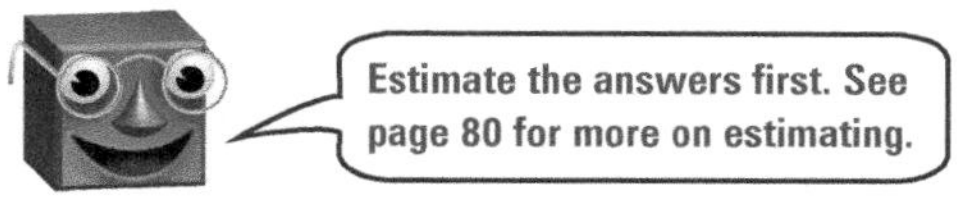
Estimate the answers first. See page 80 for more on estimating.

1 Calculate these.

a 11^2 **b** 9^2 **c** 5^2 **d** 15^2 **e** 18^2 **f** 24^2
g 91^2 **h** 87^2 **i** 113^2 **j** $(^-302)^2$ **k** $7{\cdot}9^2$ **l** $9{\cdot}3^2$
m $16{\cdot}4^2$ **n** $(^-8{\cdot}3)^2$ **o** $(^-31{\cdot}6)^2$ **p** $56{\cdot}1^2$ **q** $0{\cdot}13^2$ **r** $(^-0{\cdot}47)^2$
s $0{\cdot}85^2$ **t** $(^-0{\cdot}12)^2$

2 Calculate these. Check using inverse operations.

a $\sqrt{196}$ b $\sqrt{484}$ c $\sqrt{784}$ d $\sqrt{1156}$ e $\sqrt{1024}$ f $\sqrt{2.25}$
g $\sqrt{2.89}$ h $\sqrt{21.16}$ i $\sqrt{62.41}$ j $\sqrt{0.16}$ k $\sqrt{0.01}$ l $\sqrt{0.04}$
m $\sqrt{0.49}$ n $\sqrt{0.81}$

3 Give the answers to these to 1 decimal place.

a $\sqrt{54}$ b $\sqrt{32}$ c $\sqrt{94}$ d $\sqrt{104}$ e $\sqrt{811}$ f $\sqrt{964}$
g $\sqrt{2.6}$ h $\sqrt{5.9}$ i $\sqrt{16.9}$ j $\sqrt{27.1}$ k $\sqrt{64.7}$

4 Give the answer to these to the nearest tenth.

a $\sqrt{19}$ b $\sqrt{27}$ e $\sqrt{86}$ d $\sqrt{134}$ e $\sqrt{341}$ f $\sqrt{34.3}$
g $\sqrt{71.4}$ h $\sqrt{92.8}$ i $\sqrt{105.2}$

5 $1^2 = 1$
         $11^2 = 121$
         $111^2 = 12\ 321$

a Use your calculator to find the next line of this pattern.
b Without using your calculator, write down the answers to $11\ 111^2$ and $111\ 111^2$.

*6 a Find the answer to $20^2 - 19^2$.
Find two consecutive numbers that add to the same answer.
b Calculate $15^2 - 14^2 + 13^2 - 12^2$.
Find four consecutive numbers that add to the same answer.
c Which four consecutive numbers do you think add to give the same answer as $32^2 - 31^2 + 30^2 - 29^2$?
d What about $66^2 - 65^2 + 64^2 - 63^2$?

T **Review 1**

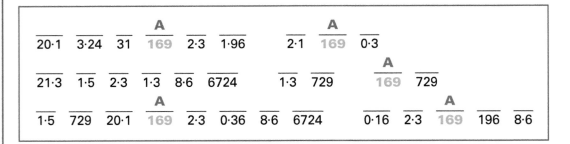

Use a copy of this box. Write the letter beside each above its answer.
Use your calculator to find these.

A $13^2 = 169$ V 14^2 N 27^2 D 82^2 O 1.8^2 Z $\sqrt{961}$
W $\sqrt{4.41}$ S $\sqrt{0.09}$ E $\sqrt{73.96}$ K 0.6^2 G 0.4^2 T 1.4^2
U $\sqrt{2.25}$ I $\sqrt{1.69}$ B $\sqrt{453.69}$ R $\sqrt{5.29}$ M $\sqrt{404.01}$

Review 2 Give the answers to these to 1 d.p.

a $\sqrt{8}$ b $\sqrt{21}$ c $\sqrt{129}$ d $\sqrt{6.8}$ e $\sqrt{17.4}$

T

Investigation

Happy numbers

Is 13 a happy number?
Step 1 Square each digit and find the sum. $1^2 + 3^2 = 1 + 9 = 10$
Step 2 Keep squaring the digits and finding the sum until you get a single digit. $1^2 + 0^2 = 1 + 0 = 1$.
If the single digit is 1, the number is a happy number. 13 is a happy number.

You could use a spread sheet to help.

Is 865 a happy number?
Step 1 $8^2 + 6^2 + 5^2 = 64 + 36 + 25 = 125$
$1^2 + 2^2 + 5^2 = 1 + 4 + 25 = 30$
Step 2 $3^2 + 0^2 = 9 + 0 = 9$
865 is not a happy number.

Is your house number a happy number?
Is your date of birth (DDMMYY) a happy number?
Is your telephone number a happy number?

Last digits

1 $4^2 = 16$. The last digit of 4^2 is 6.
   $3^2 = 9$. The last digit of 3^2 is 9.

   Find the last digits of $1^2, 2^2, 3^2, ..., 9^2$.
   Describe the pattern.

   0, 1, 4, 9, 6, 5 are the last digits of $10^2, 11^2, 12^2, 13^2, 14^2, 15^2$.
   Predict the last digits of the square numbers $16^2, 17^2, 18^2, 19^2$ and 20^2.
   Use your calculator to check your prediction.

   Predict the last digits of the square numbers from 20^2 to 40^2.

   Explain why 713, 258, 422 and 597 could not be square numbers.

Look at the last digits.

2 9604 is a square number. △ ■^{2} = 9604.   △ ■ is a 2-digit number.

   △ could be 8. What other number could ■ be?

   What could ⬡ be for these square numbers?
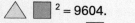
   △⬡^{2} = 2209   △⬡^{2} = 3025   △⬡^{2} = 1681   △⬡^{2} = 4356

   Khalid was asked to find what number multiplied by itself gives the answer 4489. His calculator was broken.
   He wrote $30 \times 30 = 900$ $40 \times 40 = 1600$ $50 \times 50 = 2500$
   $60 \times 60 = 3600$ $70 \times 70 = 4900$
   Khalid now knew the number must be either 63 or 67. Explain how he knew this.
   Is the number 63 or 67?

   Use Khalid's method to find out what number multiplied by itself gives these.
   2209, 3025, 1681, 4356, 1024.

Summary of key points

  We can **add and subtract integers**. Sometimes we use a number line to help.

Examples $4 + {}^-2 = 2$ $3 + {}^-5 = {}^-2$ ${}^-2 + {}^-5 = {}^-7$ ${}^-3 + 7 = 4$

$3 - {}^-4 = 7$ ${}^-6 - {}^-8 = 2$ ${}^-7 - {}^-3 = {}^-4$

  Multiplying or dividing two negative numbers gives a positive number.
Multiplying or dividing one negative and one positive number gives a negative number.

Examples ${}^-5 \times {}^-7 = 35$ ${}^-4 \times 6 = {}^-24$ $3 \times {}^-8 = {}^-24$ ${}^-32 \div {}^-4 = 8$

  A number is **divisible** by 2 if it is an even number

by 3 if the sum of its digits is divisible by 3

by 4 if the last two digits are divisible by 4

by 5 if the last digit is 0 or 5

by 6 if it is divisible by both 2 and 3

by 8 if half of it is divisible by 4

by 9 if the sum of its digits is divisible by 9

by 10 if the last digit is 0.

To check if a number is divisible by 24, check for divisibility by 3 and 8, not 4 and 6. We check using numbers that multiply to 24 but have no common factors other than 1.

D **Prime numbers** are only divisible by themselves and 1.   1 is not a prime number. The prime numbers less than 30 are 2, 3, 5, 7, 11, 13, 17, 19, 23 and 29. To test if a number is prime, try to divide it by each of the prime numbers in turn.

E 36 written as a **product of prime factors** is $2 \times 2 \times 3 \times 3$ or $2^2 \times 3^2$.
We can use a table or a factor tree to write 126 as a product of primes.

2	126
3	63
3	21
	7

or

126
 / \
2   63
   / \
  3   21
     / \
    3   7

$126 = 2 \times 3^2 \times 7$ $126 = 2 \times 3^2 \times 7$

continued ...

The **highest common factor (HCF)** of 18 and 27 is 9. This is the largest number that is a factor of both 18 and 27.

The **lowest common multiple (LCM)** of 4 and 5 is 20. This is the smallest number that is a multiple of both 4 and 5.

We can use a diagram to find the **HCF (highest common factor)** or **LCM (lowest common multiple)** of two numbers.

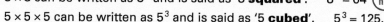

Example   To find the HCF and LCM of
360 and 288, write
$360 = 2^3 \times 3^2 \times 5$
$288 = 2^5 \times 3^2$
$HCF = 2 \times 2 \times 2 \times 3 \times 3 = 2^3 \times 3^2 = 72$
$LCM = 2 \times 2 \times 2 \times 2 \times 2 \times 3 \times 3 \times 5 = 2^5 \times 3^2 \times 5 = 1440$

> HCF is the product of the numbers in the middle. LCM is the product of all the numbers in the diagram.

 8×8 can be written as 8^2 and is said as '8 **squared**'. $8^2 = 64$

$5 \times 5 \times 5$ can be written as 5^3 and is said as '5 **cubed**'. $5^3 = 125$.

6 is the **square root** of 36 and is written as $\sqrt{36} = 6$.

4 is the **cube root** of 64 and is written as $\sqrt[3]{64} = 4$.

Squaring and finding the square root are **inverse operations.**

Cubing and finding the cube root are **inverse operations**.

 To square a number using a calculator, use the $\boxed{x^2}$ key.

Example $5 \cdot 1^2$ is keyed as $\boxed{5.1}$ $\boxed{x^2}$ $\boxed{=}$ to get $26 \cdot 01$.

$^-4 \cdot 6^2$ is keyed as $\boxed{(-)}$ $\boxed{4.6}$ $\boxed{x^2}$ $\boxed{=}$ to get $^-21 \cdot 16$.

To find the square root of a number using a calculator, use the $\boxed{\sqrt{}}$ key.

Examples $\sqrt{324}$ is keyed as $\boxed{\sqrt{}}$ $\boxed{324}$ $\boxed{=}$ to get 18.

$\sqrt{56}$ is keyed as $\boxed{\sqrt{}}$ $\boxed{56}$ $\boxed{=}$ to get $7 \cdot 5$ (1 d.p.)

Test yourself      **Except for questions 13, 14 and 16.**

1   $\boxed{2}$ $\boxed{1}$ $\boxed{0}$ $\boxed{5}$ $\boxed{^-3}$ $\boxed{1}$ $\boxed{^-2}$ $\boxed{3}$ $\boxed{^-6}$

a   Choose a number card to give the answer $^-1$. $\boxed{} + \boxed{2} = {}^-1$

b   Choose two number cards to give the answer $^-4$. $\boxed{} + \boxed{} = {}^-4$

c   Choose another two number cards that add to $^-4$.

d   Choose a number card to give the answer 3. $\boxed{^-3} - \boxed{} = 3$

e   Choose two number cards to give the answer $^-4$. $\boxed{} - \boxed{} = {}^-4$

f   Choose another two number cards that give the answer $^-4$. $\boxed{} - \boxed{} = {}^-4$

g   Choose three number cards so the answer is as small as possible. $\boxed{} + \boxed{} + \boxed{} = \underline{}$

h   Choose two number cards to give the largest possible answer. $\boxed{} - \boxed{} = \underline{}$

T **2** Use a copy of this.   ◆A ◆B

What number goes in the top circles if these instructions are followed?
a In the green circles, two numbers are added to get the numbers above.
b In the blue circles, the number on the right is subtracted from the number on the left to get the number above.
c In the yellow circles, two numbers are multiplied to get the number above.
d In the pink circles, the number on the left is divided by the number on the right to get the number above.

3 **a** How many ⁻3s are there in ⁻12?   ◆B
   b $7 \times {}^-2 = {}^-14$. Use the numbers 7, ⁻2 and ⁻14 to write one more multiplication fact and two division facts.

4 Which of 108, 117, 121, 216, 387, 600, 640 and 702 are divisible by   ◆C
   a 5   **b** 3   **c** 4   **d** 6   **e** 8   **f** 9?

5 Which of 432, 1035, 3150 and 3240 are divisible by   **a** 12   **b** 15   **c** 18?   ◆C

6 **a** Two prime numbers are added.   **b** Two prime numbers have a difference of 3.   ◆D
       The answer is 82.   What might the numbers be?
       What might the numbers be?

7 What are the missing numbers, A, B, C and D?   ◆E

                                                    232
                                            A   116
                                                2   B
                                                   C   D

8 Use a table or factor tree to write these as products of prime factors.   ◆E
   a 84   **b** 126   **c** 364   *d 7425

9 **a** What is the LCM of 4 and 6?   ◆F
   b What is the HCF of 24 and 40?

T **10** Use a copy of this.   ◆F
   a Complete this diagram for the prime factors of 300 and 96.
   b Find the HCF of 96 and 300.
   c Find the LCM of 96 and 300.

11 Find the missing numbers.   ◆G
   a ___ squared is 64   **b** 25 is the square of ___.
   c The square of ___ is 81.   **d** 64 is the cube of ___.
   e The square root of 9 is ___.   **f** The cube root of ___ is 3.

12 Calculate these.

    a 4^2 **b** $\sqrt{100}$ **c** 7^2 **d** $\sqrt{49}$ **e** 5^3 **f** 10^3 Ⓖ

    g 4^3 **h** $\sqrt[3]{8}$ **i** $\sqrt[3]{27}$ **j** $0{\cdot}5^2$ **k** $\sqrt{0{\cdot}36}$

13 Calculate these.

    a $6{\cdot}8^2$ **b** 29^2 **c** $\sqrt{8{\cdot}41}$ **d** $\sqrt{39{\cdot}69}$ Ⓗ

14 Give the answers to these to 1 decimal place.   **a** $\sqrt{21}$ **b** $\sqrt{33}$ **c** $\sqrt{109}$ Ⓗ

15 Find   **a** 60^2 **b** 200^2 **c** $\sqrt{3600}$ **d** $\sqrt{160\,000}$ Ⓖ

∗16 **a** 1996 has six factors, 1, 2, 4, 499, 998, and 1996. What is the next year   Ⓒ Ⓓ Ⓖ
           that has six factors?

      b 2025 is a square year. $45 \times 45 = 2025$.
         Which year in the 20th century was a square year?

3 Mental Calculation – Whole Numbers and Decimals

You need to know

✓ mental calculation — addition and subtraction facts to 20   page 4
                        — multiplication and division facts up to 10×10
                        — inverse operations
                        — complements in 1, 10 and 100
                        — partitioning numbers

✓ multiples and factors   page 3

Key vocabulary

add, addition, approximate, approximately, brackets, calculation, commutative, complements, decrease, difference, divide, division, double, estimate, factor, halve, increase, index, indices, multiplication, multiply, order of operations, partition, product, subtract, subtraction, sum

 ## Much ado about something

This is how Pru found the answer to 6×8.

Step 1   Subtract 6 from 10 and 8 from 10 to get 4 and 2.

Step 2   Multiply the two answers together.
              This gives the ones digit. $4 \times 2 = \mathbf{8}$

Step 3   Write down the original numbers, one under the
              other, as shown in black.
              Write down the answers from step 1, one under the
              other, as shown in red.
              Subtract diagonally to get the tens digit. $6 - 2 = \mathbf{4}$ or $8 - 4 = \mathbf{4}$

6	**4**
8	**2**

Step 4   Write down the answer, **48**

Does this work for other 1-digit numbers?

Adding and subtracting whole numbers

Discussion

Jack counted his steps on the way to and from school. He counted 750 on the way there and 680 on the way back.

Jack used a number line to explain how he worked out the total mentally.

+600   +80

750   1350   1430

Discuss Jack's way of adding 750 and 680.

How else could Jack add 750 and 680 mentally?
Discuss.

I took 1430 steps.

Here are some ways to **add and subtract mentally**.
Remember, you can add numbers in any order (commutative property).

Example   147 + 68

$147 + 68 = 140 + 7 + 60 + 8$ $= 140 + 60 + 7 + 8$ $= 200 + 15$ $= \mathbf{215}$ 40 and 60 are complements in 100. **Partitioning**	+3 +60 +5 147 150 210 **215** $147 + 68 = 147 + 3 + 60 + 5$ $= \mathbf{215}$ **Counting up**	+70 147 **215** 217 −2 $147 + 68 = 147 + 70 - 2$ $= 217 - 2$ $= \mathbf{215}$ **Adding too much then taking some away (compensation)**

Example   417 + 388

$417 + 388 = 400 + 400 + 17 - 12$ $= 800 + 17 - 12$ $= 817 - 12$ $= \mathbf{805}$ **Nearly doubles**	$417 + 388 = 420 + 390 - 3 - 2$ $= 810 - 3 - 2$ $= \mathbf{805}$ **Nearly numbers**

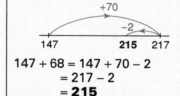

Mentally means 'in your head'. These boxes explain ways you could do it.

Example   605 – 293

$605 - 293 = 605 - 200 - 90 - 3$ $= 405 - 90 - 3$ $= 315 - 3$ $= \mathbf{312}$ **Partitioning**	+7 +300 +5 293 300 600 605 $7 + 300 + 5 = \mathbf{312}$ **Counting up**	−300 +7 305 **312** 605 $605 - 293 = 605 - 300 + 7$ $= 305 + 7$ $= \mathbf{312}$ **Taking away too much then adding some back (compensation)**

Look for **complements in 10, 50 or 100** to help you add.

Examples

$18 + 16 + 24 = 18 + 40$
$= \mathbf{58}$

$132 + 218 = 100 + 32 + 200 + 18$
$= 100 + 200 + 32 + 18$
$= 300 + 50$
$= \mathbf{350}$

$145 + 8 + 155 + 22 = 145 + 155 + 8 + 22$
$= 300 + 30$
$= \mathbf{330}$

Number

Exercise 1 **This exercise is to be done mentally.**

1 a $4 + 8 + 12 + 6 + 13$ b $9 + 7 + 13 + 11 + 15$ c $16 - 7 + 4 - 2 + 17$
   d $19 - 6 + 11 - 12 + 8$ e $15 + 12 - 8 + 16 - 17$

See *page 35* for more on adding and subtracting integers.

2 a $6 + {}^-3 + 10 + {}^-7 + {}^-3$ b $12 + {}^-6 - {}^-3 + {}^-8 - 14$
   c ${}^-7 + 10 - {}^-2 - 4 + {}^-3 - 8$ d $20 + {}^-16 - 5 - {}^-3 - 8 + 4$
   e $16 - {}^-3 - 15 + 7 - 8$ f $15 + {}^-2 + {}^-7 - {}^-4 - 8$
   g $18 - {}^-3 - 15 - 5 + {}^-3$ h $14 - {}^-6 - 5 + {}^-7 - 12$
   i $19 - {}^-6 + {}^-20 - 7 + {}^-2$ j ${}^-7 + 3 - 4 + 7 - {}^-14$
   k ${}^-21 + {}^-3 + 16 - {}^-12 - 20$

3 a $36 + 20$ b $86 - 30$ c $93 - 50$ d $37 + 80$ e $89 - 35$

Look for complements.

4 a $48 + 52 + 9$ b $63 + 37 + 16 + 4$ c $59 + 7 + 41 + 9$
   d $83 + 5 + 17 + 10$ e $7 + 72 + 18 - 3$ f $21 + 36 - 19 + 64$
   g $56 + 12 + 54 + 88$ h $147 + 9 + 53 - 4$

[T]

5 This is an arithmagon.
The number in each square is the sum of the numbers in the two circles on either side of the square.

36+35=71

Use a copy of these.
Fill them in.

a 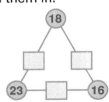   b 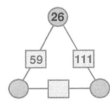   c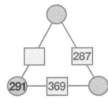

6 Use complements to find x.
   a $100 = x + 27$ b $100 - x = 71$ c $50 = 28 + x$ d $200 = 163 + x$
   e $50 - x = 17$ f $450 = 1000 - x$ g $536 = 1000 - x$

7 A German artist, Dürer, made a wood cut called 'Melancholia' which had this 4×4 magic square in it.

16	3	2	13
5	10	11	8
9	6	7	12
4	15	14	1

a What does each row, column and diagonal add to?
b If you add 2 to each of the numbers in the melancholia square, do you get a new magic square?

1	14	15	
12			6
	11	10	
13		3	16

c Copy this magic square and fill in the gaps.
In what way is this magic square the same as the melancholia square?

8 a 174 + 36　　b 326 + 42　　c 543 + 47　　d 196 – 23
　 e 285 – 32　　f 597 – 79　　g 427 – 89　　h 480 + 350
　 i 840 – 390　　j 427 + 103　　k 350 – 289　　l 925 – 479
　 m 1064 + 2387　n 7013 – 4875

You may need to use jottings for some of these.

9　Write down 3 more facts you know if these are true.
　 a　1872 + 5624 = 7496　　　　b　3850 – 1264 = 2586

10　a　Find the sum of 834 and 572 and then subtract 479.
　　 b　Decrease 1900 by 1260 and then increase this by the difference between 4860 and 3320.

11　What goes in the box?
　　 a　129 + □ = 301　　　b　211 – □ = 162　　　c　3406 – □ = 2820
　　 d　5816 + □ = 8324　　 e　□ + 3847 = 9601　　 f　□ – 8423 = 1549

*12　a　Three consecutive numbers add to 81. What are they?
　　 b　7, 9, 11, 13 are four consecutive odd numbers.
　　　　Find four consecutive odd numbers that add to 80.
　　 c　Is it possible to find four consecutive odd numbers that add to 511?
　　　　Explain your answer.
　　 d　Four consecutive even numbers add to 340. What are they?

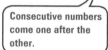

Consecutive numbers come one after the other.

*13　Use all the numbers on these cards to make two 2-digit numbers.
　　 a　How many different pairs can you make?
　　 b　What is the greatest sum you can make?
　　 c　What is the greatest difference?
　　 d　Which pair gives a difference closest to 50?

7　6　3　9

*14　Repeat question 13 a–c but make two three-digit numbers with these cards.

7　6　3　9　4　4

*15　Sandy drives his van from village to village delivering parcels.
　　 At each stop he totals his business mileage.
　　 He has to take away his personal mileage.
　　 This table gives the odometer readings as he travels.

Start of journey	86 432 km	Mileage
First stop	86 457 km	
Second stop	86 493 km	
Third stop	86 517 km	
Fourth stop	86 532 km	
Fifth stop	86 611 km	

Between the fourth and fifth stop, Sandy went to visit his sister.
This was 18 km of personal mileage he had to subtract.
What should he write down for his business mileage after each stop?

T

Review 1　Use a copy of these arithmagons.
Fill them in.

a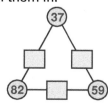

b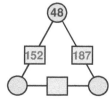

c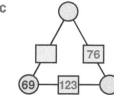

The number in each square is the sum of the numbers in the circles on either side.

Number

T | **Review 2**

			T							
$\overline{114}$	$\overline{392}$	$\overline{53}$	$\overline{4}$	$\overline{260}$		$\overline{3418}$	$\overline{431}$	$\overline{780}$	$\overline{57}$	$\overline{515}$	$\overline{3418}$

					T			**T**					
$\overline{3637}$	$\overline{57}$	$\overline{2}$	$\overline{165}$	$\overline{114}$	$\overline{4}$	$\overline{170}$	$\overline{392}$	$\overline{4}$	$\overline{260}$	$\overline{57}$	$\overline{780}$	$\overline{335}$	$\overline{114}$

		T								
$\overline{392}$	$\overline{431}$	$\overline{4}$	$\overline{260}$	$\overline{165}$		$\overline{170}$	$\overline{392}$	$\overline{780}$	$\overline{9670}$	$\overline{335}$	$\overline{114}$

$\overline{1438}$	$\overline{392}$	$\overline{9670}$	$\overline{335}$

Use a copy of this box.
Write the letter that is beside each question above its answer.

T $5 + {}^-2 + 8 + {}^-7 = 4$ N $14 + {}^-8 - {}^-3 + {}^-7$ E $78 + 87$ U $91 - 38$
S $53 + 49 + 12$ I $83 + 16 - 42$ H $186 + 74$ C $426 + 89$
R $460 + 320$ F $517 - 86$ W $540 - 370$ O $854 - 462$
D $482 - 147$ L $1430 + 8240$ G $1800 - 362$ A $4896 - 1478$
M $7823 - 4186$

Review 3 Choose numbers from the box to make these true.
In each question, use a number only once.

a ▢▢ + ▢▢ + ▢▢ + ▢ = 100
b ▢▢ + ▢▢ − ▢▢ − ▢ = 100
c ▢▢▢▢ − ▢▢▢ = 1122

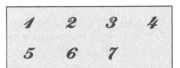

1	2	3	4
5	6	7	

? **Puzzle**

Replace the letters with digits to make these true.   a   N I N E   b   B A S E
Each letter stands for a different digit.   − T E N   + B A L L
                                                               T W O   G A M E S

Investigation

Magic Squares

1 In a magic square the numbers in each row, each column and each diagonal add to the same total.

The numbers 1, 2, 3, 4, 5, 6, 7, 8 and 9 are placed as shown to make a magic square.

Investigate other ways of placing these numbers to make magic squares.

2	9	4
7	5	3
6	1	8

2 Can a magic square be made from the first 9 even numbers, 2, 4, 6, 8, 10, 12, 14, 16, 18? **Investigate**.

What if odd numbers were used instead of even numbers?   continued...

3 Can you make a magic square with the numbers 16, 17, 18, 19, 20, 21, 22, 23, 24?

4

12
1

4	15

16	3

7
14

2	13
11	8

5	10
9	6

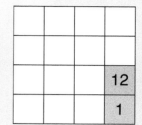

Each of the six small blocks is to be placed on the large grid to make a magic square.

Investigate ways of doing this if the

12
1

block is placed as shown.

What if the

12
1

block was put somewhere else?

Adding and subtracting decimals

Here are some ways to **add and subtract decimals mentally**.

Example   4·8 + 5·7

Partitioning	Counting up	Nearly numbers
4·8 + 5·7 = 4 + 0·8 + 5 + 0·7 = 4 + 5 + 0·8 + 0·7 = 9 + 1·5 = **10·5**	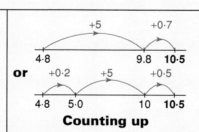	4·8 + 5·7 = 5 − 0·2 + 6 − 0·3 = 5 + 6 − 0·2 − 0·3 = 11 − 0·2 − 0·3 = 10·8 − 0·3 = **10·5**

Example   12·3 − 5·5

Partitioning	Counting up	Taking away too much then adding back (compensation)
12·3 − 5·5 = 12·3 − 5 − 0·5 = 7·3 − 0·5 = 7·3 − 0·3 − 0·2 = 7 − 0·2 = **6·8**	0·5 + 6 + 0·3 = **6·8**	12·3 − 5·5 = 12·3 − 6 + 0·5 = 6·3 + 0·5 = **6·8**

Look for **complements in 1** to help you add.

Examples   0·37 + 0·63 = 1   3·4 + 4·6 = 3 + 4 + 0·4 + 0·6   5·7 + 2·9 = 5·7 + 2·3 + 0·6
                                                  = 3 + 4 + 1   = 8 + 0·6
                                                  = 8   = **8·6**

Exercise 2 **This exercise is to be done mentally.**

1 a 0·7 + 0·8 b 0·4 + 0·6 c 0·9 + 0·8 d 0·5 + 0·7
   e 1·3 + 0·4 f 4·7 + 0·6 g 2·9 + 0·8 h 0·2 + 0·8
   i 0·7 + 1·3 j 2·7 + 1·3 k 4·3 + 6·9 l 4·5 + 4·8
   m 2·8 + 2·9 n 4·8 + 7·2 o 5·7 + 2·9 p 0·59 + 0·41
   q 0·41 + 0·57 r 0·58 + 0·74 s 0·97 + 0·68

Remember to look for complements.

2 a 0·9 − 0·5 b 0·8 − 0·3 c 1·7 − 0·8 d 1·3 − 0·7
   e 6·7 − 2·4 f 3·9 − 1·5 g 8·2 − 1·9 h 7·3 − 3·8
   i 5·6 − 3·9 j 3 − 0·7 k 7 − 5·2 l 0·78 − 0·39

3 a 0·6 + 0·7 + 0·4 b 3 + 0·3 + 4 + 0·7 c 7 + 0·7 + 5·3 + 0·3
   d 9·6 + 7 + 0·4 e 3·7 + 2 + 5·1 + 0·3

4 a 18·6 + 7·2 b 6·3 + 1·4 c 15·3 + 4·2 d 8·3 − 4·2
   e 14·3 − 5·5 f 17·8 + 3·6 g 16·4 + 2·9 h 15·3 − 5·7
   i 16·4 + 3·8 j 11·9 + 2·6 k 12·8 − 4·9 l 8·1 − 7·6
   m 5·0 − 1·5 n 16·0 − 2·7 o 18 − 4·9

You may need to use jottings for some of these.

5 a 3·4 + 6·8 = 10·2. What is the answer to 10·2 − 6·8?
   b What other three facts do you know are true if   8·63 − 4·29 = 4·34   is true?

6 a 8·6 + 8·4 + 0·8 b 0·6 + 3·4 + 8·2 c 4·7 + 5·4 − 0·2
   d 1·8 + 3·7 + 1·4 e 1·3 + 1·7 + 2·4 f 8·6 + 0·8 − 0·5
   g 5·2 + 0·4 − 0·7 h 3·9 + 2·6 − 1·3

7 a 5·23 + 0·4 + 0·07 b 6·15 + 4·21 − 3·04 c 11·56 − 3·42 + 8·9
   d 6·59 + 3·08 − 4·2 e 7·83 + 2·5 − 4·96

*8 Find ways to fill in the boxes. Put a different digit in each box.
    a □·□ + □·□ = 14·6   b □·□ − □·□ = 1·3   c □·□□ + □·□□ = 4·87

*9 Three decimal numbers add to 10. What might the three numbers be?

*10 There are five different ways Paul can go
     from home to school.
     He goes a different way each day and
     comes home the same way as he went.
     a How long is each way?
     b How far does Paul travel in total each
        week?

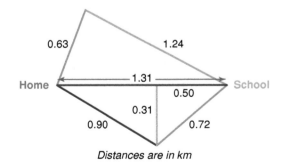

Distances are in km

Review 1
a 0·5 + 0·7 b 0·9 − 0·4 c 8·7 + 0·9 d 6·3 − 0·7 e 4·2 + 3·9
f 7·5 − 3·8 g 14·6 − 3·7 h 0·68 + 0·32 i 15·6 − 7·9 j 0·82 + 0·97
k 4·2 + 2·7 − 0·6 l 8·3 − 4·7 + 2·6

Review 2 Find ways to fill in the boxes with one digit to make these true.
a □·□ + □·□ = 9·2   b □·□ − □·□ = 2·7   c □·□□ − □·□□ = 2·01

Investigation

Decimal magic squares

1 The numbers in this magic square have been mixed up.
Change the numbers around so that the sum of each row,
each column and each diagonal is 3·3.

Can this be done in more than one way? **Investigate**.

1.5	0.9	0.3
0.7	1.3	1.7
1.1	0.5	1.9

1.9		
	1.8	
		1.7

2 **Investigate** ways of finishing this magic square.

3 Is this a magic square?
Would it be a magic square if each number had a decimal point?
Investigate.

* Make up your own magic square using decimal numbers.

155	150	250	125
450	105	95	750
850	65	550	115
350	135	145	50

Puzzle

    A·BB
+ B·AA
DC·CD

Replace A, B, C and D with numbers to make the addition
correct.

Multiplying and dividing whole numbers

Discussion

Claire had 40 beads, each 11 mm long. She wanted to know how long a necklace would
be if she threaded them all onto a nylon thread.

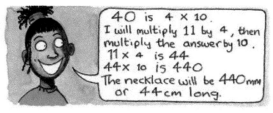

40 is 4 × 10.
I will multiply 11 by 4, then
multiply the answer by 10.
11 × 4 is 44
44 × 10 is 440
The necklace will be 440mm
or 44cm long.

Discuss Claire's method.
What other ways could you use to find the answer mentally?

Here are some **ways to multiply and divide mentally**.

Example   15 × 12

15 × 12 = 15 × (10 + 2)
      = (15 × 10) + (15 × 2)
      = 150 + 30
      = **180**
or 15 × 12 = (10 + 5) × 12
      = (10 × 12) + (5 × 12)
      = 120 + 60
      = **180**
 	
Partitioning
	15 × 12 = 3 × 5 × 4 × 3
	      = 3 × 20 × 3
	      = 60 × 3
	      = **180**
	or 15 × 12 = 15 × 2 × 6
	      = 30 × 6
	      = **180**
	
	Using factors
		15 × 12 = 30 × 6
		      = **180**
		
		Double 15 is 30
		Half of 12 is 6
		
		**Doubling one number and
		halving the other**

Example   168 ÷ 8

168 ÷ 8 = (160 ÷ 8) + (8 ÷ 8)
      = 20 + 1
      = **21**
 	
Partitioning
	168 ÷ 8 168 ÷ 2 = 84
	            84 ÷ 2 = 42
	            42 ÷ 2 = **21**
	
	168 ÷ 8 = **21**
	
	Using factors

 8 = 2 × 2 × 2

Sometimes we can multiply and then divide.

Examples   38 × 5 = **190**   38 × 10 = 380
                    380 ÷ 2 = 190

            18 × 25 = **450**   18 × 100 = 1800
                    1800 ÷ 4 = 450

5 = 10 ÷ 2

25 = 100 ÷ 4

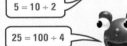

Exercise 3		**This exercise is to be done mentally.**

1 **a** 72 × 11   **b** 45 × 12   **c** 25 × 16   **d** 75 × 12
    e 23 × 11   **f** 23 × 21   **g** 42 × 25

2 Jasmine did these divisions to work out how prize money was to be shared.
    What answers should she get?
    a £84 ÷ 2   **b** £92 ÷ 2   **c** £120 ÷ 4   **d** £200 ÷ 5   **e** £420 ÷ 5
    f £208 ÷ 8   **g** £196 ÷ 4   **h** £900 ÷ 6   **i** £156 ÷ 6   **j** £940 ÷ 2
    k £340 ÷ 2   **l** £580 ÷ 2   **m** £1700 ÷ 2   **n** £5600 ÷ 2   **o** £540 ÷ 3
    p £138 ÷ 6   **q** £355 ÷ 5   **r** £268 ÷ 4   **s** £168 ÷ 12   **t** £128 ÷ 16
    u £918 ÷ 18   **v** £450 ÷ 18   **w** £112 ÷ 16

3 64 × 17 = 1088. What is the answer to 1088 ÷ 17?

4 **a** Find the product of 8 and 4 and 3.
    b Find the product of 86 and 4.

Remember we find the
product by multiplying
the numbers together.

5 Joe did these calculations to work out the wages at the council. What answers should he get?
    a 160 × 4   **b** 260 × 5   **c** 210 × 6   **d** 440 × 3   **e** 150 × 9
    f 230 × 8   **g** 124 × 4   **h** 282 × 5   **i** 514 × 6   **j** 120 × 25
    k 28 × 50   **l** 50 × 138   **m** 52 × 25

6 Sophie divided 170 ÷ 12 like this.
Use Sophie's method to find the answers to these.
  a   150 ÷ 12   b   160 ÷ 14   c   260 ÷ 11
  d   430 ÷ 13   e   470 ÷ 12   f   190 ÷ 21
  g   370 ÷ 16

100 ÷ 12 = 8 R4
+70 ÷ 12 = 5 R10
       = 13 R14
       = **14 R2**

R14 is the same
as R12 + R2.

7 Choose any three digits from 1 to 9.
Make a 2-digit number and a 1-digit number from them.
Multiply your numbers together.
Use the same digits to make other pairs of 2-digit and 1-digit numbers.
  a   Which pair gives the biggest product?
  b   Which pair gives the smallest product?
  c   Try other sets of 3 digits.
      What is the biggest product you can make?
 *d   Try this again, choosing 4-digits and making two 2-digit numbers.

```
  1   2
    3   4
  5   6
    7
  8   9
```

*8 Find ways of filling in the box and the circle.   ■ × ○ = 900

T

Review

450	484	1910		1150	814	110	80	1910	32	450
					W					
720	430	1910	32	484		*744*	24	450	1910	430
65	1910	24	36	484		48	32		48	110		36	24	110	24	600	24

Use a copy of this box. Write the letter that is beside each question above its answer.
W 62 × 12 = 744   O 74 × 11   T 18 × 25   I 96 ÷ 2
S 128 ÷ 4   G 400 ÷ 5   R 860 ÷ 2   F 180 × 4
L 230 × 5   H 121 × 4   E 382 × 5   D 24 × 25
N 660 ÷ 6   A 120 ÷ 5   C 144 ÷ 4   B 325 ÷ 5

 Puzzle

1 What digits do ▲, ■ and ✱ stand for?
  a   ▲2   b   ■■   c   2✱
    × 3   × 4   ×✱
    ────   ────   ────
    18▲   396   224

2 Chiquita made up these multiplications then replaced the numbers with letters.
What values might A, B and C have? A, B and C are different.
  a   AB   b   AB
    × B   × C
    ──   ──
    BC   CB

*3 Make up some more multiplications in which the numbers are replaced by letters.

Multiplying and dividing decimals

Here are some ways of **multiplying and dividing decimals mentally.**

Example   4·1 × 50

		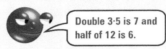
4·1 × 50 = 4 × 50 + 0·1 × 50 　　　　= 200 + 5 　　　　= **205**	4·1 × 50 = 4·1 × 10 × 5 4·1 × 10 = 41 　41 × 5　= 40 × 5 + 1 × 5 　　　　　= 200 + 5 　　　　　= **205**	50 = 100 ÷ 2 4·1 × 50　4·1 × 100 = 410 　　　　　410 ÷ 2 = **205**
Partitioning	**Using factors**	**Multiply then divide**

Sometimes we can double one number and halve the other to make the calculation easier.

Example   3·5 × 12 = 7 × 6
　　　　　　　　　= 42

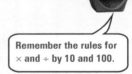

Double 3·5 is 7 and half of 12 is 6.

We can use **place value** to help us multiply.

Worked Example
a 2·1 × 40　　　**b** ☐ × 0·2 = 10

Answer
a 2·1 × 40 = 21 × 4　　　**b** We know that 5 × 2 = 10.
　　　　　　= **84**　　　　　　　So　　　**50** × 0·2 = 10.

Worked Example
a 2·4 ÷ 6　　　**b** 3·4 ÷ 2

Answer
a We know 24 ÷ 6 = 4　　　**b** We know 34 ÷ 2 = 17
　So　　　2·4 ÷ 6 = **0·4**　　　　So　　　3·4 ÷ 2 = **1.7**

Remember the rules for × and ÷ by 10 and 100.

We can use facts we already know to work out other facts.

Example   4 × 7 = 28　　so　　0·4 × 7 = 2·8　　so　　0·4 × 0·7 = 0·28

Exercise 4   **This exercise is to be done mentally.**

1 Paul did these calculations to work out the areas of some rectangles.
　　What answers should he get?
　　a　1·2 × 2　　　**b**　3·1 × 3　　　**c**　3·2 × 4　　　**d**　5·3 × 2　　　**e**　2·6 × 2
　　f　4·3 × 3　　　**g**　2·5 × 3　　　**h**　4·5 × 4　　　**i**　7·5 × 6　　　**j**　4·7 × 2
　　k　5·3 × 5　　　**l**　6·3 × 8　　　**m**　4·9 × 7　　　**n**　2·3 × 9

2 **a**　2·4 ÷ 4　　　**b**　3·6 ÷ 6　　　**c**　2·1 ÷ 7　　　**d**　1·8 ÷ 3　　　**e**　2·5 ÷ 5
　　f　3·2 ÷ 8　　　**g**　5·6 ÷ 7　　　**h**　2·7 ÷ 9　　　**i**　4·5 ÷ 5　　　**j**　4·8 ÷ 6
　　k　8·1 ÷ 9　　　**l**　3·6 ÷ 2　　　**m**　8·4 ÷ 2　　　**n**　5·4 ÷ 2　　　**o**　5·1 ÷ 3
　　p　8·4 ÷ 7　　　**q**　6·4 ÷ 4　　　**r**　9·6 ÷ 8

3 **a**　0·2 × 8　　　**b**　0·3 × 4　　　**c**　6 × 0·5　　　**d**　3 × 0·2　　　**e**　4 × 0·5
　　f　0·3 × 6　　　**g**　8 × 0·4　　　**h**　0·04 × 8　　　**i**　0·02 × 6　　　**j**　0·05 × 3
　　k　6 × 0·07　　　**l**　9 × 0·02　　　**m**　9 × 0.04　　　**n**　6 × 0·05

4 What goes in the box?
　　a　☐ × 0·2 = 1　　　**b**　☐ × 0·2 = 10　　　**c**　☐ × 0·4 = 2　　　**d**　☐ × 0·3 = 1·8
　　e　5 × ☐ = 0·5　　　**f**　3 × ☐ = 1·2　　　**g**　5 × ☐ = 4

Use what you already know.

5 **a** $1{\cdot}2 \times 20$ **b** $3{\cdot}1 \times 20$ **c** $5{\cdot}3 \times 40$ **d** $4{\cdot}2 \times 30$ **e** $5{\cdot}1 \times 50$
     f $4{\cdot}5 \times 20$ **g** $2{\cdot}6 \times 30$ **h** $3{\cdot}7 \times 60$ **i** $4{\cdot}5 \times 60$

6 **a** $0{\cdot}4 \times 0{\cdot}2$ **b** $0{\cdot}6 \times 0{\cdot}1$ **c** $0{\cdot}3 \times 0{\cdot}2$ **d** $0{\cdot}5 \times 0{\cdot}7$
     e $0{\cdot}9 \times 0{\cdot}8$ **f** $0{\cdot}6 \times 0{\cdot}7$ **g** $0{\cdot}3 \times 0{\cdot}6$ **h** $0{\cdot}2 \div 4$
     i $0{\cdot}04 \div 5$ **j** $0{\cdot}6 \div 4$ **k** $0{\cdot}07 \div 2$ **l** $0{\cdot}09 \div 2$
     m $0{\cdot}6 \div 5$ **n** $0{\cdot}03 \div 2$ **o** $0{\cdot}03 \div 5$ *p $0{\cdot}006 \div 3$
     *q $0{\cdot}004 \div 2$ *r $0{\cdot}009 \div 3$ *s $0{\cdot}007 \div 2$ *t $0{\cdot}006 \div 5$

Use what you already know.

7 **a** $3{\cdot}7 \times 12$ **b** $13 \times 1{\cdot}2$ **c** $2{\cdot}9 \times 11$ **d** $12 \times 3{\cdot}4$ **e** $4{\cdot}5 \times 19$
     f $3{\cdot}4 \times 50$ **g** $8{\cdot}6 \times 21$ **h** $9{\cdot}3 \times 11$ **i** $32 \times 2{\cdot}01$ **j** $12 \times 1{\cdot}01$
     k $18 \times 3{\cdot}02$ **l** $16 \times 2{\cdot}02$ **m** $24 \times 2{\cdot}01$ **n** $28 \times 1{\cdot}02$ *o $4{\cdot}6 \times 1{\cdot}1$
     *p $8{\cdot}4 \times 2{\cdot}5$ *q $4{\cdot}9 \times 1{\cdot}9$ *r $4{\cdot}3 \times 1{\cdot}01$

8 $3{\cdot}65 \times 7 = 25{\cdot}55$
    Write down three more facts using the numbers 3·65, 7 and 25·55.

9 $3 \times 2{\cdot}6 = 7{\cdot}8$ What does $7{\cdot}8 \div 3$ equal?

***10** Find ways of filling in the box and the circle.
     a $\square \times \bigcirc = 3{\cdot}6$ **b** $\square \times \bigcirc = 0{\cdot}06$ **c** $\square \div \bigcirc = 0{\cdot}4$

***11** **a** Copy and complete these.

$24 \div 8 = 3$	$2{\cdot}4 \div 8 = 0{\cdot}3$	$0{\cdot}24 \div 8 = \underline{\quad}$
$24 \div 0{\cdot}8 = 30$	$2{\cdot}4 \div 0{\cdot}8 = \underline{\quad}$	$0{\cdot}24 \div 0{\cdot}8 = \underline{\quad}$
$24 \div 0{\cdot}08 = 300$	$2{\cdot}4 \div 0{\cdot}08 = \underline{\quad}$	$0{\cdot}24 \div 0{\cdot}08 = \underline{\quad}$
$24 \div 0{\cdot}008 = \underline{\quad}$	$2{\cdot}4 \div 0{\cdot}008 = \underline{\quad}$	$0{\cdot}24 \div 0{\cdot}008 = \underline{\quad}$

     b Write down the next two lines of these.

$54 \div 9 = 6$	$5{\cdot}4 \div 9 = 0{\cdot}6$	$0{\cdot}54 \div 9 = 0{\cdot}06$
$54 \div 0{\cdot}9 = 60$	$5{\cdot}4 \div 0{\cdot}9 = 6$	$0{\cdot}24 \div 0{\cdot}9 = 0{\cdot}6$

     c Find the answers to these.
       i $36 \div 0{\cdot}6$ **ii** $81 \div 0{\cdot}9$ **iii** $4{\cdot}5 \div 0{\cdot}5$ **iv** $56 \div 0{\cdot}8$ **v** $72 \div 0{\cdot}08$
       vi $4{\cdot}2 \div 0{\cdot}7$ **vii** $4{\cdot}8 \div 0{\cdot}06$ **viii** $6{\cdot}3 \div 0{\cdot}07$ **ix** $0{\cdot}63 \div 0{\cdot}07$

⊤

Review 1
Use a copy of this box. Write the letter that is beside each question above its answer.

A												
$\overline{8{\cdot}6}$	$\overline{57{\cdot}2}$	$\overline{86}$	$\overline{0{\cdot}45}$	$\overline{2{\cdot}1}$	$\overline{4{\cdot}5}$	$\overline{10{\cdot}8}$	$\overline{45}$	$\overline{30}$	$\overline{0{\cdot}45}$	$\overline{2{\cdot}4}$	$\overline{108}$	$\overline{2{\cdot}4}$

									A	
$\overline{10{\cdot}8}$	$\overline{0{\cdot}45}$	$\overline{0{\cdot}8}$	$\overline{10{\cdot}8}$	$\overline{0{\cdot}7}$	$\overline{0{\cdot}7}$	$\overline{2{\cdot}4}$	$\overline{4{\cdot}5}$	$\overline{2{\cdot}1}$	$\overline{30}$ $\overline{8{\cdot}6}$	$\overline{57{\cdot}2}$

					A		
$\overline{10{\cdot}8}$	$\overline{2{\cdot}1}$	$\overline{0{\cdot}45}$	$\overline{0{\cdot}8}$	$\overline{4{\cdot}5}$	$\overline{8{\cdot}6}$	$\overline{10{\cdot}8}$	$\overline{57{\cdot}2}$

A $4{\cdot}3 \times 2 = 8{\cdot}6$ **I** $1{\cdot}8 \times 6$ **T** $0{\cdot}3 \times 7$ **S** $9 \times 0{\cdot}05$ **O** $4{\cdot}3 \times 20$
Y $3{\cdot}6 \times 30$ **N** $5{\cdot}2 \times 11$ **H** $2{\cdot}5 \times 12$ **C** $0{\cdot}9 \times 50$ **E** $4{\cdot}8 \div 2$
R $1{\cdot}5 \times 3$ **G** $5{\cdot}6 \div 8$ **B** $7{\cdot}2 \div 9$

Review 2   **a** $7 \times 0{\cdot}9$ **b** $0{\cdot}8 \times 0{\cdot}6$ **c** $0{\cdot}2 \div 5$ *d $0{\cdot}05 \times 0{\cdot}6$ *e $0{\cdot}08 \times 5$ *f $0{\cdot}027 \div 9$

Review 3   **a** $4{\cdot}8 \times 2{\cdot}5$ **b** $1{\cdot}5 \times 1{\cdot}9$ **c** $4{\cdot}8 \times 1{\cdot}2$ *d $5{\cdot}6 \times 0{\cdot}15$

Solving problems mentally

Lucy's netball team raised £86 with a cake stall and £194 selling pizza.
To find the total raised we must **add** £86 and £194.
One way to do this is   £86 + £194 = £86 + £200 − £6
                                    = £286 − 6
                                    = £280

Look for clues in the wording.

Lucy's team raised **£280** in total.

You must decide whether to add, subtract, multiply or divide to solve these problems.

Exercise 5   **This exercise is to be done mentally. You may use jottings.**

1   At birth you have 350 bones. Some fuse together so that as an adult you have only 205 bones.
    How many more bones do you have at birth?

2   Joseph could sign 20 autographs in one minute.
    How many could he sign in
    a 30 minutes   b 70 minutes   c 50 minutes?

3   The cost of hiring a bus for the school camp is £270.
    If 30 students go, how much will each have to pay?

4   In November Julian received 6000 e-mails.
    How many, on average, did he receive each day?

5   Josie stands on the scales holding her two cats.
    Each cat weighs 3 kg.
    The scales show 58 kg.
    How much does Josie weigh?

6   Aisling is exactly 4 days old.
    How many hours old is she?

7   A car uses 8 litres of petrol for every 100 km.
    How many litres will it use to travel
    a 400 km   b 2000 km   c 1500 km?

8   A pelican eats about 14·7 kg of fish each week.
    a   About how much fish would it eat each day?
    b   About how much fish would it eat in 4 days?

9   a   Naim bought 5 rolls.
        How much did this cost?
    b   Jack bought 2 packets of crisps and a pie.
        How much did this cost?
    c   Penny bought a pie and some crisps. How much did this cost?
    d   How many drinks can you buy for £5·50?
    e   Jessica has £2 to spend on lunch.
        What could she buy?
        Give at least two different answers.

Roll	£0.90
Pie	£1.50
Drink	£0.55
Crisps	£0.45

10 a Janet rode her bike the shortest way from school to the pool.
How far did she ride?

b Ramon rode his bike the longest way from school to the pool. He did not go down any road more than once and always rode in the direction of the arrows.
How far did he ride?

c How much further than Janet did Ramon ride?

d Janet rode her bike from school to the pool by the shortest way five times each week.
How far is this?

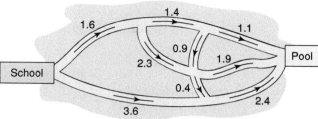

Distances are in miles

11 Dylan has a job at 'McRoberts Burgers'.
He gets £3·40 an hour.

a Last week he worked 5 hours one day and 6 hours on another.
How much did he earn?

b If Dylan works over a meal time he gets an extra £2·50.
How much would he get for working 15 hours if he worked over 4 meal times?

12 Five friends went to an outdoor concert. The total cost was £35·50.
One friend paid £10 for two of the five tickets and the others shared the rest of the cost.
How much did each of the others pay?

13 A turtle can move 26 m each minute.
How far could the turtle move in
a 2 minutes   **b** 5 minutes   **c** $\frac{1}{4}$ hour   **d** 1 hour   *∗**e** 1·4 hours?

Review 1   Chris stacked 96 bricks into 4 equal piles. How many were in each pile?

Review 2
a How much would 5 kg of apples cost?
b How much would 3 kg of oranges cost?
c How many kilograms of oranges could you buy for £18·50?

Apples £1.50 per kg

Oranges £1.85 per kg

Review 3   Eight people bought tickets for an animal park. The total cost was £25·60.
How much did each pay?

Review 4   A school bought 8 of these TVs.
They got a discount of £75 off the total price.
How much did they pay?

Ex Hire
TV and video in one
£320

Review 5   Mr Taylor took his class of 30 students on camp.
a What was the total cost?
b Each student paid £27·50.
How much did they pay altogether?
c How much money was left over?
*∗**d** How much refund did each student get?

Total Camp Costs

Bus	£75
Food	£495
Tent hire	£161

Puzzle

1 I am a two-digit number greater than 50.
I am divisible by 12.
The sum of my digits is an even number greater than 10.
What number am I?

2 I am a three-digit number less than 200.
I am divisible by 16.
The sum of my digits is an even number less than 14 but greater than 6.
What number am I?

*3 I am a three-digit number.
The sum of my digits is 7.
I am divisible by 8 but not by 16.
What number am I?

Find the two possible answers.

Order of operations

$8 + 5 \times 2 \qquad 4 + 9 \times 2 - 3 \qquad 19 - 8 \div 2$

When there is more than one operation in a calculation we do the **multiplication and division before** the **addition and subtraction**.

Examples

× first.

$7 + 3 \times 2 = 7 + 6$
$= 13$

$4 + 6 \times 5 - 3 = 4 + 30 - 3$
$= 31$

$20 \div 5 + 6 \div 2 = 4 + 3$
$= 7$
÷ first.

$17 - 6 \div 2 = 17 - 3$
$= 14$

$20 + 8 \div 4 - 2 \times 3 = 20 + 2 - 6$
$= 16$

Work from left to right, doing × and ÷ first, then work from left to right doing + and −.

Exercise 6

1 a $4 + 6 - 3$
  b $4 + 2 \times 3$
  c $2 + 3 \times 2$
  d $5 - 2 + 4$
  e $6 \times 3 - 1$
  f $7 \times 2 - 4 + 1$
  g $8 \times 3 + 8 \div 2$
  h $6 \times 4 - 7 \times 2$
  i $2 \times 3 + 5 \times 4$
  j $11 - 3 \times 2 + 1$
  k $16 - 2 \times 3 - 2$
  l $20 - 5 \times 3 + 4$
  m $24 \div 6 \div 2 + 4$
  n $4 \times 2 + 3 \times 4 - 7$
  o $3 \times 5 + 2 \times 4 + 3$
  p $8 \times 6 + 7 \times 2 + 6$
  q $12 \div 3 \div 2 - 4$
  r $12 + 8 \div 4 + 3$
  s $11 + 16 \div 4 - 1$
  t $5 \times 3 - 4 \times 4$
  u $7 \times 9 - 3 \times 10 + 4$
  v $16 \div 2 - 8 \div 1$
  w $20 \div 4 + 15 \div 3$
  x $5 + 12 \div 2 - 7 \times 5$

Review

a $5 \times 2 - 3$
  b $6 \times 7 - 5$
  c $3 + 4 \times 2$
  d $12 - 3 \times 3 + 4$
e $16 - 5 \times 2 - 7$
  f $5 \times 3 + 2 \times 4$
  g $16 \div 2 + 4 \div 2$
  h $20 - 8 \div 4 + 6$
i $4 + 3 \times 7 - 5 \times 8$
  j $40 \div 8 - 5 \div 1$
  k $5 + 6 \times 2 - 3 \times 8$

If there are **brackets** in a calculation, do the calculation inside them first.

A number immediately before a bracket means that we multiply what is inside the bracket by this number.

Examples

$5(2 + 7) = 5 \times (2 + 7)$
$= 5 \times 9$
$= 45$

$15 \div (3 + 2) = 15 \div 5$
$= 3$

Work out the brackets first.

In $\frac{7 + 3}{2}$, the horizontal line acts as a bracket.

$\frac{7 + 3}{2} = \frac{(7 + 3)}{2}$
$= \frac{10}{2}$
$= 5$

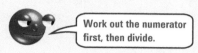

Work out the numerator first, then divide.

Examples

$\frac{18 - 3}{5} = \frac{(18 - 3)}{5}$
$= \frac{15}{5}$
$= 3$

$\frac{16 + 4}{8 - 3} = \frac{(16 + 4)}{(8 - 3)}$
$= \frac{20}{5}$
$= 4$

Work out both the numerator and denominator, then divide.

Exercise 7

1
a $3(7 + 2)$ b $5(6 - 4)$ c $7(8 - 3)$ d $5(6 - 1)$
e $4(8 - 3)$ f $6(5 + 6)$ g $(15 - 3) \times 2$ h $(6 - 3) \times 4$
i $(14 - 6) \times 3$ j $5(20 - 8)$ k $(8 + 6) \times 2$ l $13(20 - 19)$
m $5 \times (20 + 5 - 16)$ n $(18 + 3 - 14) \times 2$ o $2(10 - 2 \times 3)$ p $3(8 - 2 \times 3)$
q $21 - 3(4 - 1)$ r $2(4 + 3) - 7$ s $18 - 2(3 \times 2)$ t $14 - 3(2 \times 2)$
u $(10 - 3) \times (12 - 4)$ v $(20 - 5) \div (8 - 3)$ w $(18 - 6) \div (9 - 6)$

2
a $\frac{16 + 4}{2}$ b $\frac{8 + 12}{4}$ c $\frac{5 + 9}{7}$ d $\frac{24 - 3}{7}$ e $\frac{15 - 9}{6}$
f $\frac{24}{3 \times 8}$ g $\frac{100}{2 \times 5}$ h $\frac{60}{5 \times 6}$ i $\frac{81}{3 \times 3}$ j $\frac{120}{2 \times 15}$
k $\frac{100}{5 \times 5}$ l $\frac{14}{6 - 4}$ m $\frac{7 + 9}{10 - 6}$ n $\frac{25 + 5}{7 - 2}$

3
a $3 \times 4 - 2(5 - 1)$ b $5 \times 4 - 3(6 - 3)$ c $9 \times 8 + 3(2 \times 2)$ d $5 \times 4 + 6(3 \times 0)$
e $6 \times 2 + 4(3 + 0)$ f $5 \times 8 - 3(8 - 2 \times 2)$ g $6 \times 2 - 2(4 \times 3 - 6)$
h $15 \div (7 - 4) - 9$ i $45 \div (12 - 6 \div 2) - 6$ j $5 \times (16 - 18 \div 3) + (12 - 3 \times 3)$

4 $2 + 3 \times 6 = 20$ $2(6 - 3) = 6$

a   What other answers can you make using 2, 3 and 6?
    You may use $+$, $-$, $\times$, $\div$ and brackets.
b   What is the biggest answer you can make?
*c   What is the second biggest?

*5   Choose any four different numbers from 1 to 9.
a   What answers can you make using your four numbers together with $+$, $-$, $\times$, $\div$ or brackets?
b   What is the biggest answer you can make?
c   What is the second biggest?
d   Repeat b and c for five different numbers from 1 to 9.

Number

[T] | **Review 1**

3	⁻5	3	5	35		6	4	10	5	7		36	⁻5	81	

$$\overline{42} \quad \overset{I}{\overline{21}} \quad \overline{35} \quad \overline{81} \quad \overline{32} \qquad \overline{14} \quad \overline{81} \quad \overline{5} \quad \overline{28} \qquad \overset{I}{\overline{21}} \quad \overline{10} \qquad \overline{{}^-10} \quad \overline{18} \quad \overset{I}{\overline{21}} \quad \overline{10} \quad \overline{{}^-5}$$

Use a copy of this box. Write the letter that is beside each question above its answer.

I $3(5 + 2) = 21$ **W** $4(5 + 4)$ **F** $6(9 - 2)$ **U** $7(8 - 6)$ **D** $(8 + 6) \times 2$
T $(6 - 2) \times 8$ **S** $9(12 - 4 + 1)$ **R** $5(1 + 2 \times 3)$ **C** $30 - 2(4 \times 5)$ **H** $5 \times 6 - 3(8 - 4)$
M $\frac{8 + 4}{2}$ **Y** $\frac{20 + 8}{4}$ **O** $\frac{32 - 4}{7}$ **N** $\frac{80}{2 \times 4}$ **E** $\frac{26 + 4}{2 \times 3}$
A $3 \times 9 - 4(5 \times 4 - 12)$ **P** $60 \div (12 - 12 \div 2) - 7$

Review 2   Use the digits 2, 3 and 7 together with $+$, $-$, $\times$, $\div$ or brackets.
What is the second largest answer you can make?

We work out **squares (indices) before multiplication and division but after brackets**.

The order in which we do operations is   **B**rackets
Squares (**I**ndices)
Division and **M**ultiplication
Addition and **S**ubtraction.

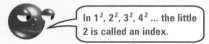

In $1^2, 2^2, 3^2, 4^2$ ... the little 2 is called an index.

To help you remember this you could use the word BIDMAS.

Example $3 \times 4^2 = 3 \times 16$
$= 3 \times 10 + 3 \times 6$
$= 30 + 18$
$= \textbf{48}$

$(3^2 - 2^2)^2 = (9 - 4)^2$
$= 5^2$
$= \textbf{25}$

A square root sign acts a bracket. We work out the answer to $8^2 - 28$ *before* we take the square root.

$* \sqrt{8^2 - 28} = \sqrt{64 - 28}$
$= \sqrt{36}$
$= \textbf{6}$

Exercise 8

1 **a** 2×3^2 **b** $5^2 - 2$ **c** 4×2^2 **d** $3(2^2 - 1)$ **e** $5(3^2 + 1)$
   f $4^2 \times 2 - 3 \times 4$ **g** $4^2 \div 4 + 2$ **h** $5^2 - 12 \div 3$

2 **a** $6(5 - 2)^2$ **b** $2(3 + 1)^2$ **c** $3(7 - 2)^2$ **d** $4(5 - 3)^2$
   e $8 - 3 \times 2^2$ **f** $7 + 2 \times 3^2$ **g** $20 + 4 \times 5^2$ **h** $2 \times 4^2 + 4$
   i $4 \times 3^2 - 10$ **j** $(8 - 3 \times 2)^2$ **k** $(1 + 2 \times 5)^2$ **l** $(16 - 3 \times 4)^2$
   m $(5 \times 3 - 8)^2$ **n** $7(7 - 3 \times 2)^2$ **o** $2(20 - 5 \times 3)^2$

3 **a** $(5 - 8)^2$ **b** $(11 - 15)^2$ **c** $(4 - 9)^2$ **d** $2(5 - 11)^2$
   e $5(2 - 5)^2$ **f** $4(1 - 2 \times 5)^2$ **g** $2(3 - 3 \times 4)^2$

*4 **a** $\frac{4^2}{8}$ **b** $\frac{6^2}{9}$ **c** $\frac{(3 \times 2)^2}{12}$ **d** $\frac{24}{5^2 - 1}$
   e $\frac{(4 \times 2 - 1)^2}{7}$ **f** $\frac{(3 - 5 \times 2)^2}{7}$ **g** $\frac{(3 + 1)^2}{(3 - 1)^2}$ **h** $\frac{(2 + 4)^2}{(6 - 4)^2}$
   i $\frac{(2 \times 4 - 2)^2}{2^2}$ **j** $\frac{(1 + 2 \times 4)^2}{(2 + 1)^2}$ **k** $\frac{(7 + 5)^2}{(15 - 3)^2}$ **l** $\frac{(2 \times 5)^2}{4 \times 5}$

*5 **a** $\sqrt{25 - 9}$ **b** $\sqrt{25 + 11}$ **c** $\sqrt{37 - 12}$ **d** $\sqrt{10 \times 7 - 6}$ **e** $\sqrt{4 + 5 \times 9}$
   f $\sqrt{20 \times 6 + 1}$ **g** $\sqrt{5^2 + 24}$ **h** $\sqrt{3^2 - 5}$ **i** $\sqrt{2^2 - 3}$ **j** $\sqrt{23 + 11^2}$
   k $\sqrt{3^2 + 4^2}$ **l** $\sqrt{10^2 - 6^2}$ **m** $\sqrt{5^2 - 4^2}$

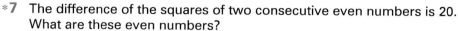

*6 Mr Jones was a farmer. He had five square grassed pens.
Find the length of one side if the area was
   a 900 m²   b 324 m²   c 576 m²   d 225 m²   e 441 m²

> Write as the product of two square numbers.

*7 The difference of the squares of two consecutive even numbers is 20.
What are these even numbers?

Review 1

a $(7 + 2)^2$ b $(11 - 14)^2$ c $(15 - 4 \times 2)^2$ d $5(6 - 2)^2$

e $3(22 - 6 \times 2)^2$ f $\frac{(1-7)^2}{4}$ g $\frac{(3+2)^2}{3 \times 2 - 1}$ h $\frac{(5-2)^2}{(2+1)^2}$

*Review 2

a $\sqrt{19 + 6}$ b $\sqrt{6 \times 5 - 14}$ c $\sqrt{9^2 - 17}$ d $\sqrt{40 - 2^2}$

e $\sqrt{10^2 - 8^2}$ f $\sqrt{256}$ g $\sqrt{729}$

? Puzzle

Choose numbers from the ring to make these true.
In each question use a number only once.

1 ☐ – ☐ + ☐ = 4   2 ☐ + ☐ – ☐ = 6

3 ☐ × ☐ – ☐ = 7   4 (☐ + ☐) ÷ 2 = 4

5 (☐ + ☐) ÷ (☐ + ☐) = 3   6 ☐ × ☐² – ☐ × ☐ = 38

7 $\sqrt{☐^2 + ☐^2} = 5$ 8 $\sqrt{(☐^2 - ☐^2)} = 3$

(ring of numbers: 1 2 3 4 5)

🎲 Calculation full house – a game for a group

To prepare

Each player writes down the numbers 1 to 20.
Beside each number, write a calculation that has this number as the answer.

Example $1 = 3 + 4 + 5 - 9 + 1 - 3$
           $2 = 8 \times 3 - 22$
           $3 = 21 - 18$
           $4 = 14 - 2 \times 5 \ldots$

> These will be used if you are the leader.

To play

● Choose a leader.
● Everyone except the leader, writes down five numbers from 1 to 20. Give your five numbers to the person on your left.
● The leader calls out one of the calculations he or she wrote when preparing for the game.
● If the answer to the calculation is one of the five numbers you have, cross it out.

Example   Pete had the numbers 2, 3, 8, 10 and 18.
             The leader called the calculation $7 + 9 - 6$.
             The answer is 10.
             Pete crossed out 10.

● The first student to cross out all 5 numbers is the winner.

Investigation

Number Chains

1 Begin with any number.

If the number is even, divide it by 2.
If the number is odd, multiply it by 3 and add 1.
Repeat with the new number until you get the digit 1.

Example   If we begin with 20 we get

$20 \div 2 = 10$
$10 \div 2 = 5$
$5 \times 3 + 1 = 16$
$16 \div 2 = 8$
$8 \div 2 = 4$
$4 \div 2 = 2$
$2 \div 2 = 1$

The number chain is 10, 5, 16, 8, 4, 2, 1. There are seven numbers in this chain.

Find all the pairs of numbers from 2 to 20 that have number chains of the same length. For example, 3 and 20 have number chains of length 7.

2 Begin with any 2-digit number less than 40.
Multiply the ones digit by 4. Add the tens digit to this.
Repeat until you get back to the number you started with.

Example $32 \rightarrow 11 \rightarrow 5 \rightarrow 20 \rightarrow 2 \rightarrow 8 \rightarrow 32$

What is the longest number chain you can make this way?

Making estimates

Sometimes we don't need an exact answer to a question.

An **estimate is good enough**.

Examples   How many people were at the school fair?
How many burgers does a typical 11-year old eat in a year?

Example   Amanda was asked to estimate the number of times her sister said 'um' in a 24 minute phone call.

She counted the number of times her sister said 'um' in 1 minute. She got 16.

She multiplied this by 24 to get 384.

We would probably say she said 'um' **about 400** times.

Exercise 9

1 For which of the following would an exact number, rather than an estimate, be needed?
   a A reporter wants to know how many people were at a football match.
   b The school finance committee wants to know how much profit was made at the school fair.
   c A doctor wants to know how much water a patient drinks in a typical day.
   d A journalist wants to know the cost of compensating the farmers in Cumbria after the foot-and-mouth outbreak.
   e A football club manager wants to know the number of wins, draws and losses for his team.

Review Would an exact number or an estimate be needed by the organisers for the number of children at a school camp?

 Practical

Estimate these. Explain how you got your estimate.

1 The number of litres of water your household uses in a day to the nearest 10 litres.

2 The distance your family car travels in one year to the nearest 1000 miles.

3 The distance you travel to school and back in one year to the nearest 10 miles.

4 The number of phone calls your family makes in one year to the nearest 100.

5 The number of leaves on a bush.

6 The number of people in this picture.

Estimating answers to calculations

Jenna's school is extending the library.
There are 580 books that need extra shelves.
Each shelf holds about 28 books.
Jenna wants to **estimate** how many extra shelves will be needed.
To **estimate the answer to a calculation** we round the numbers.

≈ means 'is approximately equal to'.

Jenna worked out her estimate like this.
$$580 \div 28 \approx 600 \div 30$$
$$= 20$$

There is often more than one possible estimation.

Examples
$562 \times 2 \cdot 4 \approx 600 \times 2 \quad = 1200$
or $562 \times 2 \cdot 4 \approx 600 \times 2 \cdot 5 = 1500$
or $562 \times 2 \cdot 4 \approx 560 \times 2 \quad = 1120$

$273 \div 28 \approx 300 \div 30 = 10$
or $273 \div 28 \approx 270 \div 30 = 9$

When rounding numbers we try to round to 'nice numbers' which are easy to calculate mentally.

Examples $\frac{58}{7} \approx \frac{56}{7} = 8$

Approximate $\frac{58}{7}$ to $\frac{56}{7}$ rather than $\frac{60}{7}$.
56 is a *multiple* of 7.

$334 \div 3 \cdot 5 \approx 300 \div 3 = 100$

$\frac{89 \times 32}{18} \approx \frac{90 \times 30}{20} = \frac{2700}{20} \approx \frac{2800}{20} = 140$

Round to 2800 to get a multiple of 20.

Worked Example
Estimate the answers to
a $\frac{186 \times 32}{21}$
b $(18 \cdot 6 - 5 \cdot 2) \div (2 \cdot 6 + 8 \cdot 2)$

Answers
a $\frac{186 \times 32}{21} \approx \frac{200 \times 30}{20} = \frac{6000}{20} = \mathbf{300}$
b $(18 \cdot 6 - 5 \cdot 2) \div (2 \cdot 6 + 8 \cdot 2) \approx (19 - 5) \div (3 + 8)$
$$= 14 \div 11$$
$$\approx 14 \div 10$$
$$= \mathbf{1 \cdot 4}$$

Exercise 10

1 Choose the best approximation.

		A	B	C
a	33×57	A 30×60	B 40×50	C 40×60
b	82×78	A 80×70	B 100×80	C 80×80
c	$942 \div 32$	A $1000 \div 30$	B $900 \div 40$	C $900 \div 30$
d	748×26	A 800×30	B 700×30	C 700×20
e	$324 \div 53$	A $300 \div 60$	B $300 \div 50$	C $400 \div 50$
f	$642 \div 43$	A $600 \div 50$	B $700 \div 40$	C $600 \div 40$
g	$20 \cdot 9 \times 39 \cdot 8$	A 20×39	B 20×30	C 20×40
h	$9 \cdot 18 \times 4 \cdot 81$	A 10×5	B 9×4	C 9×5
i	$17 \cdot 8 \div 4 \cdot 7$	A $18 \div 4$	B $20 \div 5$	C $178 \div 47$

2 Estimate the answers to these. Show how you found your estimate.
- **a** 125×76
- **b** 341×77
- **c** 234×178
- **d** 594×312
- **e** $234 \div 19$
- **f** $527 \div 48$
- **g** $378 \div 37$
- **h** $583 \div 26$
- **i** 211×82
- **j** 762×47
- **k** $724 \div 34$
- **l** $513 \div 24$

3 Estimate the answers to these.
- **a** 81.6×5
- **b** $21.6 \div 3$
- **c** $48.4 \div 5$
- **d** $64.2 \div 8$
- **e** 5.6×3.2
- **f** 8.9×7.6
- **g** $10.3 \div 9.2$
- **h** 4.85×3.2
- **i** 4.83×7.2
- **j** $16.3 \div 4.2$
- **k** $287 \div 5.3$
- **l** 24.6×5.7
- **m** $204 \div 4.1$
- **n** $196 \div 5.3$
- **o** 8.1×19.6

4 Write down a calculation you could do to find an estimate for these.
Find the estimate.
- **a** $\frac{126 \times 32}{18}$
- **b** $\frac{87 \times 91}{8}$
- **c** $\frac{53 \times 79}{22}$
- **d** $\frac{18.6 \times 23.2}{4.9}$
- **e** $\frac{898}{3.1 \times 2.8}$
- **f** $\frac{682}{5.3 \times 2.17}$
- **g** $\frac{104 + 82}{78 - 19}$
- **h** $\frac{137 + 69}{86 - 17}$
- **i** $\frac{16.9 + 11.2}{3.8 - 1.1}$
- **j** $\frac{18.6 + 11.9}{9.7 - 6.84}$
- **k** $12.1 \div (3.7 + 4.2)$
- **l** $8.3 \div (4.2 - 2.3)$
- **m** $(17.6 - 5.2) \div (0.8 + 9.6)$
- **n** $(29.7 - 5.4) \div (11.4 - 5.27)$
- **o** $(8.3 - 4.9) \div (2.68 + 1.37)$

5 For each of these write down a calculation you could do to estimate the answer.
- **a** A club spent £24 862 on a new building and £6842 on landscaping.
How much was spent altogether?
- **b** On Saturday 8642 people visited an exhibition. On Sunday 11 542 visited it. How many more people visited the exhibition on Sunday than on Saturday?
- **c** Each class at Hendley School had 32 students.
How many students were in this school if there were 17 classes?
- **d** 46 students each made 23 masks to be sold at the school fête.
How many masks did the students make altogether?
- **e** In a fire drill, the students at a school lined up in rows of 38.
How many rows of students would there be if there were 798 students at school on the day of a fire drill?
- **f** There are 18 classes at a school. Each class was given raffle tickets to sell.
How many raffle tickets must each class sell if the school has to sell 630 tickets?
- **g** Robert made 18 identical wooden statues. He used 16·65 m of wood altogether.
How much did he use for each statue?
- **h** Carly frames pictures of pop stars. Each frame uses 1·35 m of wood.
How much wood does she need for 21 of these pictures?
- **i** Elastic is £0·85 per metre. Elaine bought 6·2 m.
How much did this cost?
- **j** One of the highest rainfalls in 1 minute happened in Maryland, USA in 1956. 1·23 inches fell in 1 minute.
If 1 inch = 2·54 cm, how many centimetres fell?
- **k** Jupiter takes 11·862 Earth years to orbit the Sun. How many Earth years does it take to orbit the Sun 24 times?
- ***l** Mars takes 1·881 Earth years to orbit the Sun.
How many times does Mars orbit the Sun in 78 Earth years?

*6 Firaz is finding the lengths of these rectangles from the areas and widths.
Estimate the answers he should get.
- **a** area = 560 m² width = 40 m
- **b** area = 400 cm² width = 18 cm
- **c** area = 1864 cm² width = 59 cm

This links to measures

Review 1 Choose the best approximation for each calculation.
a 343×22 A 400×20 B 300×20 C 300×30
b $782 \div 38$ A $700 \div 40$ B $800 \div 30$ C $800 \div 40$
c $128 \div 19$ A $100 \div 20$ B $200 \div 20$ C $100 \div 10$
d $3{\cdot}86 \div 2{\cdot}4$ A $4 \div 2$ B $3 \div 2$ C $4 \div 3$
e $29{\cdot}23 \times 16{\cdot}72$ A 30×20 B 30×10 C 20×20

Review 2 Estimate the answer to these calculations. Show how you found your estimate.
a 189×126 b 583×650 c $598 \div 26$ d $888 \div 32$
e $209 \div 19$ f $396 \div 22$ g $684 \div 36$ h $578 \div 34$

Review 3 Estimate the answers to these.
a $19{\cdot}7 \times 3$ b $56{\cdot}2 \div 8$ c $93{\cdot}6 \times 4$ d $72{\cdot}8 \div 9$
e $51{\cdot}7 \times 3{\cdot}6$ f $80{\cdot}7 \div 4{\cdot}2$ g $285 \div 3{\cdot}1$ h $6 \times 24{\cdot}3$

Review 4 Write down a calculation you could do to find an estimate for these.
Find the estimate.
a $\frac{139 \times 52}{19}$ b $\frac{394}{5{\cdot}7 + 1{\cdot}9}$ c $\frac{21{\cdot}7 + 9{\cdot}3}{8{\cdot}6 - 3{\cdot}2}$
d $(16{\cdot}4 - 8{\cdot}9) \div (3{\cdot}64 + 4{\cdot}27)$

Review 5 For each of these, write down a calculation you could do to estimate the answer.
a   A train to Blackpool had 23 carriages.
    Each carriage had 88 people inside.
    How many people were on the train?
b   Martha used 8 bottles of sauce. Each bottle contained $16{\cdot}75$ mℓ.
    How much sauce did she use altogether?
c   A bus company charges £$48{\cdot}50$ to take a class of 25 on a trip.
    How much is this per pupil?
d   Dipesh bought 23 tickets to a school concert at £$7{\cdot}99$ per ticket.
    How much did they cost altogether?
e   778 football fans travelled by coach from Derby to Liverpool. Find the number of coaches
    needed if each could carry 43 passengers.

Summary of key points

Some ways of **adding and subtracting mentally** are
partitioning
counting up
adding too much then taking some away (compensation)
recognising complements
nearly numbers
subtracting too much then adding back (compensation).

Some ways of **multiplying and dividing mentally** are
partitioning
using factors
doubling one number and halving another
multiplying then dividing

C The order in which we do operations is

Brackets
⇓
Squares (**I**ndices)
⇓
Division and **M**ultiplication
⇓
Addition and **S**ubtraction

Remember this with the word BIDMAS.

Work from left to right doing brackets, then left to right doing indices, then left to right doing multiplication and division, then left to right doing addition and subtraction.

Examples $7 + 20 \div 5 = 7 + 4$ $3 \times 8 + 2 \times 5 = 24 + 10$
                         $= 11$ $= 34$

$\dfrac{80}{4 \times 5} = \dfrac{80}{(4 \times 5)}$ $\dfrac{4^2 + 8}{3^2 - 1} = \dfrac{16 + 8}{9 - 1}$

The horizontal line acts as a bracket.
                         $= \dfrac{80}{20}$ $= \dfrac{24}{8}$
                         $= 4$ $= 3$

 D Sometimes **an estimate** is a good enough answer to a question.

Examples   How many people attended a concert?
                     How many animals does a vet see each year?

 E To **estimate the answer to a calculation** we round the numbers.
There is often more than one possible estimation.

Example $382 \times 24 \approx 400 \times 20 = 8000$
             or $382 \times 24 \approx 400 \times 25 = 10\,000$
             or $382 \times 24 \approx 380 \times 20 = 7600$

Example $217 \div 25 \approx 200 \div 20 = 10$
             or $217 \div 25 \approx 210 \div 30 = 7$
             or $217 \div 25 \approx 200 \div 25 = 8$

Try and **round to 'nice numbers'**.

Examples   Approximate $\frac{61}{7}$ to $\frac{63}{7}$ rather than $\frac{60}{7}$.

                     Approximate $\frac{183 + 297}{26}$ to $\frac{500}{25}$ rather than to $\frac{500}{30}$.

Test yourself   **Find the answers mentally.**

1 a $15 + 3 - 11 + 16 - 2$ **b** $100 - 50 + 20$ **c** $7 + {}^-2 + 3 - {}^-4 + 8$ **A**
  d $14 - {}^-3 - 8 + {}^-4 + 2$ **e** $36 + 14$ **f** $159 + 58$
  g $411 - 379$ **h** $5200 - 384$

2 Find x.   **a** $100 - x = 63$ **b** $100 = 56 + x$ **c** $752 = 1000 - x$ **A**

3 Copy this magic square.
   Fill it in.

4 a $3 \times 5 \times 4$ b 65×2 c 37×4 d 82×5 e 150×3
   f $78 \div 2$ g $128 \div 4$ h 40×20 i 120×30 j 82×11
   k 35×12 l $390 \div 3$ m 36×50 n $1850 \div 2$

5 a $1\cdot3 \times 2$ b $4\cdot3 \times 5$ c $5\cdot6 \times 7$ d $4\cdot8 \div 2$ e $6\cdot3 \div 9$
   f $0\cdot3 \times 8$ g $9 \times 0\cdot6$ h $5 \times 0\cdot04$ i $1\cdot8 \div 3$ j $4\cdot2 \div 6$
   k $5\cdot1 \times 11$ l $4\cdot3 \times 30$ m $5\cdot3 \times 50$ n $4\cdot7 \times 13$

6 a Find the sum of $8\cdot6$ and $2\cdot4$ and subtract $1\cdot5$.
   b Find the difference between 784 and 396.
   c Find the product of 5 and 97.

7 a $2\cdot4 \times 50$ b $6\cdot6 \times 40$ c $4\cdot7 \times 60$ d $3\cdot9 \times 30$

8 a $7 \times 0\cdot7$ b $0\cdot8 \times 0\cdot3$ c $0\cdot4 \div 8$ d $0\cdot6 \div 5$ *e $0\cdot018 \div 6$ *f $0\cdot036 \div 0\cdot6$

9 a $650 \div 5$ b $464 \div 4$ c $36\cdot3 \div 3$ d $750 \div 15$

10 The most overdue library book ever, was taken out in 1823 by a man and returned
   in 1968 by his great grandson.
   How many years overdue was the book?

11 Tricia wanted to buy 9 kg of potatoes which cost £8·10 in total.
   She only had £4·39.
   a How much more did Tricia need?
   b How much does one kilogram of potatoes cost?

12 a $5 \times 3 - 4$ b $6 + 2 \times 4$ c $15 - 3 \times 4$ d $15 \times 2 - 3 \times 4$
   e $5 + 12 \div 2$ f $6(8 + 2)$ g $\frac{9+6}{5}$ h $\frac{24}{3 \times 2}$
   i 2×5^2 j $3(4^2 - 8)$ k $\frac{3^2 - 1}{2}$

13 a The organisers of a concert want to know how many people a concert hall
      seats. Does this need to be an exact number rather than an estimate?
   b Estimate, to the nearest 100, how many phone calls you have made in the last year.
      Show how you found your estimate.

14 Estimate the answers to these. Show how you found your estimate.
   a 186×24 b $781 \div 36$ c $31\cdot6 \times 7\cdot2$ d $392 \div 4\cdot5$
   e $\frac{4\cdot2 \times 3\cdot6}{3\cdot7}$ f $\frac{3\cdot8 + 7\cdot2}{5\cdot1 - 2\cdot7}$ g $5\cdot4 \div (8\cdot6 - 2\cdot7)$

15 Write down a calculation you could do to find an approximate answer to this.
   Renée bought 24 chocolate bears. Each one has a mass of $16\cdot97$ g.
   What is their total mass?

4 Written and Calculator Calculations

You need to know

✓ written calculation — adding and subtracting whole numbers   page 4
— adding and subtracting decimals with up to 2 decimal places
— multiplying 2- and 3-digit numbers by a single digit
— multiplying 3-digit by 2-digit numbers
— dividing numbers by a single digit

✓ estimating   page 4

Key vocabulary

add, adding, brackets, calculate, calculator, change, currency, difference, divide, dividend, division, divisor, equivalent calculation, estimate, inverse, key, multiplication, multiply, operation, product, quotient, remainder, subtract, subtraction, sum, total

Kaprekar's number

Choose any 4-digit number, such as, 3259.

Step 1   Write the digits from largest to smallest.   9532
Write the digits from smallest to largest.   −2359
Subtract   7173

Step 2   Take the answer you get and keep repeating step 1.

7731	6543	8730	8532
−1377	−3456	−0378	−2358
6354	3087	8352	6174

6174 is **Kaprekar's number**.

What happens if you repeat *Step 1* with Kaprekar's number?

Do you get Kaprekar's number if you begin with *any* 4-digit number?

Adding and subtracting

Rosie loved stamps. She had collected 7654.
She bought her brother's collection of 278 stamps.
She wanted to know how many she had then.

We **add and subtract whole numbers** by lining up the units column.

```
  7654          ⁵¹⁴₁
+  278         7̶6̶5̶4̶
 ────          −  278
 7932          ────
  ₁₁           7376
```

Rosie has 7932 stamps now. She had 7376 more stamps than her brother before he sold her his collection.

To **add and subtract decimals** we line up the decimal points.

Worked Example
Calculate   a $862 \cdot 3 + 9 \cdot 87$ b $283 \cdot 6 - 27 \cdot 57$

Answer

a $862 \cdot 3$
   $+ \quad 9 \cdot 87$
   ─────────
     $872 \cdot 17$
       ₁₁

 The decimal points are lined up.

b   ⁷₁ ⁵₁
      $283 \cdot 6̶0̶$
    $- \quad 27 \cdot 57$
    ─────────
      $256 \cdot 03$

 Fill the gaps with zeros.

Exercise 1

1 a $84 + 37 + 586$
   c $369 + 52 + 1043 + 2$
   e $53 + 862 + 1957 - 841$
   g $4279 - 1302 + 48 - 361$

   b $47 + 392 + 487$
   d $4327 + 964 + 53 - 270$
   f $3427 - 589 - 36$

2 a 156 and 432 are 3-digit numbers that add to 588.
       Use the digits 1, 2, 3, 4, 5 and 6 to make two other
       3-digit numbers that add to 588.
       How many ways are there of doing this?

   b Use the digits 1 to 7 to make a 4-digit and a 3-digit number that add to 7777.
      Find at least six ways of doing this.

```
   156
+  432
 ─────
   588
```

3 Choose two different numbers from the cloud.
   Find the sum and the difference of the numbers.
   Make all the sums and differences it is possible to make
   by choosing two different numbers from the cloud.
   How many are there?

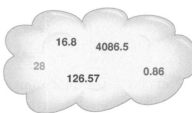

16.8   4086.5
28
126.57   0.86

4 a $5 \cdot 3 + 27 \cdot 96 + 4 \cdot 7$
   c $56 \cdot 56 + 412 \cdot 2 + 6 \cdot 07$
   e $864 \cdot 39 + 2 \cdot 7 + 89 \cdot 07$
   g $96 \cdot 4 - 3 \cdot 75 + 2 \cdot 86 + 31 \cdot 47$
   i $93 - 37 \cdot 68 - 2 \cdot 1 + 13 \cdot 75$

   b $186 \cdot 04 + 3 \cdot 7 + 86 \cdot 94$
   d $83 \cdot 75 + 1842 \cdot 9 + 3 \cdot 57$
   f $4 \cdot 27 + 0 \cdot 8 + 18 \cdot 46 + 42 \cdot 3$
   h $576 \cdot 3 - 28 \cdot 47 + 3 \cdot 81 + 16$

5 ☐☐☐ + ☐☐ = 213   ☐☐☐ − 123 = ☐☐☐   ☐☐·☐ + ☐☐·☐☐ = 186·24
   Find at least three ways to fill in these boxes. Put one digit in each box.

6 **a** Find the total of 86·3 and 5·27.
   b Find the difference between 298·03 and 36·4.
   c Find the sum of 8·68, 19·24 and 0·07.
   d The difference between two numbers is 3·45.
       What might the numbers be?

7 Anthony wanted to fill in his car log book. He made four short journeys in his car.
   These were 6·8 km, 10·7 km, 9 km and 4·4 km.
   The reading on the speedometer at the end of these journeys was 69 435·1 km.
   What was the reading before the journeys?

8 Here is a table of record throws.

Frisbee	190·07 m	Boomerang	121 m
Gumboot	52·73 m	Cricket ball	128·6 m
Sling shot	437·13 m	Haggis	55·11 m
Rolling pin	53·4 m	Brick	44·54 m

Find out some more
records for throwing.

   i Find the difference between the throws of
       a a haggis and a brick   **b** a frisbee and a cricket ball
       c a rolling pin and a sling shot   **d** a boomerang and a gumboot.
   ii Find the difference between the longest and shortest throws given.

9 4·86 mm, 3·08 mm and 6·4 mm of rain fell in Manchester one bank holiday weekend.
   a How much rain fell altogether?
   b The same weekend, 16 mm of rain fell in Minehead.
       How much more rain fell in Minehead?

10 Thomas pours 1·38 litres then 2·6 litres out of a 5 litre container of milk.
    How much milk is left in the container?

11 Prada pays £32·65 for groceries and £16·70 for dinner.
    How much does she have left out of £100?

12 Which is greater, the sum of 18·76 and 12·9 or the difference between 112 and 79·68?

13 Five people were in the final of an archery contest.
    They each had six shots.
    Only one bullseye was scored. The scores are in order of size.
    What are the missing scores?

							TOTAL
Rajiv	19	19	39	39	45	45	206
Peter		39					243
Toby					45		186
Frederick	19						226
Edward	19						244

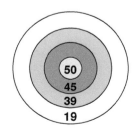

14 Use the number cards Billie is holding to make two 4-digit decimal numbers with at least one digit before and after the decimal point.
Find
   a   the largest possible sum
   b   the smallest possible difference
  *c   the third largest sum
  *d   the second smallest difference.

Review 1

		E					E		
11·09	3905	12 426		76·38	223·43	6·6	12 426	14·09	11·09

								E
18·21	76·38	105·74	6·6		25·03	4·86		11·09	3905	12 426

E
12 426   4·86   29·42   223·43   25·03   14·09   3905

							E			
223·43	1906	4·86	29·42	3·54	1906	29·42	12 426		25·03	14·09

11·09   76·38   18·21   4·86

Use a copy of this box. Write the letter that is beside each question above its answer.

E   46 + 8327 + 582 + 3471 = 12 426
I   17 + 3·8 + 4·23
U   5·36 + 2·9 − 4·72
L   52·6 + 147·2 + 23·63
N   15 − 3·8 − 6·34

A   4862 − 3042 + 86
R   98·71 + 2·8 + 4·23
D   16·48 − 5·2 − 4·68
O   64·7 + 33·4 − 21·72
W   53·76 − 40·25 + 4·7

H   4371 − 586 + 204 − 84
S   5·07 + 3 + 0·82 + 5·2
G   57 + 3·84 − 31·42
T   16 + 0·82 − 5·73

Review 2   Guy had £118·62 in the bank. He earned another £78·96 mowing lawns.
How much more did he need to buy a stereo for £211·86?

Review 3   Joe bought some soccer cards for £14·63. He bought a pouch for them for £2·58.
He sold them both for a total of £15·73.
How much did he lose?

Review 4   Put the digits 1 to 8 in the boxes to make each true. Use each digit only once.
a   ☐☐ ☐☐☐ + ☐☐☐ = 83 979
b   ☐☐ ☐☐☐ − ☐☐☐ = 70 666
c   ☐☐ + ☐·☐ + ☐☐·☐☐ = 110·25
d   ☐☐☐·☐ − ☐☐·☐ = 664·15

?   Puzzle

36 + 12 = 372   136 − 4 = 96   281 + 12 − 145 = 31

Where should the decimal points be placed to make these true?

Is there more than one answer?

Investigation

Ten pounds and eighty-nine pence

Follow these steps, **writing all numbers as decimals with two decimal places**.

Example

1 Choose an amount of money less than £10. **£6·72**
2 Reverse the digits and write down the new amount. **£2·76**
3 Find the difference. £6·72 − £2·76 = **£3·96**
4 Reverse the digits in your answer to **3**. **£6·93**
5 Add the answers to **3** and **4** together. £3·96 + £6·93 = **£10·89**

Repeat **1** to **5**, starting with different amounts.
Do you always get £10·89? **Investigate**.

Multiplication

Multiplying by a 1-digit number

Marie wanted to find the cost of 3·74 m of plastic tube for her mouse house at £8 per metre.
3·74 × 8 is approximately 4 × 8 = 32.
Two ways of finding the answer are shown.

Always estimate the answer first.

3·74 × 8				3·74 × 8 is equivalent to 374 × 8 ÷ 100.
×	**3**	**0·7**	**0·04**	374
8	24	5·6	0·32	× 8
Answer 24 + 5·6 + 0·32 = **29·92**				2992 2992 ÷ 100 = **29·92** ₅₃

Use the estimate to check that your answer is about the right size.

Worked Example 562·7 × 4

Answer
Two ways of finding the answer are shown.

562·7 × 4				562·7 × 4 is equivalent to 5627 × 4 ÷ 10.
×	**500**	**60**	**2**	**0·7**	5627
4	2000	240	8	2·8	× 4
Answer 2000 + 240 + 8 + 2·8 = **2250·8**				22508 22508 ÷ 10 = **2250·8** ₂₁₂

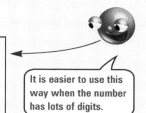
It is easier to use this way when the number has lots of digits.

Exercise 2 Estimate the answers, then calculate.

1
a 3·42 × 3	b 4·61 × 2	c 8·17 × 3	d 3·76 × 4
e 5·93 × 6	f 12·7 × 3	g 16·4 × 4	h 21·3 × 5
i 2·46 × 7	j 25·62 × 8	k 19·34 × 5	l 17·21 × 6
m 121·8 × 3	n 115·6 × 4	o 135·2 × 6	p 83·61 × 4
q 152·7 × 9	r 309·5 × 4	s 443·5 × 8	

2
a 462 × 0·6	b 383 × 0·8	c 729 × 0·03	d 594 × 0·05
e 52·3 × 0·3	f 89·6 × 0·7	g 43·2 × 0·09	h 4·86 × 0·03
i 1·82 × 0·7	j 82·6 × 0·04	k 41·3 × 0·06	

3 Johnny bought 4 m of wood at £11·59 per metre.
How much did it cost?

4 Shabir measured one of her paces as 78·3 cm.
Her bedroom is 6 paces long and 4 paces wide.
 a What is the length and width of Shabir's bedroom?
 b What is the area of Shabir's bedroom?

Link to measures page 355.

5 A table tennis ball travels 47·22 m in one second.
A squash ball travels 64·64 m in one second.
How much further would the squash ball travel in five seconds than the table tennis ball?

6 A chair is 42·63 cm wide. Nine of these chairs are joined side by side.
 a How long is the row of nine chairs?
 b How long is a row of 27 chairs, if they are put in three sets of nine chairs, with a space of 0·89 m between each set?

7
Find four different ways to fill in the boxes.

8 A banana cake has 0·46 grams of dietary fibre per 100 grams.
How much dietary fibre does a 600 gram piece have?

Gosh, I'm tired.

*9 Baxter the snail crawls 0·043 m per hour. If he keeps up this pace, how far does he crawl in
 a 8 hours b a day (use your answer to **a**)?

Review 1 Multiply these.
a 5·61 × 4	b 14·6 × 5	c 1·64 × 7	d 24·69 × 9
e 11·35 × 8	f 187·3 × 4	g 196·7 × 3	h 352·4 × 9
i 486 × 0·7	j 51·3 × 0·04	k 1·46 × 0·9	

Review 2
a Pat bought 6 ℓ of this juice.
How much did it cost?
b Peter bought 9 ℓ of this juice.
How much did it cost?

Fill your own
Juice
£1.84 per litre

* Review 3 The instructions on a packet of fertiliser say to spread 0·75 kg per square metre. How much is needed for this lawn?

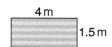
4 m
1.5 m

Multiplying by a 2-digit number

Martine invited 165 people to her parents' 50th wedding anniversary. The meal for each guest cost £12.
The total cost was £165 × 12.
We always **estimate** the answer first.
165 × 12 is approximately 200 × 10 = 2000.

```
    165
×    12
  1650   165 × 10
   330   165 × 2
  1980
```

Worked Example
21·6 × 4·5

Answer
21·6 × 4·5 is approximately 20 × 5 = 100.
Two ways of finding the answer are shown.

Method 1				
×	**20**	**1**	**0·6**	**Check**
4	80	4	2·4	86·4
0·5	10	0·5	0·3	+10·8
	90	+ 4·5	+ 2·7 =	**97·2**

We add the totals of the columns to get the answer.
90 + 4·5 + 2·7 = **97·2**
We check this by adding the totals of the rows.
86·4 + 10·8 = 97·2

Method 2
21·6 × 4·5 is equivalent to 21·6 × **10** × 4·5 × **10** ÷ **100**
or 216 × 45 ÷ 100

```
   216
×   45
  8640   216 × 40
  1080   216 × 5
  9720
    1
```

9720 ÷ 100 = **97·2**

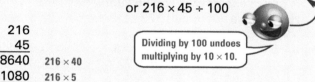
Dividing by 100 undoes multiplying by 10 × 10.

Worked Example
2·37 × 27

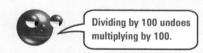
Dividing by 100 undoes multiplying by 100.

Answer
2·37 × 27 is approximately 2 × 30 = 60.
2·37 × 27 = 2·37 × 100 × 27 ÷ 100
      = 237 × 27 ÷ 100

```
   237
×   27
  4740   237 × 20
  1659   237 × 7
  6399
    1
```

6399 ÷ 100 = **63·99**

Exercise 3 **Estimate the answers then calculate.**

T

1 Use a copy of this crossnumber.
Fill it in.

Across		**Down**	
1.	341 × 38	1.	59 × 19
3.	235 × 21	2.	211 × 45
5.	133 × 18	3.	128 × 35
7.	769 × 89	4.	447 × 86
8.	15 × 11	6.	118 × 31
9.	11 × 23	10.	131 × 40
12.	33 × 19	11.	28 × 13
14.	99 × 27	12.	102 × 62
15.	266 × 34	13.	214 × 33
16.	43 × 22	15.	12 × 8
17.	103 × 14		

2 There are 687 Earth days in a year on Mars.
How many Earth days are there in these Mars years?
a 28   **b** 16   **c** 34

Challenge: How old are you in Mars years?

3 A pencil can draw a line 56 kilometres long.
How long a line could 128 of these pencils draw?

4
a 6·2 × 17	**b** 3·2 × 13	**c** 5·6 × 23	**d** 8·2 × 16
e 14 × 19·3	**f** 25 × 27·4	**g** 684 × 7·3	**h** 892 × 9·4
i 534 × 4·6	**j** 72 × 5·21	**k** 39 × 7·32	**l** 94 × 3·27
m 88 × 4·62	**n** 37 × 5·09	**o** 53 × 6·04	

5
a 6·3 × 8·4	**b** 5·2 × 7·8	**c** 6·4 × 3·7	**d** 4·2 × 10·7
e 26·4 × 2·4	**f** 46·3 × 2·5	**g** 19·9 × 1·6	**h** 2·7 × 37·4
i 5·37 × 2·6	**j** 4·71 × 8·2	**k** 9·63 × 5·7	**l** 5·3 × 8·39
m 6·4 × 3·26	**n** 5·82 × 3·8	**o** 4·5 × 3·99	

6 Find the cost of these. You may have to round your answer to
the nearest penny.
a Margaret bought 18 m of silk.
b Thomas bought 8·4 m of cotton.
c Wasim bought 9·4 m of linen.
d Raewyn bought 5·2 m of velvet.
e Allanah bought 6·3 m of linen and 2·7 m of silk.

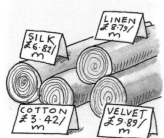

SILK £6.82/m
LINEN £8.79/m
COTTON £3.42/m
VELVET £9.89/m

7 Rohan bought 21·6 m of garden edging at £2·30 per metre and 18·6 kg of compost at £2·45
per kilogram.
a How much did this cost?
b How much change did he get from £100?

8 Make up word problems for these calculations.
a 36 × 41 = 1476   **b** 862 × 21 = 18 102   *c 15 × 21 × 33 = 10 395

*9 Find two consecutive numbers with a product of
a 552   **b** 1406.

To find the *product* we multiply numbers.

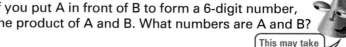

*10   Find three consecutive numbers with a product of 24 360.

*11   2·4 × 6·3 = 15·12   4·2 × 3·6 = 15·12
       4·2 is 2·4 with the digits reversed.
       3·6 is 6·3 with the digits reversed.
       Find other numbers between 0 and 10, with 1 d.p., that this works for.

*12   A and B are two 3-digit numbers. If you put A in front of B to form a 6-digit number,
       this 6-digit number is three times the product of A and B. What numbers are A and B?

This may take some time

T   **Review 1**

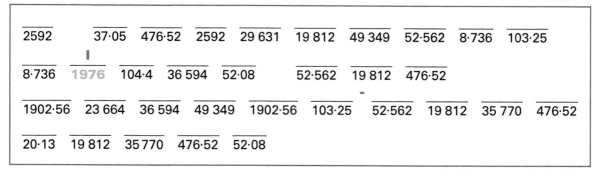

| 2592 | | 37·05 | 476·52 | 2592 | 29 631 | 19 812 | 49 349 | 52·562 | 8·736 | 103·25 |

| 8·736 | **1976** | 104·4 | 36 594 | 52·08 | | 52·562 | 19 812 | 476·52 |

| 1902·56 | 23 664 | 36 594 | 49 349 | 1902·56 | 103·25 | 52·562 | 19 812 | 35 770 | 476·52 |

| 20·13 | 19 812 | 35 770 | 476·52 | 52·08 |

Use a copy of this box. Write the letter beside each question above its answer.
I   52 × 38 = 1976   A   96 × 27   O   381 × 52   U   730 × 49   N   809 × 61
G   357 × 83   E   642 × 57   W   986 × 24   V   3·6 × 29   S   14 × 3·72
H   18·3 × 1·1   T   82·72 × 23   Y   41·3 × 2·5   L   3·64 × 2·4   R   83·6 × 5·7
D   24·7 × 1·5   F   6·41 × 8·2

Review 2   Find the cost of these. You may have to round your answer to the nearest penny.
a   Rose bought 15 m of rope.
b   Peter bought 26 m of hose.
c   Rhian bought 3·5 m of chain.
d   Habib bought 17·2 m of plastic.
e   Jake bought 3·2 m of rope and 15·6 m of plastic.

rope £1·86/m   plastic £1·20/m
hose £3·92/m   chain £8·20/m

Review 3   Find two numbers with a product of 875 and a difference of 10.

Review 4   The currency exchange rate between British pounds and American dollars is
£1 = US $1·62. How many dollars would you get for £18?

Puzzle

1   What values might A, B and C have?
   a   A·B   b   A·A
        × B A   × B B
        ─────   ─────
         BC·B   BC·B

*2   Make up some more decimal multiplications where the numbers have been
     replaced by letters. Make sure your multiplication has at most, three different
     digits. Some of the digits in the answer should be the same as the digits in the
     numbers being multiplied. Give them to someone else to solve.

Division

Dividing by a 1–digit whole number

Worked Example $26·28 \div 6$

Answer $26·28 \div 6$ is approximately $24 \div 6 = 4$

24 is the closest number that divides easily by 6.

```
  6 ) 26·28
    −24·00     6 × 4
      2·28
    − 1·80     6 × 0·3     because 6 × 3 = 18
      0·48
    − 0·48     6 × 0·08    because 6 × 8 = 48 and 6 × 0·8 = 4·8
      0·0
```

or

```
      4·38
      ²⁴
  6 ) 26·28
```

Notice that the decimal points line up.

Answer   **4·38**

Exercise 4   Estimate the answers, then calculate.

1
a 5.5 ÷ 5	b 6·78 ÷ 2	c 4·84 ÷ 4	d 45·5 ÷ 5
e 12·72 ÷ 6	f 9·38 ÷ 7	g 34·5 ÷ 5	h 74·88 ÷ 6
i 69·02 ÷ 7	j 740·4 ÷ 6	k 393·4 ÷ 7	l 510·3 ÷ 9
m 438·9 ÷ 3	n 826·36 ÷ 4	o 762·88 ÷ 8	p 499·44 ÷ 6
q 457·24 ÷ 7	r 589·95 ÷ 9	s 496·85 ÷ 5	*t 13·2 ÷ 5
*u 20·4 ÷ 8	*v 6·09 ÷ 6	*w 0·42 ÷ 5	*x 0·049 ÷ 2
*y 4·036 ÷ 8	*z 0·351 ÷ 4		

Hint: The answer to the ones with a * will have more d.p. than the number you divided into (dividend).

2   Eight equal-sized books were stacked on top of one another. The stack was 49·6 cm high. How thick was each book?

3   Nine people shared the cost of a meal equally. The meal cost £239·22. How much did each pay?

4   Five people stood on a bridge. Their total mass was 384·25 kg.
   a   What was the average mass of each person?
   b   The bridge mass limit is 700 kg.
      Can nine people with the average mass you found in **a**, safely stand on the bridge?

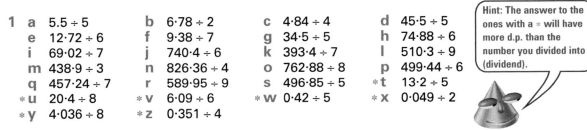

5   Which is the better buy?
   a   8 kg of apples for £14·32 or 5 kg of apples for £8·90.
   b   2 m of tape for £3·64 or 5 m of tape for £8·90.
   *c   2 kg of cereal for £11·34 or $\frac{1}{2}$ kg of cereal for £2·85.
   *d   3 ℓ of juice for £3·95 or $\frac{1}{4}$ ℓ of juice for 35p.

T| **Review 1**

| $\overline{0\cdot8}$ | $\overline{23\cdot4}$ | $\overline{73\cdot49}$ | | $\overline{0\cdot8}$ | $\overline{59\cdot7}$ | $\overline{89\cdot4}$ | $\overline{1\cdot2}$ | $\overline{73\cdot49}$ | $\overline{73\cdot49}$ | $\overline{0\cdot8}$ | $\overline{5\cdot6}$ | $\overline{70\cdot4}$ | $\overline{56\cdot7}$ |

| $\overline{23\cdot4}$ | $\overline{1\cdot2}$ | | $\overline{73\cdot49}$ | $\overline{89\cdot73}$ | $\overline{56\cdot7}$ | $\overline{56\cdot7}$ | $\overline{78\cdot58}$ | $\overline{56\cdot7}$ | | $\overline{97\cdot3}$ | $\overline{0\cdot8}$ | $\overline{23\cdot4}$ | $\overline{58\cdot6}$ |

| $\overline{3\cdot03}$ | $\overline{1\cdot2}$ | $\overline{4\cdot1}$ | $\overline{8\cdot3}$ | | $\overline{56\cdot7}$ | $\overline{3\cdot03}$ | $\overline{56\cdot7}$ | $\overline{73\cdot49}$ | | $\overline{1\cdot2}$ | $\overline{89\cdot4}$ | $\overline{56\cdot7}$ | $\overline{89\cdot73}$ |

Use a copy of this box. Put the letter that is beside the division above its answer.

I $6\cdot4 \div 8 = 0\cdot8$	O $10\cdot8 \div 9$	B $39\cdot2 \div 7$	R $49\cdot8 \div 6$
U $32\cdot8 \div 8$	Y $21\cdot21 \div 7$	T $210\cdot6 \div 9$	H $351\cdot6 \div 6$
W $486\cdot5 \div 5$	M $238\cdot8 \div 4$	L $563\cdot2 \div 8$	P $268\cdot2 \div 3$
E $226\cdot8 \div 4$	N $448\cdot65 \div 5$	S $587\cdot92 \div 8$	Z $314\cdot32 \div 4$

* **Review 2** A block of apricot cheese is to be divided into eight equal pieces. The block has a mass of $98\cdot2$ g.
What is the mass of each piece?

* **Review 3** Which is the better buy?
9 kg of potatoes for £6·57 or 8 kg for £5·76

Dividing by a 2-digit whole number

Worked Example
472 books are put onto shelves. Each shelf holds 23 books.
a How many shelves are filled?
b How many books are on the shelf that is not filled?

Answer
We must divide 472 by 23 and find the remainder.

$472 \div 23$ is approximately $500 \div 20 = 25$

$$23 \overline{)472}$$
$$\underline{-460} \quad 23 \times 20 \qquad \text{because } 23 \times 2 = 46$$
$$12$$

> 23 goes into 47 just more than twice so 23 will go into 472 just more than 20 times.

Answer 20 R12
a **20 shelves** can be filled.
b **12 books** are on the shelf that is not full.

Worked Example
$59\cdot5 \div 14$

Answer
$$14 \overline{)59\cdot5}$$
$$\underline{-\ 56} \quad 14 \times \mathbf{4}$$
$$3\cdot5$$
$$\underline{-2\cdot8} \quad 14 \times \mathbf{0\cdot2} \qquad \text{because } 14 \times 2 = 28$$
$$0\cdot7$$
$$\underline{-0\cdot7} \quad 14 \times \mathbf{0\cdot05} \qquad \text{because } 14 \times 5 = 70 \text{ and } 14 \times 0\cdot5 = 7$$
$$0\cdot0$$

Answer **4·25**

Exercise 5 **Always estimate your answer first.**

1   What is the remainder when
    a   747 is divided by 27
    b   864 is divided by 33
    c   573 is divided by 47?

2   680 eggs are packed into cartons which each hold 12 eggs.
    a   How many cartons can be packed?
    b   How many eggs are left over?

3   784 boxes of baked beans were given to 15 shops in a supermarket chain. Each shop got
    the same number of boxes.
    a   How many boxes did each shop get?
    b   How many boxes were left over?

4   A box of chocolate bars with 181 bars was given to a children's holiday camp.
    The 14 children got the same number each.
    a   How many bars did each child get?
    b   How many were left over?

5   Give the answer to these as a decimal.
    a $192 \div 15$ b $296 \div 16$ c $402 \div 15$ d $570 \div 24$ e $378 \div 28$
    f $238 \div 28$ g $318 \div 24$ h $70 \cdot 3 \div 19$ i $55 \cdot 2 \div 23$ j $41 \cdot 6 \div 26$
    k $58 \cdot 4 \div 16$ l $22 \cdot 5 \div 18$ m $44 \cdot 8 \div 14$ n $72 \cdot 8 \div 28$ o $88 \cdot 2 \div 21$
    p $78 \cdot 2 \div 17$ q $45 \cdot 5 \div 14$ r $58 \cdot 5 \div 13$

6   Dividend = $35 \cdot 1$ Divisor = 13   Find the quotient.

7   Twelve friends shared these costs.
    How much did each pay for
    a   the train   b   meals?

| train | £75.60 total |
| meals | £98.40 total |

*8   296 people live in an area of 16 square miles.
     What is the mean number of people per square mile?

*9 $\boxed{}\boxed{}\cdot\boxed{} \div \boxed{}\boxed{} = 5 \cdot 8$
     Put the numbers 1, 2, 6, 8 and 9 in the boxes to make this true.

Review 1   Voting papers are packed in bundles of 25.
a   How many bundles can be made from 873 voting papers?
b   How many would be left over?

Review 2   Give the answer to these as a decimal.
a $252 \div 15$ b $132 \div 16$ c $89 \cdot 6 \div 14$

Dividing by a decimal

> ## Discussion
>
>
>
> Jill   Madhu
>
> Is Madhu right?   **Discuss**.
>
> How would you find the answer to these?
>
> $81 \div 0.4$ $35 \div 0.08$ $686 \div 4.9$ $616 \div 3.2$ $162 \div 3.6$

To **divide by a decimal** we do an **equivalent calculation** which has a whole number as the divisor.

Worked Example
$21 \div 0.04$

Answer

$21 \div 0.04$ is approximately equal to $20 \div 0.04 = 2000 \div 4$

$= 500$

$21 \div 0.04$ is equivalent to $2100 \div 4$.

```
4)2100
 −2000    4 × 500
   100
  −100    4 × 25
     0
```

Answer   **525**

Exercise 6

        Always estimate your answer first.

1   Which of A, B, C or D is equivalent to the calculation given?
    a $42 \div 0.6$ A $42 \div 0.06$ B $4.2 \div 6$ C $42 \div 6$ D $420 \div 6$
    b $58 \div 0.4$ A $58 \div 4$ B $580 \div 0.4$ C $580 \div 40$ D $580 \div 4$
    c $300 \div 0.8$ A $300 \div 8$ B $3000 \div 8$ C $30 \div 8$ D $30 \div 0.8$
    d $642 \div 0.03$ A $642 \div 3$ B $64.2 \div 3$ C $64\,200 \div 3$ D $6420 \div 3$
    e $836 \div 0.05$ A $836 \div 5$ B $83\,600 \div 5$ C $83.6 \div 5$ D $8360 \div 50$
    f $542 \div 0.08$ A $542 \div 8$ B $54\,200 \div 8$ C $542 \div 0.8$ D $5420 \div 8$

2   Write an equivalent division you could do to find the answer to these.
    a $82 \div 0.2$ b $93 \div 0.3$ c $72 \div 0.04$ d $841 \div 0.09$

3 Calculate these.

   a 18 ÷ 0·2   b 24 ÷ 0·8   c 32 ÷ 0·4   d 81 ÷ 0·3
   e 42 ÷ 0·5   f 22 ÷ 0·04   g 273 ÷ 0·3   h 482 ÷ 0·4
   i 405 ÷ 0·9   j 301 ÷ 0·07   k 255 ÷ 0·06

*4 The currency exchange rate between British pounds and Australian dollars is about
   A$1 = £0·3. How many dollars would you get for £255?

Review 1 Write an equivalent division you could do to find the answer.
a 86 ÷ 0·3   b 914 ÷ 0·07

Review 2 Calculate these.
a 16 ÷ 0·4   b 72 ÷ 0·03   c 585 ÷ 0·5

Mixed calculations

Discussion

● Roy worked out the cost of 26 kg of apples at £1·99 per kg like this.

$$199$$
$$\times 26$$

Sarah worked out the cost like this.

$$26 \times £2 - 26 \times 1p = £52 - 26p$$
$$= £51·74$$

Which way is easier? **Discuss**.
Who is less likely to make a mistake? **Discuss**.

−6p   −20p

£51·74   £51·80   £52

$$199$$
$$\times 26$$
$$\overline{3980}$$
$$1194$$
$$\overline{5174}$$

$$5174 ÷ 100 = £51·74$$

In the next exercise you will need to decide whether to **add**, **subtract**, **multiply** or **divide**.
In most problems you will need to do more than one of these.

Exercise 7

1 What goes in the box, +, −, × or ÷ ?
   a 990 ☐ 120 = 1110   b 990 ☐ 120 = 118 800   c 990 ☐ 120 = 870
   d 990 ☐ 120 = 8·25

2 Mr Bascand prints books.
  Claire pays £6·05 to have her book printed, including the cover.
  How many pages are there in her book?

Book Printing
Each page
Cover
4p
85p

3 Eight friends each bought the 'Todays Special'.
   a How much did this cost altogether?
   b How much change did they get from £75?
   c Maddison bought a pudding for £3·75 and a 'Todays Special'.
      How much change did she get from £20?

TODAYS
SPECIAL
£8.65

4 A drink and a chocolate bar cost 80p.
  Two drinks and a chocolate bar cost £1·25.
  How much does the chocolate bar cost?

5 A mini-bus can carry eight people.
   a How many minibuses will be needed to take 187 people to a game?
   b On the return journey, one minibus didn't turn up. A taxi which can take four passengers was called. Was one taxi enough? Explain.

6 Choco Treats sells chocolates.
   Saskia pays £11·75 for a gift-wrapped box of chocolates.
   How many chocolates were in Saskia's box?

Chocolates 85p each
Gift wrapped in a box
an extra £1.55 per box

7 Ross, the snail, crawled 335 cm in 8 hours.
   a How far did he crawl, to the nearest cm, each hour?
   b How far did he crawl in 0·25 of a day? Round you answer to the nearest cm.

8 Mark worked in a mail order seed warehouse.
   Some of the weights are shown in the table.
   a Work out the weight of these if they are put in a bag and then a stamped envelope.
      i 4 packets of pansies
      ii 3 packets of pansies and 2 packets of mixed seeds
      iii 12 packets of pansies and 7 packets of mixed seeds
   b How much would 12 stamps weigh?
   c A sheet of stamps weighed 2 g. How many stamps made up the sheet?

Weights	
Bag	2.76 g
Envelope	3.82 g
Stamp	0.08 g
Pansies	8.53 g
Mixed Seeds	12.87 g

9 Make up some word problems for these calculations.
   a $186 \times 7 = 1302$ **b** $19·4 \times 3 = 58·2$ **c** $183·4 \div 4 = 45·85$ **d** $9·6 + 3·7 - 2·4 = 10·9$

Review 1 Which operation, +, −, × or ÷, does * stand for?
a $306 * 85 = 26\,010$ **b** $306 * 85 = 221$ **c** $306 * 85 = 3·6$ **d** $306 * 85 = 391$

Review 2
a In a church wall there are eight small windows side by side. Each window is 1·14 m wide. How long is the row of eight windows?
b The church wall is 15·5 m long. What length of wall has no window?

Review 3 In olympic diving, each competitor is given a score by seven judges. The highest and lowest scores are eliminated. The rest are added and the total divided by 5. This gives the final score. What was the final score for these competitors?
a Yanus   7·0, 8·0, 7·5, 6·5, 7·5, 7·0, 7·5
b Beck   8·6, 8·7, 9·1, 7·8, 8·2, 7·4, 8·3
c Ewitt   4·3, 5·9, 6·1, 6·4, 5·8, 7·7, 7·2

Checking if answers are sensible

Gareth calculated the distance between his classroom and the school gate.
He got an answer of 2·6 km.
This is not sensible because 2·6 km is much too far.

Always **check** that **your answer** to a calculation is **sensible**.
Ask yourself 'Is the answer sensible and about what I would expect?'

Worked Example
Jason worked out how many 24p pens he could buy for £5.
His answer was 20.
Do you think Jason was right or wrong? Why?

Answer
We check if the answer is sensible.
He can buy 4 pens for just under £1.
So he can buy 4 × 5 = 20 for under £5.
His answer is about right.

You can check the answer is sensible in other ways.

Examples   The sum of two odd numbers should be an even number.
If the last digits of two numbers you have multiplied together are 6 and 7, then the last digit of the answer should be **2** because 6 × 7 = **42**.
If you multiply by 3, the answer should be divisible by 3.
If you multiply 1·59 × 24, the answer should be between 1 × 24 = 24 and 2 × 24 = 48.

Exercise 8

Answer these questions **without** doing the calculation.

1   Adam worked out the cost of two chocolate bars at 48p each.
His answer was £8·16. Is he correct or incorrect? Say why.

2   Annabel added up these amounts to find the total length of wood needed for a garden border.

   3·7 m   3·8 m   2·1 m   3·4 m

She got an answer of 57·2 m.
Is this answer about what you would expect? Say why or why not.

3   Rebecca worked out how much she had earned doing jobs for her parents.

   £4·50   £5·25   £8·70   £9·85

She added these up and got £82·50.
Is she correct or incorrect? Say why.

4   David worked out that he could buy 25 of these easter eggs for £5.
Could this answer be right? Explain you answer.

5   Mrs Street worked out how many litres of paint were needed to paint the school pool.
She got 0·57 ℓ.
Is this answer sensible? Say why or why not.

6   Hitesh worked out £3·52 + 86p on the calculator. He got 89·52 as his answer. How can we tell that the answer on the calculator is wrong?

7   Mikey can run 100 m in 15 seconds.
He worked out it would take him 16 minutes 15 seconds to run 650 m.
Could this answer be right? Explain your answer.

8 Maha bought a pie for £2·68, a muffin for £1·20 and a chocolate square for 84p.
She gave £10 and got £2·26 change.
Is this correct or incorrect? Say why.

9 Pete multiplied 864 × 529 and got 447 534.
How can you tell his answer is wrong?

10 Nick multiplied 783 × 467 and got 36 810.
How can you tell this answer is wrong?

Review 1 Morgan worked out the cost of four model aeroplanes.
He got £42·16.
Is he correct or incorrect? Say why.

Review 2 Nick added up these times he had spent watching TV.

8 min   27 min   53 min   35 min   15 min   42 min

He got 6 hours and 32 minutes.
Is this answer about what you would expect? Say why or why not.

Review 3 Raj worked out that he needed 864 m of wood to make a picture frame.
Is this answer sensible? Say why or why not.

Review 4
a Pam gave a shop assistant a £10 note and got £3·86 change.
   She bought items costing 89p, £1·08, £2·03, 26p, 45p and 28p.
   Did she get the correct or incorrect change? Say why.
b Pam bought six stamps for 19p each. She worked out this would cost £1·15.
   How can you tell she is wrong?

Checking answers using inverse operations

We can check the answer to a calculation using **inverse operations**.

Addition and subtraction are inverse operations.
Multiplication and division are inverse operations.

Example   check 2·4 × 6 = 14·4   using 14·4 ÷ 6 or 14·4 ÷ 2·4
            check 42·4 × 28·5 = 1208·4   using 1208·4 ÷ 42·4 or 1208·4 ÷ 28·5
            check 96·3 + 5·89 = 102·19   using 102·19 − 5·89 or 102·19 − 96·3
            check 6 ÷ 7 = 0·8571428 ...   using 7 × 0·8571428 ...

> Remember:
> Inverse operations
> undo each other.

Sometimes we check the answer to a calculation using **estimating** *and* **inverse operations**.

Examples

6783 ÷ 19 = 357
appears to
be about
right because
350 × 20 =
7000.

9357 − 4216 =
5141
appears to be
about right
because
5000 + 4000 =
9,000.

Exercise 9

1 Find the missing numbers.
  a Check 3864 − 1258 = 2606 using 2606 + ☐ = 3864.
  b Check 9432 + 4176 = 13 608 using 13 608 − ☐ = 4176.
  c Check 563 × 14 = 7882 using 7882 ÷ ☐ = 563.
  d Check 3705 ÷ 65 = 57 using ☐ × 65 = 3705.
  e Check 152·75 + 81·06 = 233·79 using 233·79 − ☐ = 81·06.
  f Check 8·75 × 25·3 = 221·375 using ☐ ÷ 25·3 = ☐.
  g Check 3 ÷ 7 = 0·4285714 ... using ☐ × 7 = ☐.
  h Check 96·483 + 2·79 = 99·273 using ☐ − ☐ = 96·483.
  i Check 1·31² = 1·7161 using ☐

2 i Find the answer to these calculations.
  ii Write down a calculation using inverse operations that you could do to check your answer.
  a 9613 + 9423   b 5873 − 1659   c 321 × 89   d 2080 ÷ 65
  e 20·7 ÷ 9   f 834·62 + 93·1 g 8·5 ÷ 2·5   h 97 − 3·842
  i 150·4 ÷ 16   j 92·3 × 4·1   k 8 ÷ 14   l 4·3²
  m 2·81²   n √19·36   o √23·7

3 What calculation could you do to check these? Use estimation **and** inverse operations.
  a 80 231 − 4975 = 75 256   b 5421 ÷ 65 = 83·4   c 89·6 − 30·5 = 59·1
  d 242·73 ÷ 2·9 = 83·7   e 581 × 62 = 36 022   f 9842 + 3754 = 13 596
  g 78·72 ÷ 3·2 = 24·6

4 Neroli wanted to know how much 18 bags of crisps would cost. Each bag costs £1·49.
  a What calculation does she need to do?
  b How much would 18 bags of crisps cost?
  c What calculation could she do to check her answer?

5 Rosalie had 86·4 m of rope. She wanted to cut it into 16 equal pieces to use as guy ropes for her tent.
  a What calculation would she do to find out how long each piece would be?
  b How long would each piece be?
  c What calculation could she do to check her answer?

Review 1 Find the missing numbers.
a Check 96 423 + 82 471 = 178 894 using 178 894 − ☐ = 96 423.
b Check 83 × 96 = 7968 using ☐ ÷ ☐ = 83.
c Check 11 ÷ 7 = 1·5714286 ... using ☐ × ☐ = 11.
d Check 89·06 − 5·79 = 83·27 using ☐ + ☐ = 89·06.

Review 2 Check, using inverse operations, which of these answers are wrong.
a 5·6 × 3·7 = 20·72   b 8·64 ÷ 5 = 1·86   c 187·9 − 3·02 = 184·88

Review 3 What calculation could you do to check these?
Use estimation **and** inverse operations.
a 50 964 − 3827 = 47 137   b 237·38 ÷ 8·3 = 28·6

Puzzle

Using the numbers in the squares write

four divisions such as 16 ÷ 2 = 8
four multiplications such as 8 × 3 = 24
four additions such as 16 + 2 = 18
and   four subtractions such as 12 – 2 = 10

Each square must be used only once.
The 16 calculations will use all of the numbers.

16	8	2	2	8	6
5	6	18	16	4	40
6	18	2	6	10	32
32	2	24	3	3	4
8	4	6	8	24	8
18	32	5	8	2	3
12	18	40	6	10	24
12	24	2	9	6	12

Checking answers using an equivalent calculation

We can check the answer to a calculation by doing an **equivalent calculation.**

Example   494 × 5 = 2470 can be checked by doing one of these calculations.
    (500 – 6) × 5 = 500 × 5 – 6 × 5
  or 494 × 10 ÷ 2
  or 400 × 5 + 90 × 5 + 4 × 5

Example   20·8 ÷ 8 = 2·6 can be checked by doing one of these calculations.
    20·8 ÷ 4 ÷ 2
  or 20 ÷ 8 + 0·8 ÷ 8

Exercise 10

1 Check the answers to these by doing an equivalent calculation. If the answers are wrong
   give the correct answer.
   a   304 × 5 = 1520   b   893 × 6 = 5358   c   52 × 40 = 2080   d   705 × 50 = 35 250
   e   388 ÷ 4 = 48·5   f   495 ÷ 5 = 99   g   86 × 20 = 1270   h   45 × 98 = 4410
   j   √1600 = 40

Review   Check the answer to these by doing an equivalent calculation.
a 594 × 9 = 5346   b 38 × 24 = 912   c 729 ÷ 9 = 81   d 54 × 80 = 4320

Number

Using a calculator

Jan did a calculation on her calculator. The answer was displayed as $\boxed{45.2}$.

How we write down the answer depends on what the question was.

Question
Pip bought five pictures for £226.
How much did each cost?

Answer
£45·20

Madhu took 226 minutes to run five laps.
How long did she take to run each lap?

45·2 minutes is 45 minutes and 0·2 of a minute.
There are 60 seconds in a minute.
0·2 × 60 = 12 seconds.
The answer is **45 minutes and 12 seconds**.

Peter bought five lengths of fencing.
Each length was 9·04 m long.
What is the total length he bought?

45·2 m is 45 m and 0·2 of a metre.
There are 100 cm in a metre.
0·2 × 100 = 20 cm
The answer is **45 m and 20 cm**.

6 m and 74 cm = 6·74 m
We key 6 m 74 cm into the calculator as $\boxed{6.74}$

3 hours 15 min = $3\frac{1}{4}$ hours
= 3·25 hours

We key 3 hours 15 min into the calculator as $\boxed{3.25}$

Exercise 11

Always check your answer.

1   Robert paid £164 for five tickets to a show.
    How much did each ticket cost?

2   Chloe bought 164 cm of ribbon. She cut it into five equal pieces. How long, in cm and mm, was each piece?

3   Freda took 164 minutes to walk round the block five times.
    How long did she take to walk round once? Give the answer in minutes and seconds.

4   Helen did a calculation and got the answer 16·4.
    What should she write down if the answer is to be in
    a £   b cm and mm   c minutes and seconds
    d hours and minutes   e m and cm?

5   Rosemary took 2 hours and 15 minutes to do an experiment. How long would she take to do the same experiment five times? Assume it takes the same time for each repeat.

6   Matthew took 5 minutes and 24 seconds to run round the block once.
    How long would he take to run round it four times?

7 Pritesh is 7 years 3 months old. His mother is four times as old.
How old is his mother?

8 Tiffany measured one of her pet stick insects. It was 3 cm and 4 mm long. If eight of these stick insects lined up one behind the other, how long would the line be?

Review 1 Samuel did a calculation and got the answer 42·6.
What should he write down if the answer is to be in
a £   b cm and mm   c minutes and seconds   d hours and minutes?

Review 2 Beth spends 3 hours and 45 minutes at work each day.
How long does she spend at work each week (5 days)?

Discussion

$51 \div 8 = 6$ remainder 3 or 6 R 3
The calculator gives the answer to $51 \div 8$ as 6·375.
Sally said she could get the remainder 3 on the calculator.
She subtracted 6 from 6·375 to get 0·375 on the screen.
Then she multiplied by the divisor, 8.
Does this give the remainder, 3? **Discuss.**

What do you need to key to give the remainder in these? $\frac{34}{8}$ $\frac{71}{4}$ $\frac{102}{5}$.
Discuss.

Worked Example
Find the remainder.   a $\frac{128}{5}$ b $\frac{60}{7}$ c $\frac{123}{9}$

Answer
a **Key** (128) (÷) (5) (=) (−) (25) (=) (×) (5) (=) to get remainder **3**.
b **Key** (60) (÷) (7) (=) (−) (8) (=) (×) (7) (=) to get remainder **4**.
  Some calculators show the remainder 4 as 3·99999997.
c **Key** (123) (÷) (9) (=) (−) (13) (=) (×) (9) (=) to get remainder **6**.
  Some calculators show the remainder 6 as 6·00000003.

*Worked Example
Write 259 hours in days and hours.

Answer
Key (259) (÷) (24) (=) to get 10·79166667 days. There are 10 whole days.

Now Key (−) (10) (=) (×) (24) (=) to get remainder 19. There are 19 hours left over.
259 hours = **10 days 19 hours**.

Discussion

Discuss how to write   259 minutes in hours and minutes
                            259 weeks in years and weeks
                            259 months in years and months.

Exercise 12

1 Find the remainder.

a $\frac{154}{5}$ b $\frac{261}{8}$ c $\frac{342}{4}$ d $\frac{411}{6}$

e $\frac{279}{6}$ f $\frac{326}{3}$ g $\frac{249}{7}$ h $\frac{178}{19}$

i $\frac{473}{26}$ j $\frac{806}{35}$ k $\frac{903}{41}$

*2 Convert
- a 227 hours to days and hours
- b 1862 hours to days and hours
- c 527 minutes to hours and minutes
- d 613 minutes to hours and minutes
- e 426 weeks to years and weeks
- f 379 weeks to years and weeks
- g 195 months to years and months.

*3 Simone took 569 minutes to play the same CD six times. In minutes and seconds, how long is the CD?

*4 Wasim spent the same amount of time at the gym each week. At the end of the year he worked out that he had spent a total of 4108 minutes there.
- a How many hours and minutes is this?
- b How many hours and minutes did Wasim spend at the gym each week?

Review 1 Find the remainder.   a $\frac{261}{4}$ b $\frac{307}{8}$ c $\frac{247}{6}$ d $\frac{834}{17}$

Review 2 Write
a 417 minutes in hours and minutes
b 375 hours in days and hours
c 572 months in years and months.

Brackets on a calculator

Discussion

- $\frac{24}{6+2}$

  Enid and Tom did this on the calculator.

  Enid keyed 24 ÷ 6 + 2 = and got the answer 6.

  Tom keyed 24 ÷ (6 + 2) = and got the answer 3.
  Who is right? **Discuss**.

- $\frac{12+6}{3-1}$

  Janita, Asad, Mel and Peter did this on the calculator.
  Janita got 5.
  Asad got 9.
  Mel got 13.
  Peter got 15.
  Asad got the right answer. He keyed

  (12 + 6) ÷ (3 − 1) =

  What did the others key? **Discuss**.

Sometimes, to find the answer to a calculation, we need to use the brackets on a calculator.

Worked Example

Calculate

a $(46 + 15) \times (26 - 7)$

b $3 \cdot 7 - (2 \cdot 09 - 1 \cdot 6)$

c $46 - 3 \cdot 2(4 \cdot 1 - 2 \cdot 7)$

d $\frac{8 \cdot 2}{3 \cdot 2 - 1 \cdot 8}$

e $\frac{9 \cdot 6 - 4 \cdot 7}{3 \cdot 8 + 2 \cdot 1}$

Remember
BIDMAS

Remember:
the horizontal line
acts as a bracket.

Answer

a Key (46 + 15) × (26 − 7) = to get **1159**

b Key 3·7 − (2.09 − 1.6) = to get **3·21**.

c Key 46 − 3.2 (4.1 − 2.7) = to get **41·52**.

   Note On some calculators you need to insert a ⊗ before the first bracket.

d Key 8.2 ÷ (3.2 − 1.8) = to get 5·857142857 or **5·9 (1 d.p.)**

e Key (9.6 − 4.7) ÷ (3.8 + 2.1) = to get 0·830508474 or **0·8 (1 d.p.)**

   Both the numerator and denominator must have brackets.

Exercise 13

1 a $7 + 3(14 + 8)$ b $10 + 4(27 + 9)$ c $48 - 2(3 + 11)$ d $124 - 5(14 + 9)$

   e $47 \times (396 - 72)$ f $52 \times (851 - 36)$ g $(7 + 14) \times (23 + 28)$ h $(83 + 27) \times (94 - 39)$

   i $(52 + 81) \times (37 - 19)$ j $314 - (827 - 619)$ k $816 - (327 - 214)$ l $\frac{512}{63 + 65}$

   m $\frac{475}{82 - 57}$ n $\frac{16 + 92}{112 - 94}$ o $\frac{183 + 293}{89 - 61}$

2 Calculate these.

   a $5 \cdot 2 \times (9 \cdot 6 - 3 \cdot 85)$ b $7 \cdot 2 \times (15 \cdot 73 - 8 \cdot 6)$ c $(5 \cdot 2 + 1.89) \times (12 - 3 \cdot 8)$

   d $(8 \cdot 1 - 2 \cdot 97) \times (53 \cdot 4 - 48 \cdot 63)$ e $364 \cdot 7 - (27 \cdot 92 - 8 \cdot 3)$ f $8 \cdot 2 + 5(6 \cdot 4 + 1 \cdot 04)$

   g $15 \cdot 2 - 2(3 \cdot 6 - 1 \cdot 8)$ h $57 \cdot 3 - 4(19 \cdot 7 - 14 \cdot 32)$ i $4 \cdot 2 \times 3 - (5 \cdot 6 - 2 \cdot 39)$

   j $3 \cdot 7 \times 8 \cdot 6 + (8 \cdot 3 - 0 \cdot 75)$ k $4 \cdot 2 \times 6 \cdot 8 - (3 \cdot 2 - 1 \cdot 6)$

3 Calculate these. Give the answers to 1 d.p.

   a $\frac{86}{24 + 17}$ b $\frac{133 + 89}{416 - 329}$ c $\frac{12 \cdot 7}{5 \cdot 2 + 1 \cdot 6}$ d $\frac{7 \cdot 2 + 8 \cdot 6}{5 \cdot 2 - 3 \cdot 4}$

   e $\frac{16 \cdot 5 + 4}{13 - 4}$ f $\frac{1 \cdot 4 + 3 \times 8}{6}$ g $\frac{2 \cdot 1 + 3 \times 6}{9 - 1 \cdot 4}$ h $6 \times (4 \cdot 31)^2$

   i $\sqrt{(18^2 - 12^2)}$ j $\frac{(2 + 3)^2}{(12 - 3)^2}$

4 $4 \cdot 2 \times 6 + 3 - 3 \times 4 + 1 \cdot 8$

   Put one set of brackets in this calculation to make

   a the largest possible answer   b the smallest possible answer

   *c the second largest possible answer.

Review Calculate these. Give the answer to 1 d.p. if you need to round.

a $53 \times (584 - 79)$ b $(46 + 81) \times (187 - 152)$ c $4 \cdot 2 \times (3 \cdot 8 - 1 \cdot 26)$

d $\frac{368}{90 - 67}$ e $\frac{8 \cdot 4 + 2 \cdot 6}{9 \cdot 6 - 7 \cdot 4}$ f $4 \cdot 2 \times 5 - (6 \cdot 4 + 2 \cdot 3)$

g $18 \cdot 7 - 3(1 \cdot 89 - 1 \cdot 5)$ h $7 \cdot 2 \times 3 \cdot 6 - (5 \cdot 2 - 4 \cdot 85)$ i $\frac{6 \cdot 4 + 5 \times 2}{7 - 2 \cdot 6}$

j $\frac{(4 + 3)^2}{(13 - 4)^2}$ k $\sqrt{(20^2 - 14^2)}$

Summary of key points

We **add and subtract** by first lining up the units (ones) column.

Examples

```
  85247          8·73        ¹³ ¹⁵ ¹
+  3952       +16·059       14·06̶0̶
  89199        24·789      −  8·327
                              5·733
```

We add and subtract decimals by lining up the decimal points.

Example 53·8 − 6·24 + 3·8

```
  ⁴ ₁⁷ ₁
  5̶3̶·8̶0̶          47·56
−  6·24         + 3·8
  47·56          51·36
                  ₁ ₁
```

B Always **estimate the answer** first.

C **Multiplication**

By a 1-digit number

Example 4·79 × 6

4·79 × 6 is approximately 5 × 6 = 30 and is equivalent to 479 × 6 ÷ 100.

×	4	0·7	0·09
6	24	4·2	0·54

Answer 24 + 4·2 + 0·54 = **28·74**

or

```
    479
  ×   6
   2874     2874 ÷ 100 = 28·74
    ₄ ₅
```

By a 2-digit number

Example 83·2 × 3·4

83·2 × 3·4 is approximately equal to 100 × 3 = 300.

Method 1

×	80	3	0·2
3	240	9	0·6
0·4	32	1·2	0·08

272 + 10·2 + 0·68 = **282·88**

Check

```
  249·6
+ 33·28
 282·88
```

Method 2

83·2 × 3·4 is equivalent to
83·2 × 10 × 3·4 × 10 ÷ 100 or
832 × 34 ÷ 100.

```
    832
  ×  34
  24960    30 × 832
   3328     4 × 832
  28288
```

Answer 28 288 ÷ 100 = **282·88**

Check the answer is about the right size by looking at the estimate.

D **Division**

By a 1-digit whole number

Example 194·88 ÷ 6

194·88 ÷ 6 is approximately 180 ÷ 6 = 30

180 is the closet number that divides easily by 6.

```
 6)194·88              or          32·48
  −180·00   6 × 30            6)194·²8̶⁴8
    14·88
   −12·00   6 × 2
     2·88
    −2·40   6 × 0·4
     0·48
    −0·48   6 × 0·08
     0·00
```

Answer **32·48**

By a 2-digit whole number

Examples   646 ÷ 19

646 ÷ 19 is approximately 600 ÷ 20 = 30

```
19)646
  −570    19 × 30
   76
  −76     19 × 4
    0
```

Answer **34**

37·6 ÷ 16

37·6 ÷ 16 is approximately 40 ÷ 20 = 2

```
16)37·6
  −32·0    16 × 2
    5·6
   −4·8    16 × 0·3
    0·8
   −0·8    16 × 0·05
      0
```

Answer **2·35**

 When we **divide by a decimal** we do an equivalent division. We make the divisor a whole number.

Example   54 ÷ 0·9 is equivalent to 540 ÷ 9.

20 ÷ 1·6 is equivalent to 200 ÷ 16.

 Always check that your **answer is sensible**.

Example   Beth buys items for 89p, £1·24, £1·06, 35p and 92p.

She gives £10 and gets £2·58 change.

This is not correct because the items would add to less than £5.

G We can check the answer to a calculation using **inverse operations**.

Addition and **subtraction** are inverse operations.

Multiplication and **division** are inverse operations.

Examples   1184 − 397 = 787 so 787 + 397 = 1184

197 × 31 = 6107 so 6107 ÷ 31 = 197 and 6107 ÷ 197 = 31

H We can check the answer to a calculation by doing an **equivalent** calculation.

Example   56·4 × 5 = 282 can be checked by doing one of these calculations.

56·4 × 10 ÷ 2 = 564 ÷ 2   50 × 5 + 6 × 5 + 0·4 × 5 = 250 + 30 + 2

= 282   = 282

I Sometimes we need to **interpret** the answer given by a calculator.

Examples | 5.6 | might be £5·60 or 5 hours 36 minutes or 5 metres 60 centimetres and so on.

 To find the **remainder** when dividing using a calculator, subtract the whole number part of the answer and then multiply by the divisor.

Examples   387 ÷ 63

Key [387] [÷] [63] [=] to get 6·142857143.

Key [−] [6] [=] to get 0·142857143.

Key [×] [63] [=] to give the remainder 9.

Sometimes we use **brackets** on the calculator.

Example   4·2 − (3·6 − 1·2)

Key 4.2 − ((3.6 − 1.2)) = to get **1·8**.

In a fraction, the horizontal line acts as a bracket.

Example $\frac{8·6 + 3·2}{4·5 − 3·7}$

Key ((8.6 + 3.2)) ÷ ((4.5 − 3.7)) = to get **14·8 (1 d.p.)**

Test yourself   **You may use a calculator for questions 18, 19 and 24.**

1 Calculate these.
  a 1723 + 589   b 76 423 + 8659   c 7·96 + 8·3   d 15·01 + 9·63
  e 36·8 + 5·709   f 9806 − 394   g 4509 − 3689   h 20·6 − 19·24
  i 113·5 − 24·68   j 314·7 − 19·64   k 800·5 − 3·69

2 a 8·36 + 52·9 + 4 + 0·23   b 16 − 5·83 − 4·27 + 14.9

3 Amy bought trainers for £79·60, laces for £0·58 and socks for £4·72. How much
change did she get from £100?

4 Find the answer to these. Estimate first.
Write down your estimate and your answer.
  a 4·6 × 7   b 18·3 × 5   c 1·68 × 9   d 4·83 × 7   e 16·42 × 5
  f 18·24 × 6   g 142·7 × 8   h 523 × 0·6   i 42·7 × 0·04

5 Calculate these.
  a 56 × 34   b 56 × 78   c 345 × 45   d 678 × 34   e 569 × 89   f 789 × 56
  g 457 × 38 h 987 × 34 i 289 × 67   j 678 × 73   k 569 × 34

6 a 5·2 × 18   b 38 × 4·63   c 5·74 × 54   d 3·2 × 11·4

7 Akila needed 4·8 m of lace for a tablecloth.
  a How much did it cost, to the nearest penny?
  b How much change did she get from £10?

Lace
£1.89
per m

8 796 tennis balls are packed in tubes of 15.
  a How many tubes can be packed?
  b How many balls are left over?

9 Give the answer to these as a decimal.
  a 207 ÷ 15   b 342 ÷ 24

10 For each of the 52 weeks in a year, a supermarket is open for 86 hours.
What is the total number of hours it is open?

11 Estimate the answers, then calculate.
  a 5·6 ÷ 7   b 13·8 ÷ 6   c 58·4 ÷ 4   d 170·4 ÷ 8   e 212·04 ÷ 9

12 Shirley bought three toothbrushes, four bars of soap, one deodorant and one sunscreen.
- **a** How much did this cost?
- **b** How much change did she get from £25?

13 Seventeen people buy tickets for a concert.
The total bill is £92·65
What is the cost of each ticket?

14 **a** 189 ÷ 0·3   **b** 426 ÷ 0·06

15 A magazine, 'World' sells for £3.75. An opposition magazine 'Us', sells for £4.20 but there is a discount of £20 for orders over £100.
- **a** How much would it cost to buy a class set of 30 copies of World?
- **b** How much would it cost to buy a class set of 30 copies of Us?
- **c** Which magazine is it cheaper to buy 23 copies of?
- **d** Which magazine is it cheaper to buy 24 copies of?

16 Bev's garden needed 128·78 m of edging.
On Saturday she edged 25·6 m and on Sunday she edged 36·83 m.
How many more metres does she need to edge?

17 **a** The ⑧ key on your calculator doesn't work.
How could you make your calculator display 858?
- **b** The ②, ④, ⑥ and ⑧ keys on your calculator do not work.
How could you make your calculator display 846?

18 Find the remainder when 396 is divided by 27.

19 Calculate these. Give the answers to **c** and **d** to 1 d.p.
- **a** (53 + 21) × 75   **b** 4·2 + (6·8 − 0·87)   **c** $\frac{17\cdot6}{4\cdot2+3\cdot7}$ **d** $\frac{8\cdot3+4\cdot2}{8\cdot3-5\cdot7}$

20 Answer these questions **without** doing the calculation.
- **a** Verity added up these amounts to find the total amount of money made on a sponsored walk.
£8·69   £12·70   £9·50   £13·64   £6·35   £8·24
She got an answer of £167·12.
Is she correct or incorrect? Say why.
- **b** Annabel worked out that she would need to order 39 coaches to take 204 people to a concert. Each coach holds 52 people.
Is she correct or incorrect? Explain your answer.

21 Write down a calculation, using inverse operations, that you could do to check the answer to each of these.
- **a** 836 × 42 = 35 112   **b** 364·8 ÷ 1·9 = 192   **c** 8·36 + 19·72 = 28·08

22 Write down an equivalent calculation that you could do to check these.
- **a** 603 × 27 = 16 281   **b** 42 × 59 = 2478   **c** 632 ÷ 4 = 158

23 185   56   589   423   156   86
Put one of the numbers above in each box to make a true statement.
- **a** ☐ − ☐ + ☐ = 323
- **b** ☐ + ☐ − ☐ = 490

24 Convert
- **a** 186 hours to days and hours   **b** 373 seconds to minutes and seconds.

5 Fractions

You need to know

✓ fractions — numerator and denominator   page 5
- know that $\frac{3}{5}$ means 3 out of 5
- fraction and division
- equivalent fractions
- cancelling a fraction to its lowest terms
- fraction of
- proper and improper fractions and mixed numbers

Key vocabulary

cancel, convert, decimal, denominator, equivalent fraction, fraction, improper fraction, lowest terms, mixed number, numerator, proper fraction, recurring decimal, simplest form, terminating decimal, unit fraction

 Bits and pieces

About $\frac{1}{4}$ of all meat eaten in the world is meat from pigs.

Four out of five teenagers worry about exams.

Ninety-seven hundredths of water on earth is in the ocean.

Find some more real-life fraction facts.
Make a poster or collage of your facts.
Show your fractions on a number line or diagram.

Fractions of shapes

One out of nine **parts** of this diagram is shaded.
One ninth or $\frac{1}{9}$ is shaded.

Three out of eight pieces of cake have been iced.
$\frac{3}{8}$ have been iced.

Worked Example

This pie chart shows the ages of people at a rock concert.
Estimate the fraction of people that were
a 16–30   b 50+

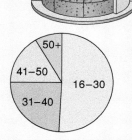

Answer

a **About $\frac{1}{2}$**.
b About three 50+ pieces would fit into $\frac{1}{4}$ of the pie chart.
So we could fit about twelve 50+ sections into the whole pie chart. 50+ is **about $\frac{1}{12}$**.

Exercise 1

1   What fraction of each of these is coloured?
a   b   c   d   e

2   The pie chart shows the ingredients of a muesli bar.
Estimate the fraction of the muesli bar that is
a   sugar
b   fat.

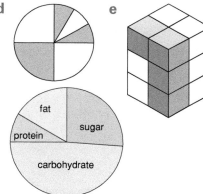

3
| chicken | veg | beef | hawaiian |

This strip shows the proportion of each type of pizza at a party.
Estimate the fraction that is
a   chicken   b   beef.

Review 1   What fraction of each shape is shaded?   a   b

Review 2   The pie chart shows the time of day that burglaries
took place in one year.
Estimate the fraction that took place
a at night   b in the afternoon.

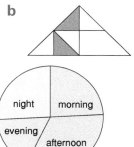

Unit fractions have numerator 1.

Examples $\frac{1}{5}$, $\frac{1}{8}$ $\frac{1}{15}$ are unit fractions.

$\frac{2}{3}$, $\frac{3}{7}$, $\frac{5}{12}$ are not.

Investigation

Unit Fractions

1   This 8×3 rectangle is divided into three parts.
    $\frac{1}{2}$ is blue, $\frac{1}{3}$ is red, $\frac{1}{6}$ is yellow

Divide an 8×3 rectangle into four parts that are $\frac{1}{2}$, $\frac{1}{3}$, $\frac{1}{8}$, $\frac{1}{24}$ of the rectangle.

There is another way of dividing the rectangle into four parts so that each part is a different unit fraction. **Investigate.**

Divide an 8×3 rectangle into five parts so that each part is a different unit fraction.
Hint Two of the parts are $\frac{1}{2}$ and $\frac{1}{8}$.

An 8×3 rectangle can be divided into six parts, with each part a different unit fraction. **Investigate.**

2   A 3×4 rectangle can be divided into three parts, with each part a different unit fraction. **Investigate.**

Can you divide this rectangle into four parts with each part a different unit fraction?
What about more than four parts?

*3 Draw 30 squares in a rectangular shape.
    Investigate the number of ways you can divide your rectangle so that each part is a different unit fraction.
    What if you drew 16 squares?
    What if you drew 15 or 18 or 40 squares?
    What if ...

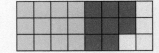

Hint: Find factors.

One number as a fraction of another

Remember
A fraction is in its **lowest terms** if the numerator and denominator have no common factors.
We **cancel** a fraction to its lowest terms by dividing the numerator and denominator by the highest common factor.

Examples $\frac{^3\cancel{15}}{_4\cancel{20}} = \frac{3}{4}$ We divide 15 and 20 by 5.
                  5 is the highest common factor of 15 and 20.

$\frac{^2\cancel{160}}{_3\cancel{240}} = \frac{2}{3}$ We divide 160 and 240 by 80.
                  80 is the highest common factor of 160 and 240.

If you find it hard to find the highest common factor, you can cancel in steps.

Example $\frac{36}{80} = \frac{18}{40}$ We divide 36 and 80 by 2.

$= \frac{9}{20}$ We divide 18 and 40 by 2 again.

Worked Example
Write 43 centimetres as a fraction of a metre.

Answer
1 metre = 100 centimetres
43 cm as a fraction of a metre = $\frac{43}{100}$.

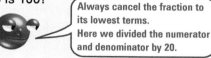
When writing one number as a fraction of another, the numerator and denominator must have the same units.

Worked Example
A small serving of toasted muesli has 120 Calories. A small serving of light toasted muesli has 100 Calories. What fraction of 120 is 100?

Answer
We write this as $\frac{100}{120} = \frac{5}{6}$
100 is $\frac{5}{6}$ of 120.

Always cancel the fraction to its lowest terms.
Here we divided the numerator and denominator by 20.

Worked Example
Five friends shared a pizza.
This bar chart shows the number of slices each friend ate.
What fraction of the slices did Pete eat?

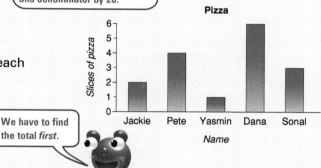

Pizza

Answer
Total number of slices = 2 + 4 + 1 + 6 + 3
= 16
Fraction that Pete ate = $\frac{4}{16}$
= $\frac{1}{4}$

We have to find the total *first*.

Worked Example
In Louise's class, 18 pupils had brothers and six had no brothers.
What fraction had brothers?

Answer
Total in class = 18 + 6
= 24
Fraction who had brothers = $\frac{18}{24}$
= $\frac{3}{4}$

Exercise 2

1 Cancel these to their simplest form.
a $\frac{3}{6}$ b $\frac{3}{9}$ c $\frac{5}{15}$ d $\frac{3}{12}$ e $\frac{4}{20}$ f $\frac{6}{18}$
g $\frac{4}{6}$ h $\frac{10}{15}$ i $\frac{9}{12}$ j $\frac{15}{45}$ k $\frac{16}{20}$ l $\frac{24}{40}$
m $\frac{20}{25}$ n $\frac{12}{18}$ o $\frac{24}{36}$ p $\frac{35}{42}$ q $\frac{56}{64}$ r $\frac{48}{64}$
s $\frac{56}{72}$ t $\frac{36}{96}$ u $\frac{48}{108}$ v $\frac{80}{144}$

2 What fraction of
a 1 metre is 51 centimetres   b 1 metre is 89 centimetres
c 1 kilogram is 37 grams   d 1 kilogram is 59 grams
e 1 hour is 13 minutes   f 1 hour is 47 minutes
g 1 metre is 17 centimetres   h 1 minute is 17 seconds
i £2 is 83p   j £3 is 61p
k 1 yard is a foot?

Link to measures.

3 Write these as fractions of an hour in their simplest form.
   a 30 minutes   b 20 minutes   c 45 minutes   d 27 minutes.

4 What fraction of a turn does the minute hand turn through between
   a 8 p.m. and 8:25 p.m.   b 2:20 a.m. and 2:40 a.m.
   c 11:10 a.m. and 11:38 a.m.   d 7:35 p.m. and 8:20 p.m.
   e 5:35 a.m. and 7:25 a.m.?

5 What fraction of a turn takes you from facing
   a north to facing south   b south to facing west
   c north to facing north-west   d south to facing south-west
   e north to facing south-west?

Turn this way.

Remember: there are 360° in a full turn.

6 What fraction of a turn is
   a 90°   b 180°   c 120°   d 36°   e 450°?

7 What fraction of the large shape is the small one?
   a   b   c

8 a What fraction of 1 kg is 250 g?   b What fraction of 2 ℓ is 500 mℓ?
   c What fraction of £2 is 75p?   d What fraction of 20 cm is 10 mm?
   e What fraction of 1 m is 350 mm?

9 Write these as fractions of 2 hours.
   a 30 minutes   b 20 minutes   c 50 minutes   d 36 minutes   e 12 minutes

10 a What fraction of 12 is 8?   b What fraction of 24 is 16?
    c What fraction of 72 is 54?   d What fraction of 120 is 30?
    e Give 30 as a fraction of 45   f Give 20 as a fraction of 120.
    g Give 32 as a fraction of 56.

11 In Jasmine's class, eight pupils wear glasses and 12 do not wear glasses.
    a How many pupils are in Jasmine's class?
    b What fraction wear glasses?

12 At a meeting, 23 people voted in favour and 37 voted against sending pupils on a tour.
    a How many people voted?
    b What fraction voted for sending the pupils on the tour?
    c What fraction voted against sending them?
   *d For the tour to take place, at least $\frac{3}{5}$ have to vote for it.
       Will the tour take place?

13 Annette made a bar chart to show how much she saved
    each month for six months.
    What fraction did she save in
    a February
    b April
    c January **and** February together?

14 Sue was doing a survey on cars.
    Three trucks and 147 cars went past her house.
    a What fraction were trucks?
    b What fraction were cars?

*15 In Nazir's class there are 14 girls and 18 boys
What fraction are boys?

*16 Of 2400 videos at Video Village, 80 were new releases.
What fraction were not new releases?

*17 In one litre of air there is about
       210 mℓ oxygen
       780 mℓ nitrogen
        10 mℓ argon.
About what fraction of the air is oxygen?

Review 1   What fraction of
a 1 metre is 27 centimetres   b 1 hour is 48 minutes?

Review 2   What fraction of a turn does the minute hand turn through between
a 2:15 p.m. and 2:35 p.m.   b 12:45 a.m. and 1:20 p.m.   c 4:36 a.m. and 5:50 a.m.?

Review 3   What fraction of the large shape is the small one?
a

b

Review 4   Write these as fractions of a complete circle.
a 90°   b 270°   c 162°   d 234°

Review 5
a   What fraction of 36 is 24?
b   What fraction of 81 is 45?
c   Give 45 as a fraction of 150.

Review 6   Pure gold is 24 carat gold.
What fraction of pure gold is 18 carat gold?

*Review 7**   Rowena asked her friends which colour they would like the new school tracksuit
to be.
This bar chart shows her results.

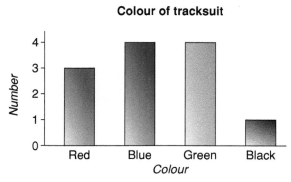

What fraction wanted the tracksuit to be
a red   b blue   c black   d green or black?

Practical

Link to collecting data page 395.

1 Write some fractions for your class.
For example, the fraction that are left-handed, like pizza, have blue eyes, have curly hair, wear glasses, walk to school, ...

2 Write some fractions for your friends, family or school.

3 Choose a topic you are interested in. Write some fractions about it. For example, you might choose cars and write '$\frac{1}{3}$ of all cars are made in Japan'.

Practical

You will need 4 dice (2 red dice and 2 green dice).

a Toss the two **red** dice.
Add the numbers.
This gives you the numerator.

Example The dice shown give the numerator 7.

b Toss the two green dice.
Add the numbers.
This gives you the denominator.

Example The dice shown give the denominator 3.

c Make the fraction, $\dfrac{\text{red dice total}}{\text{green dice total}}$.

Example The dice shown give the fraction $\frac{7}{3}$.

Find the following.

i the smallest fraction it is possible to toss
ii the largest improper fraction it is possible to toss
iii the largest proper fraction it is possible to toss
iv the fractions that can be written as whole numbers
v the number of different fractions that can be tossed.

Notes Count $\frac{2}{2}$ as a fraction.
Treat $\frac{2}{4}$ as the same fraction as $\frac{3}{6}$.

Investigation

Equivalent Fractions

1 $\frac{1}{2} = \frac{8}{16}$ $\frac{8}{16}$ has three digits 8, 1 and 6.

Which other 3-digit fractions are equal to $\frac{1}{2}$?
Have you found them all? How do you know?

$\frac{1}{3} = \frac{4}{12}$

Find all the 3-digit fractions equal to $\frac{1}{3}$.
Have you found them all? How do you know?

What if you began with $\frac{1}{4}$?
What if you began with $\frac{1}{5}$?
What if ... ?

2 $2\frac{1}{2} = \frac{5}{2} = \frac{10}{4}$ $\frac{10}{4}$ is a 3-digit fraction equal to $2\frac{1}{2}$.

Find all the 3-digit fractions equal to $2\frac{1}{2}$.

What if you began with $1\frac{1}{2}$?
What if you began with $3\frac{1}{2}$?
What if you began with $4\frac{1}{2}$?
What if ... ?

3 $1\frac{1}{3} = \frac{4}{3} = \frac{12}{9}$

$\frac{12}{9}$ is the only 3-digit fraction equal to $1\frac{1}{3}$.

What if you began with $2\frac{1}{3}, 3\frac{1}{3}, ...$?
What if you began with $1\frac{3}{4}, 2\frac{3}{4}, 3\frac{3}{4}, ...$?
What if ... ?

*4 $\frac{7293}{14\,586}$ is made using each of the digits 1 to 9 once.

It is equal to $\frac{1}{2}$.

Find other ways of arranging the digits 1 to 9 to make a fraction equal to $\frac{1}{2}$.
What if the fraction was $\frac{1}{5}$?

Fractions and decimals

To convert a **decimal to a fraction** write it with denominator 10, 100 or 1000.

Examples $0.54 = \frac{54}{100}$ $0.839 = \frac{839}{1000}$ $0.375 = \frac{375}{1000}$

$= \frac{27}{50}$ $= \frac{75}{200}$

$= \frac{3}{8}$

Cancel the fraction to its lowest terms.

Exercise 3

1 Write these as fractions in their lowest terms.

   a 0·60   b 0·82   c 0·04   d 0·157   e 0·931
   f 0·250   g 0·800   h 0·125   i 0·555   j 0·625
   k 0·085   l 0·056   m 0·028   n 0·005   o 0·008
   p 1·7   q 1·4   r 2·43   s 7·28   t 3·65
   u 4·032   v 3·005

2 Which of these does *not* show 0·6?

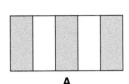

   A   B   C   D

T

Review

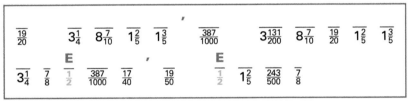

$\frac{19}{20}$ $3\frac{1}{4}$ $8\frac{7}{10}$ $1\frac{2}{5}$ $1\frac{3}{5}$ $\frac{387}{1000}$ $3\frac{131}{200}$ $8\frac{7}{10}$ $\frac{19}{20}$ $1\frac{2}{5}$ $1\frac{3}{5}$

E

$3\frac{1}{4}$ $\frac{7}{8}$ $\frac{1}{2}$ $\frac{387}{1000}$ $\frac{17}{40}$ $\frac{19}{50}$ $\frac{1}{2}$ $1\frac{2}{5}$ $\frac{243}{500}$ $\frac{7}{8}$

Use a copy of this box.
Write these as fractions in their lowest terms. Write the letter that is beside the decimal above the answer.

E 0·50 = $\frac{1}{2}$ A 0·95   S 0·387   T 0·380   H 0·486   O 0·875
N 0·425   K 1·6   U 8·7   C 1·4   D 3·25   Q 3·655

To convert a **fraction to a decimal** we can

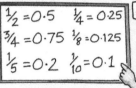

1 use known facts

   Example We know that $\frac{1}{8}$ = 0·125

   $\frac{5}{8}$ = 0·125 × 5

   = **0·625**

   $\frac{1}{2}$ = 0·5 $\frac{1}{4}$ = 0·25
   $\frac{3}{4}$ = 0·75 $\frac{1}{8}$ = 0·125
   $\frac{1}{5}$ = 0·2 $\frac{1}{10}$ = 0·1

2 make an equivalent fraction with denominator 10 or 100 then divide

   Examples $\frac{9}{25} = \frac{36}{100}$ $\frac{12}{30} = \frac{4}{10}$ $\frac{17}{20} = \frac{85}{100}$ $6\frac{4}{5} = 6\frac{8}{10}$

   = **0·36**   = **0·4**   = **0·85**   = **6·8**

Remember: $\frac{5}{8}$ is the same as 5 ÷ 8.

3 divide the numerator by the denominator.

   Examples $\frac{5}{8}$ $\frac{31}{5}$ $\frac{3}{7}$

   8)5·0   5)31   7)3·0
   −4·8   8 × 0·6   − 30   5 × 6   −2·8   7 × 0·4
   ‾‾‾‾   ‾‾‾‾   ‾‾‾‾
   0·20   1·0   0·20
   −0·16   8 × 0·02   −1·0   5 × 0·2   −0·14   7 × 0·02
   ‾‾‾‾   ‾‾‾‾   ‾‾‾‾‾
   0·04   0·0   0·060
   −0·04   8 × 0·005   −0·056 7 × 0·008
   ‾‾‾‾   ‾‾‾‾‾
   0·0   0·004

   Answer **0·625**   *Answer* **6·2**   *Answer* 0·428 R 0·004
                                                               0·43 (2 d.p.)

Fractions convert to either a **terminating** or a **recurring** decimal.
0·625 and 6·2 are **terminating** decimals because they end and have a finite number of digits.
A recurring decimal has one or more digits that repeat.

Examples $\frac{1}{3} = 0·3333 ...$ $\frac{2}{3} = 0·6666 ...$ $\frac{3}{11} = 0·272\ 727 ...$

          $0 = 0·\dot{3}$ $= 0·\dot{6}$ $= 0·\dot{2}\dot{7}$

> The dot above the 3
> shows that 3 repeats.

> The dots above the 2 and 7
> show that these both repeat.

If we continued to divide $\frac{3}{7}$ we would get $\frac{3}{7} = 0·$**428571428571**.
When the repeating pattern is longer than 1 or 2 digits we usually round the decimal. In the
example on the previous page we rounded $\frac{3}{7}$ to 0·43 (2 d.p.)

For harder divisions we can use the calculator.

Example $\frac{21}{29}$

       key 21 ÷ 29 = to get 0·72 (2 d.p.)

Exercise 4   **Except for questions 6, 7, 8, 10 and Review 2.**

1   Convert these fractions to decimals. Find the answers mentally.
  a $\frac{1}{2}$ b $\frac{1}{4}$ c $\frac{3}{4}$ d $\frac{3}{5}$ e $\frac{4}{8}$ f $\frac{3}{12}$ g $1\frac{1}{2}$ h $\frac{9}{12}$
  i $\frac{1}{8}$ j $2\frac{1}{4}$ k $2\frac{1}{5}$ l $\frac{12}{16}$ m $3\frac{3}{6}$ n $5\frac{9}{12}$

2   Write these as decimals by first writing them with denominators of 10 or 100.
  a $\frac{4}{20}$ b $\frac{4}{5}$ c $\frac{4}{50}$ d $\frac{2}{5}$ e $\frac{14}{20}$ f $\frac{3}{20}$ g $\frac{7}{50}$ h $\frac{40}{200}$ i $\frac{9}{20}$
  j $\frac{8}{40}$ k $\frac{36}{60}$ l $\frac{36}{200}$ m $\frac{9}{30}$ n $\frac{24}{80}$ o $\frac{16}{40}$ p $\frac{104}{50}$ q $\frac{48}{40}$ r $\frac{175}{50}$

3   Mandy made this fraction poster.
  Write each fraction as a decimal.

USELESS FACTS

a $\frac{3}{8}$ of my family wear contact lenses.

b $\frac{17}{25}$ of my friends like red better than green.

c $\frac{18}{60}$ of Year 7 pupils go home for lunch.

d $\frac{27}{50}$ of my music school play the piano.

e $\frac{23}{25}$ of my friends have a computer.

f My class used $1\frac{3}{5}$ rolls of paper in art last term.

g My friends ate $4\frac{5}{8}$ pizzas at my birthday party.

4   Use division to write these as decimals. Don't use a calculator.
  Round to 2 d.p. if you need to round.
  a $\frac{1}{7}$ b $\frac{5}{9}$ c $\frac{4}{7}$ d $\frac{26}{5}$ e $\frac{21}{4}$
  f $\frac{13}{7}$ g $\frac{7}{6}$ h $\frac{35}{8}$ i $\frac{19}{7}$

5   Which of these, when written as a decimal, gives a terminating decimal?
  a $\frac{3}{7}$ b $\frac{5}{9}$ c $1\frac{1}{3}$ d $\frac{7}{8}$

6   Use your calculator to write these as decimals. Round to 2 d.p.
  a $\frac{23}{27}$ b $\frac{14}{17}$ c $\frac{15}{23}$ d $\frac{25}{41}$ e $\frac{104}{163}$ f $\frac{37}{29}$ g $\frac{147}{109}$

7 What fraction of these shapes is shaded?
Write your answer as a recurring decimal.

a 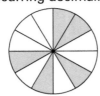   b   c   d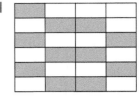

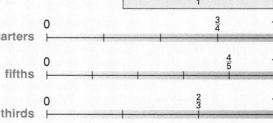 **8** Write these as recurring decimals.

a $\frac{1}{9}$ b $\frac{2}{3}$ c $\frac{5}{11}$ d $\frac{1}{6}$ e $\frac{5}{6}$ f $\frac{2}{11}$

g $\frac{17}{6}$ h $\frac{47}{9}$ i $\frac{26}{3}$ j $1\frac{4}{9}$ k $2\frac{1}{3}$

*9 Write down the recurring decimals for $\frac{1}{3}$ and $\frac{2}{3}$.
Predict what answer you would get if you divided

a 4 by 3   b 5 by 3   c 6 by 3   d 7 by 3.

* **10 a** Write $\frac{1}{11}$, $\frac{2}{11}$, $\frac{3}{11}$ and $\frac{4}{11}$ as decimals.

   b Describe the pattern.

   c Predict what answer you would get if you divided

   i 5 by 11   ii 6 by 11   iii 7 by 11   iv 8 by 11.

Review 1 Write these as decimals.

a $\frac{4}{5}$ b $\frac{7}{20}$ c $\frac{11}{25}$ d $\frac{13}{20}$ e $\frac{19}{25}$

f $\frac{31}{50}$ g $1\frac{7}{10}$ h $4\frac{7}{8}$ i $\frac{65}{50}$ j $\frac{108}{50}$

k $\frac{27}{5}$ l $\frac{23}{4}$ m $\frac{15}{7}$ n $\frac{12}{7}$

Round the answers to *m* and *n* to 1 d.p.

Review 2 Use your calculator to write these as decimals. Round to 2 d.p.

a $\frac{24}{31}$ b $\frac{18}{29}$ c $\frac{26}{38}$ d $\frac{8}{53}$

* **Review 3**

a Write these as decimals
$\frac{1}{9}$, $\frac{2}{9}$, $\frac{3}{9}$, $\frac{4}{9}$, $\frac{5}{9}$

b Describe the pattern.

c Use the pattern to write the decimals for $\frac{6}{9}$, $\frac{7}{9}$ and $\frac{8}{9}$.

Comparing fractions

A diagram is a good way to **compare fractions**.

We can see from this diagram that

$\frac{1}{2} > \frac{1}{3} > \frac{1}{4} > \frac{1}{5} > \frac{1}{6} > \frac{1}{7} > \frac{1}{8}$.

We could also use lines divided into parts to compare fractions.

Worked Example
Write these fractions in order, smallest to largest.

$\frac{3}{4}$, $\frac{4}{5}$, $\frac{2}{3}$

A line 60 mm long is easy to divide up.

Answer
We can see from the lines that the fractions in order from smallest to biggest are $\frac{2}{3}$, $\frac{3}{4}$, $\frac{4}{5}$.

quarters

fifths

thirds

We can also **compare fractions** by writing them as **decimals**.

Worked Example
Which is largest, $\frac{5}{8}$, $\frac{3}{5}$ or $\frac{5}{9}$?

Answer

$\frac{5}{8} = 5 \times 0{\cdot}125$ $\frac{3}{5} = 3 \times 0{\cdot}2$ $\frac{5}{9} = 0{\cdot}56$ (2 d.p.)   because

$\quad = 0{\cdot}625$ $\quad = 0{\cdot}6$

```
        9)5·0
         -4·5      9 × 0·5
         ────
          0·50
          0·45     9 × 0·05
         ────
          0·05
          0·045    9 × 0·005
         ─────
          0·005
```

Answer   0·56 (2 d.p.)

$0{\cdot}625 > 0{\cdot}6 > 0{\cdot}56$
$\frac{5}{8}$ is the largest.

T | **Exercise 5**

1   Which of $<$ or $>$ goes in the box?
   a $\frac{1}{2}\square\frac{2}{3}$ b $\frac{4}{5}\square\frac{1}{2}$ c $\frac{2}{7}\square\frac{1}{2}$ d $\frac{3}{8}\square\frac{1}{2}$
   e $\frac{2}{3}\square\frac{4}{5}$ f $\frac{5}{6}\square\frac{7}{8}$ g $\frac{6}{7}\square\frac{3}{4}$

2   Compare these by writing the fractions as decimals.
   Which of $<$ or $>$ goes in the box?
   a $0{\cdot}82\square\frac{5}{7}$ b $0{\cdot}76\square\frac{7}{9}$ c $\frac{5}{6}\square 0{\cdot}82$ d $\frac{4}{9}\square\frac{5}{11}$
   e $\frac{7}{19}\square\frac{4}{15}$ f $\frac{5}{7}\square\frac{19}{27}$ g $\frac{7}{11}\square\frac{29}{45}$

3   Paul and Rani each bought a quiche.
   Paul ate $\frac{4}{5}$ of his. Rani ate $\frac{7}{8}$ of hers.
   Who had the most left?

4   In a survey it was found that $\frac{5}{12}$ of those surveyed have holidays
   north of where they live and $\frac{4}{9}$ have holidays south of where they
   live. The rest have no holiday.
   Did more people have holidays north or south of where they live?

5   Sam and Rebecca are in different classes for French.
   In the last test Sam got 12 out of 20. Rebecca got 7 out of 12.
   Who got the better mark?

6   Write these fractions in ascending order.
   a $\frac{1}{2}, \frac{1}{4}, \frac{3}{8}$ b $\frac{5}{8}, \frac{2}{3}, \frac{3}{4}$ c $\frac{5}{6}, \frac{6}{7}, \frac{2}{3}$ d $\frac{7}{8}, \frac{3}{4}, \frac{4}{5}$
   e $\frac{2}{3}, \frac{5}{6}, \frac{3}{5}$ f $\frac{1}{4}, \frac{2}{5}, \frac{3}{8}$

7   Write these fractions in order, smallest first.
   a $1\frac{1}{4}, 1\frac{1}{2}, 1\frac{2}{3}$ b $3\frac{1}{4}, 3\frac{3}{8}, 3\frac{2}{5}, 3\frac{1}{5}$ c $2\frac{3}{10}, 1\frac{7}{8}, 2\frac{3}{5}, 1\frac{3}{5}, 1\frac{5}{6}$ d $4\frac{1}{4}, 4\frac{1}{5}, 4\frac{2}{7}, 4\frac{3}{8}, 4\frac{1}{3}$
   e $1\frac{1}{2}, 1\frac{2}{3}, 1\frac{5}{6}, 1\frac{3}{4}, 1\frac{7}{9}$ f $1\frac{5}{6}, 1\frac{11}{12}, 1\frac{7}{9}, \frac{13}{8}$

*8 Which of these fractions is closer to 1? Explain how you can tell.
  a $\frac{3}{4}$ or $\frac{4}{3}$ b $\frac{7}{9}$ or $\frac{9}{7}$

Review 1 Write these fractions in descending order.
a $\frac{3}{4}, \frac{2}{3}, \frac{5}{6}$ b $\frac{1}{2}, \frac{7}{8}, \frac{1}{5}, \frac{1}{6}$ c $4\frac{1}{10}, 4\frac{1}{2}, 3\frac{1}{5}, 3\frac{3}{4}, 3\frac{1}{2}$ d $2\frac{1}{4}, 2\frac{3}{7}, 2\frac{5}{8}, 2\frac{1}{3}, 2\frac{1}{5}$

Review 2 A charity concert was held to raise money for refugees.
Two out of seven pupils from Caro's school went. Four out of 15 pupils from Stella's school went.
Whose school sent the greater proportion?

 Puzzle

I am bigger than $\frac{2}{3}$ but smaller than $\frac{7}{9}$. My denominator is 18.
What fraction am I?

Adding and subtracting fractions

We can **add and subtract simple fractions mentally**.

$\frac{1}{4} + \frac{1}{2} = \frac{3}{4}$

$\frac{3}{4} + \frac{3}{4} = 1\frac{1}{2}$

$\frac{1}{8} + \frac{1}{8} = \frac{1}{4}$

It is easy to **add and subtract fractions when the denominators are the same**.

Examples

$\frac{2}{5} + \frac{1}{5} = \frac{2+1}{5}$
$= \frac{3}{5}$

$\frac{5}{8} - \frac{3}{8} = \frac{5-3}{8}$
$= \frac{2}{8}^{1}_{4}$
$= \frac{1}{4}$

Cancel the fraction to its lowest terms.
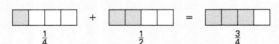

$\frac{3}{10} + \frac{2}{10} + \frac{4}{10} + \frac{3}{10} = \frac{3+2+4+3}{10}$
$= \frac{12}{10}$
$= 1\frac{2}{10}^{1}_{5}$
$= 1\frac{1}{5}$

Change improper fractions to mixed numbers.

Exercise 6 **This exercise is to be done mentally.**

1 a $\frac{1}{2}+\frac{1}{4}$ b $\frac{3}{4}+\frac{1}{2}$ c $\frac{1}{2}-\frac{3}{8}$ d $\frac{3}{4}-\frac{1}{2}$ e $\frac{1}{8}+\frac{1}{8}+\frac{1}{8}$

   f $\frac{1}{2}-\frac{1}{8}$ g $\frac{1}{8}+\frac{1}{8}+\frac{1}{4}$ h $\frac{3}{4}+\frac{1}{8}$ i $\frac{5}{8}-\frac{1}{4}$

2 a $\frac{1}{3}+\frac{1}{3}$ b $\frac{1}{4}+\frac{1}{4}$ c $\frac{1}{8}+\frac{2}{8}$ d $\frac{2}{5}+\frac{2}{5}$ e $\frac{6}{7}-\frac{4}{7}$

   f $\frac{3}{10}+\frac{2}{10}$ g $\frac{5}{8}+\frac{3}{8}$ h $\frac{9}{12}-\frac{3}{12}$ i $\frac{11}{20}+\frac{3}{20}$ j $\frac{7}{12}-\frac{5}{12}$

   k $\frac{13}{15}-\frac{8}{15}$ l $\frac{3}{10}+\frac{5}{10}$ m $\frac{7}{12}-\frac{1}{12}$ n $\frac{17}{20}-\frac{13}{20}$

> **Remember**
> $\frac{3}{3} = 1$

3 a $\frac{7}{10}+\frac{3}{10}+\frac{1}{10}+\frac{4}{10}$ b $\frac{3}{12}+\frac{5}{12}+\frac{7}{12}+\frac{4}{12}$ c $\frac{5}{20}+\frac{17}{20}+\frac{8}{20}+\frac{6}{20}$

   d $\frac{7}{10}+\frac{2}{10}-\frac{4}{10}$ e $\frac{4}{20}+\frac{13}{20}-\frac{8}{20}$ f $\frac{3}{12}+\frac{5}{12}+\frac{4}{12}-\frac{3}{12}$

   g $\frac{19}{20}+\frac{3}{20}-\frac{7}{20}$ h $\frac{8}{10}+\frac{3}{10}-\frac{5}{10}$

4 $\frac{1}{8}$ of the pupils at Jake's school walked to school. $\frac{3}{8}$ came by bus.
What fraction either walked for came by bus?

5 $\frac{1}{5}$ of a stall at a fair is used for sweets and $\frac{2}{5}$ for cakes. The rest is to
be used for biscuits.
What fraction of the stall is left for biscuits?

Review 1

a $\frac{1}{2}+\frac{1}{8}$ b $\frac{1}{2}+\frac{3}{8}$ c $\frac{1}{2}-\frac{1}{4}$ d $\frac{1}{4}+\frac{1}{8}$ e $\frac{3}{4}-\frac{1}{8}$ f $\frac{1}{4}+\frac{1}{4}+\frac{1}{8}$

g $\frac{3}{5}+\frac{3}{5}$ h $\frac{5}{8}+\frac{3}{8}+\frac{1}{8}$ i $\frac{4}{10}+\frac{7}{10}+\frac{5}{10}-\frac{1}{10}$ j $\frac{15}{20}-\frac{5}{20}+\frac{7}{20}$

Review 2 $\frac{5}{9}$ of a class went on a trip in the morning. $\frac{2}{9}$ went in the afternoon.
What fraction of the class went on the trip?

We **add and subtract fractions with different denominators** using equivalent fractions.

Example $\frac{1}{3}+\frac{5}{6}=\frac{2}{6}+\frac{5}{6}$
           $=\frac{2+5}{6}$ [$\frac{1}{3}=\frac{2}{6}$]
           $=\frac{7}{6}$
           $=1\frac{1}{6}$

Example $\frac{7}{8}-\frac{1}{4}=\frac{7}{8}-\frac{2}{8}$
           $=\frac{7-2}{8}$ [$\frac{1}{4}=\frac{2}{8}$]
           $=\frac{5}{8}$

Exercise 7

1 a $\frac{1}{2}+\frac{3}{10}$ b $\frac{1}{3}+\frac{1}{6}$ c $\frac{1}{3}+\frac{4}{9}$ d $\frac{3}{8}+\frac{1}{4}$ e $\frac{3}{7}+\frac{5}{14}$ f $\frac{7}{10}-\frac{1}{5}$ g $\frac{7}{8}-\frac{1}{4}$

  h $\frac{9}{10}-\frac{1}{2}$ i $\frac{5}{12}-\frac{1}{6}$ j $\frac{2}{3}+\frac{1}{4}$ *k $\frac{5}{6}-\frac{3}{4}$ *l $\frac{1}{3}+\frac{1}{5}$ *m $\frac{7}{8}-\frac{5}{6}$

2 a $\frac{1}{2}+\frac{1}{4}+\frac{3}{8}$ b $\frac{3}{10}+\frac{2}{5}+\frac{1}{10}$ c $\frac{3}{4}+\frac{1}{8}-\frac{1}{4}$ d $\frac{5}{9}+\frac{1}{3}-\frac{2}{9}$ e $\frac{3}{4}+\frac{2}{3}+\frac{5}{12}$ f $\frac{7}{20}-\frac{1}{5}+\frac{1}{4}$

  g $\frac{13}{15}-\frac{1}{5}+\frac{1}{3}$ *h $\frac{7}{10}+\frac{1}{2}-\frac{2}{3}$ *i $\frac{7}{10}-\frac{2}{5}+\frac{1}{3}$ *j $\frac{5}{12}+\frac{3}{4}-\frac{2}{3}$ *k $\frac{3}{4}+\frac{5}{6}-\frac{1}{8}$

3 Fran bought $\frac{2}{5}$ m of ribbon and then another $\frac{3}{10}$ m. What fraction of a metre did she buy altogether?

4 $\frac{1}{4}$ of a park is grass and $\frac{3}{8}$ is a playground. What fraction of the park is this altogether?

5 Olivia won some money in a competition. She gave $\frac{1}{2}$ to her mother and $\frac{1}{3}$ to her brother. What fraction did she give away?

6 A bottle had $\frac{9}{10}$ ℓ of water. James poured $\frac{3}{5}$ ℓ into a jug. How much was left in the bottle?

7 Two different fractions add to 1. What might they be?

8 Two different fractions subtract to give $\frac{1}{2}$. What might they be?

*9 Three fractions with different denominators add to 1. What might they be?

*10 What digit might ▲ be? It need not be the same in each case.

　a $\frac{▲}{▲} + \frac{▲}{4} = \frac{7}{▲▲}$ 　　b $\frac{▲}{▲▲} + \frac{▲}{6} = \frac{7}{▲▲}$ 　c $\frac{▲▲}{▲▲} + \frac{▲}{9} = \frac{13}{▲▲}$

Review 1 　a $\frac{1}{4} + \frac{3}{8}$ 　b $\frac{3}{4} + \frac{1}{12}$ 　c $\frac{2}{3} - \frac{2}{9}$ 　d $\frac{1}{8} + \frac{3}{16}$ 　e $\frac{7}{20} - \frac{1}{4}$

Review 2 　a $\frac{1}{4} + \frac{7}{8} + \frac{1}{2}$ 　b $\frac{9}{10} - \frac{3}{5} + \frac{1}{2}$ 　c $\frac{11}{12} - \frac{3}{4} + \frac{5}{6}$

Review 3 Thomasina filled a glass which holds $\frac{1}{4}$ ℓ from a flask with $\frac{7}{8}$ ℓ in it. How much was left in the flask?

'Fraction of' and multiplying an integer by a fraction

Remember $\frac{1}{4}$ of 16, $\frac{1}{4} \times 16$, $16 \times \frac{1}{4}$ and $16 \div 4$ are all equivalent.

Worked Example
a $\frac{1}{5} \times 4$ 　　b $\frac{4}{5} \times 6$ 　　c $\frac{7}{25}$ of 29

In maths of means multiply.

Answer
a $\frac{1}{5} \times 4 = \frac{4}{5}$

b $\frac{4}{5} \times 6 = 4 \times \frac{1}{5} \times 6$
$= \frac{24}{5}$
$= 4\frac{4}{5}$

c $\frac{7}{25}$ of $29 = 7 \times \frac{1}{25} \times 29$
$= \frac{203}{25}$
$= 8\frac{3}{25}$

Worked Example
Lucy made nine large cakes for her sister's 21st birthday. $\frac{4}{5}$ of them were eaten. How many were eaten?

Answer
$\frac{4}{5}$ of $9 = \frac{4}{5} \times 9$
$= 4 \times \frac{1}{5} \times 9$
$= \frac{36}{5}$
$= 7\frac{1}{5}$
$7\frac{1}{5}$ of them were eaten.

Sometimes when multiplying an integer by a fraction we can cancel.

Example **a** $\frac{7}{12} \times 16 = \frac{7}{12^3} \times \frac{16^4}{1}$

$= \frac{7}{3} \times \frac{4}{1}$

$= \frac{28}{3}$

$= 9\frac{1}{3}$

16 can be written as $\frac{16}{1}$.

b $1\frac{3}{4} \times 20 = \frac{7}{4_1} \times \frac{20^5}{1}$

$= \frac{35}{1}$

$= \mathbf{35}$

Exercise 8

1 Write true or false for these.

a $\frac{5}{9} \times 30 = 5 \times \frac{1}{9} \times 30$ **b** $\frac{3}{8} \times 13 = 13 \times \frac{3}{8}$ **c** $24 \times \frac{7}{16} = \frac{7}{16} \times \frac{24}{1}$

d $\frac{5}{6} \times 9 = \frac{54}{5}$ **e** $\frac{3}{8} \times 12 = \frac{3}{8} \times \frac{1}{12}$ **f** $\frac{3}{5} \times 6 = \frac{18}{5}$

g $\frac{2}{3}$ of $18 = 18 \div 3 \times 2$

2 Calculate these.

a $\frac{3}{5} \times 20$ **b** $\frac{3}{8} \times 16$ **c** $\frac{7}{10} \times 40$ **d** $\frac{5}{6} \times 24$

e $\frac{2}{3} \times 60$ **f** $\frac{3}{4} \times 48$ **g** $\frac{7}{8} \times 56$ **h** $\frac{3}{7} \times 210$

i $45 \times \frac{2}{5}$ **j** $32 \times \frac{5}{8}$ **k** $48 \times \frac{5}{6}$ **l** $100 \times \frac{3}{4}$

m $120 \times \frac{2}{3}$ **n** $1\frac{1}{2} \times 4$ **o** $1\frac{1}{2} \times 10$ **p** $1\frac{1}{4} \times 20$

q $2\frac{1}{2} \times 6$ **r** $2\frac{1}{4} \times 16$ **s** $1\frac{3}{4} \times 24$

3 a $\frac{1}{3} \times 2$ **b** $\frac{1}{5} \times 3$ **c** $\frac{1}{6} \times 5$ **d** $\frac{1}{8} \times 5$

e $\frac{1}{4} \times 3$ **f** $\frac{2}{5} \times 3$ **g** $\frac{3}{8} \times 5$ **h** $\frac{5}{6} \times 5$

i $\frac{3}{5} \times 4$ **j** $1\frac{1}{2} \times 3$ **k** $1\frac{1}{4} \times 10$

4 a $0{\cdot}25 \times 12$ **b** $0{\cdot}75 \times 16$ **c** $0{\cdot}2 \times 40$ **d** $0{\cdot}4 \times 50$

e $0{\cdot}8 \times 25$ **f** $0{\cdot}1 \times 70$ **g** $0{\cdot}3 \times 120$ **h** $1{\cdot}1 \times 60$

*i $1{\cdot}3 \times 40$ ***j** $2{\cdot}7 \times 20$ ***k** $3{\cdot}3 \times 30$

Change the decimals to fractions first.

5 Give the answers as mixed numbers.

a $\frac{3}{4}$ of 11   **b** $\frac{5}{8}$ of 17   **c** $\frac{2}{3}$ of 37   **d** $\frac{5}{6}$ of 19

e $\frac{7}{9}$ of 80   **f** $\frac{4}{7}$ of 52   **g** $\frac{4}{5}$ of 29   **h** $\frac{7}{8}$ of 9 m

i $\frac{5}{6}$ of 20 m   **j** $\frac{3}{7}$ of 12 cm   **k** $\frac{4}{9}$ of 21 mm   **l** $\frac{2}{3}$ of 130 g

m $\frac{3}{25}$ of 38 ℓ **n** $\frac{7}{20}$ of 82 km   **o** $\frac{5}{6}$ of 51 m

6 a Jillian bought 16 m of rope to make a swing.
She used $\frac{5}{6}$ of it. How much rope did she use?

b Askra's school painted a 48 m long mural around the walls of the hall.
She worked out that her class had painted $\frac{3}{7}$ of it.
What length did her class paint?

c A flea can long jump 33 cm. It can high jump $\frac{3}{5}$ of this distance.
How high can a flea jump?

d Tania knitted a 120 cm long scarf.
$\frac{2}{9}$ of the scarf was blue.
What length of scarf was *not* blue?

7 Write the next two lines for these.

a $\frac{1}{7} \times 1 = \frac{1}{7}$

$\frac{1}{7} \times 2 = \frac{2}{7}$

$\frac{1}{7} \times 3 = \frac{3}{7}$

$\frac{1}{7} \times 4 = \frac{4}{7}$

b $\frac{2}{7} \times 1 = \frac{2}{7}$

$\frac{2}{7} \times 2 = \frac{4}{7}$

$\frac{2}{7} \times 3 = \frac{6}{7}$

$\frac{2}{7} \times 4 = \frac{8}{7}$

c $\frac{4}{5} \times 1 = \frac{4}{5}$

$\frac{4}{5} \times 2 = \frac{8}{5}$

$\frac{4}{5} \times 3 = \frac{12}{5}$

$\frac{4}{5} \times 4 = \frac{16}{5}$

8 Use cancelling to find the answer to these.

a $\frac{5}{6}$ of 12

b $\frac{2}{3}$ of 12

c $\frac{7}{8}$ of 16

d $\frac{5}{6}$ of 9

e $\frac{7}{9}$ of 24

f $\frac{9}{24}$ of 16

g $\frac{3}{15} \times 50$

h $\frac{3}{12} \times 42$

i $36 \times \frac{5}{18}$

j $48 \times \frac{4}{30}$

k $72 \times \frac{8}{45}$

l $64 \times \frac{3}{40}$

m $1\frac{1}{5} \times 20$

n $1\frac{1}{8} \times 6$

o $1\frac{3}{10} \times 25$

p $2\frac{1}{4} \times 10$

9 What goes in the gap, increases or decreases?

a Multiplying a number by a fraction bigger than 1 _____ the number.

b Multiplying a number by a fraction smaller than 1 _____ the number.

Review 1 Give the answers as mixed numbers.

a $\frac{3}{4}$ of 19

b $\frac{7}{8}$ of 50

c $\frac{2}{7}$ of 93

d $\frac{1}{3} \times 4$

e $\frac{2}{3} \times 8$

f $\frac{3}{5} \times 9$

g $1\frac{1}{2} \times 5$

h $2\frac{1}{4} \times 3$

Review 2 a 0.6×50 b 0.8×35 c 4.3×20

Review 3 Peter took seven hours to drive from his home to London. On the way home he took $\frac{7}{8}$ as long.
How long did Peter take to drive home? Give your answer as fraction.

Review 4 Use cancelling to find the answers to these.

a $\frac{5}{12}$ of 18

b $\frac{7}{24}$ of 54

c $\frac{3}{20}$ of 48

d $\frac{7}{24} \times 32$

e $64 \times \frac{7}{12}$

f $2\frac{3}{10}$ of 8

g $1\frac{3}{4} \times 22$

Puzzle

* A coin collection was to be divided amongst three children so that the eldest got half of the coins, the youngest got a ninth of them and the other child got a third. There were 17 coins in this collection. The eldest child decided the collection could be shared like this.

Borrow one coin from a friend to make a total of 18 coins.

The eldest takes $\frac{1}{2}$ of 18 = 9 coins.

The youngest takes $\frac{1}{9}$ of 18 = 2 coins.

The other child takes $\frac{1}{3}$ of 18 = 6 coins.

Now that the 17 coins have been shared the borrowed coin is returned to the friend.

Does each child get the correct number of coins? Explain.

Dividing an integer by a fraction

$2 \div \frac{1}{3}$ means 'How many $\frac{1}{3}$s are there in 2?' The answer is 6.
We could write this as $2 = \blacksquare \times \frac{1}{3}$.

Discussion

- $30 \times \frac{1}{5} = 6$.
  How could you use this to find the answer to $6 \div \frac{1}{5}$? **Discuss**.

-

  Each circle has been divided into thirds.
  How could we use this to find the answer to $4 \div \frac{1}{3}$? **Discuss**.

  How would you divide these circles to help
  you find the answer to $2 \div \frac{1}{4}$? **Discuss**.

  Remember:
  $4 \div \frac{1}{3}$ means "How
  many $\frac{1}{3}$s are in 4?"

  Draw circles and divide them into parts to find the answers to these.
  $3 \div \frac{1}{7}$ $4 \div \frac{1}{5}$ $5 \div \frac{1}{6}$

-

  The blue part is $\frac{2}{5}$ of a circle.
  How many $\frac{2}{5}$s are in four whole circles?
  What is the answer to $4 \div \frac{2}{5}$? **Discuss**.
  Draw circles and divide them into parts to find the answers to these.
  $6 \div \frac{3}{5}$ $9 \div \frac{3}{4}$ $8 \div \frac{2}{3}$

Worked Example
a $3 \div \frac{1}{6}$ b $4 \div \frac{2}{3}$

Answer
a We read this as 'How many $\frac{1}{6}$s in 3?'
   From the diagram we can see that there are
   18 sixths in 3.
   $3 \div \frac{1}{6} = \mathbf{18}$

   Three circles divided into sixths

b We read this as 'How many $\frac{2}{3}$s in 4?'
   From the diagram we can see that there are
   six two-thirds in 4.
   $4 \div \frac{2}{3} = \mathbf{6}$

131

Number

Exercise 9

1 $60 \times \frac{1}{4} = 15$

$30 \times \frac{2}{4} = 15$

$20 \times \frac{3}{4} = 15$

Use this to find the answers to these.

a $15 \div \frac{1}{4}$ **b** $15 \div \frac{2}{4}$ **c** $15 \div \frac{3}{4}$

2 Write a division for each of these.
The answer to **a** is $3 \div \frac{1}{4}$.

   a How many quarters are there in three oranges?
   b How many thirds are there in seven pizzas?
   c How many fifths are there in four pies?
   d How many eighths are there in six apples?

3 **a** $4 \div \frac{1}{4}$ **b** $6 \div \frac{1}{5}$ **c** $8 \div \frac{1}{3}$ **d** $5 \div \frac{1}{8}$ **e** $7 \div \frac{1}{6}$

   f $12 \div \frac{1}{3}$ **g** $20 \div \frac{1}{5}$ **h** $16 \div \frac{1}{2}$ **i** $24 \div \frac{1}{4}$

4 What goes in the box?

   a $32 \times \frac{3}{4} = 24$ so $24 \div \frac{3}{4} = \square$

   b $55 \times \frac{3}{5} = 33$ so $33 \div \frac{3}{5} = \square$

   c $60 \times \frac{2}{15} = 8$ so $8 \div \frac{2}{15} = \square$

5 Use the diagrams to help find the answers to these.

   a $6 \div \frac{2}{3}$

   b $6 \div \frac{3}{4}$

6 Find the answers to these.

   a $4 \div \frac{4}{5}$ **b** $10 \div \frac{5}{6}$ **c** $6 \div \frac{2}{5}$

   d $12 \div \frac{2}{3}$ **e** $9 \div \frac{3}{4}$

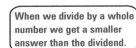

When we divide by a whole number we get a smaller answer than the dividend.

7 What goes in the gap, increases or decreases?

Dividing a number by a fraction smaller than 1 _____ the number.

Review 1 Write a division for each of these.
a How many quarters in 6?   **b** How many thirds in 8?

Review 2   **a** $5 \div \frac{1}{6}$ **b** $4 \div \frac{1}{8}$ **c** $3 \div \frac{1}{6}$ **d** $7 \div \frac{1}{9}$

Review 3 What goes in the box?
a $\frac{2}{5}$ of $45 = 18$ so $18 \div \frac{2}{5} = \square$
b $\frac{3}{7}$ of $56 = 24$ so $24 \div \frac{3}{7} = \square$

Review 4 Use the diagrams to help find the answers.

a $8 \div \frac{2}{3}$

b $9 \div \frac{3}{4}$

Summary of key points

  Unit fractions have a numerator of 1.

Examples $\frac{1}{8}$, $\frac{1}{12}$, $\frac{1}{16}$

 When writing **one number as a fraction of another**, make sure the numerator and denominator have the same units.

Example   To write 37 minutes as a fraction of an hour we change one hour to 60 minutes.

37 minutes as a fraction of an hour = $\frac{37}{60}$.

To write a fraction in its **lowest terms** we divide the numerator and denominator by the highest common factor.

Example   80 as a fraction of 240 = $\frac{80}{240} = \frac{1}{3}$

> Numerator and denominator have been divided by 80.

If it is hard to find the HCF, you can cancel in steps.

 To write a **decimal as a fraction**, write it with denominator 10, 100 or 1000.

Examples $0.82 = \frac{82}{100}$ $0.785 = \frac{785}{1000}$

$= \frac{41}{50}$ $= \frac{157}{200}$

> Always cancel the fraction to its simplest form.

 To write a **fraction as a decimal** we can

1 use known facts   *Example* $\frac{3}{5} = 3 \times \frac{1}{5} = 3 \times 0.2 = 0.6$

2 make the denominator 10 or 100   *Example* $\frac{7}{25} = \frac{28}{100} = 0.28$

3 divide the numerator by the denominator   *Example* $\frac{4}{7}$ 7)4·0

4 use a calculator

Example $\frac{36}{42}$

Key [36] [÷] [42] [=] to get 0·86 (2 d.p.)

```
         3·5      7 × 0·5
         0·50
         0·49     7 × 0·07
         0·010
         0·007    7 × 0·001
         0·003
```

Answer   0·571 R 0·003
           0·57 (2 d.p.)

 A **terminating** decimal has a finite number of digits.

Example $\frac{3}{8} = 0.375$

A **recurring decimal** is a decimal which has one or more repeated digits.

Examples $\frac{2}{3} = 0.6666 \ldots$ $\frac{5}{11} = 0.454\ 545 \ldots$

$= 0.\dot{6}$ $= 0.\dot{4}\dot{5}$

> The dots above the 4 and 5 show that these repeat.

Sometimes we round recurring decimals to 1 or 2 d.p. when 2 or more digits repeat.

> The dot above the 6 shows that the 6 repeats.

F We can use this diagram to **compare fractions**.

We can also divide lines of the same length into parts.

Example

We can also compare fractions by converting them to decimals.

Example Compare $\frac{3}{4}$ and $\frac{5}{7}$.

$\frac{3}{4} = 0\cdot75$ $\frac{5}{7} = 0\cdot71$ (2 d.p.)

$0\cdot75 > 0\cdot71$

$\frac{3}{4} > \frac{5}{7}$

 G To **add (or subtract) fractions** that have **different denominators**, change to equivalent fractions with the same denominator.

Example $\frac{1}{2} + \frac{3}{10} = \frac{5}{10} + \frac{3}{10}$

$= \frac{5+3}{10}$

$= \frac{8}{10}$

$= \frac{4}{5}$

$\frac{1}{2} = \frac{5}{10}$

 H $\frac{1}{8}$ of 24, $\frac{1}{8} \times 24$, $24 \times \frac{1}{8}$, $24 \div 8$ are all equivalent.

Examples $3 \times \frac{5}{8} = 3 \times 5 \times \frac{1}{8}$ $\frac{2}{3}$ of $7 = 2 \times \frac{1}{3} \times 7$ $1\frac{2}{3} \times 12 = \frac{5}{3_1} \times \frac{\overset{4}{12}}{1}$

$= \frac{15}{8}$ $= \frac{14}{3}$ $= \frac{20}{1}$

$= 1\frac{7}{8}$ $= 4\frac{2}{3}$ $= 20$

We have cancelled.

 I $8 \div \frac{1}{4}$ means 'How many $\frac{1}{4}$s are there in 8?'

The answer is 32.

Test yourself

1 The pie chart shows the proportion of each colour in a scarf.
   Estimate the fraction of the scarf that is
   a red   b yellow.

 A

2 a What fraction of £5 is 75p?   b What fraction of 150 is 75?

 B

3 At a dance class there were 20 girls and 5 boys.
   What fraction of the class were girls?

 B

4 Write these as fractions in their lowest terms.
   a 0·855   b 0·036

 C

5 Write these as decimals. Round c to 2 d.p.
   a $\frac{13}{25}$ b $\frac{12}{30}$ c $\frac{6}{7}$ d $\frac{27}{6}$ e $3\frac{2}{5}$

 D

6 What fraction of this shape is shaded?
Give the answer to 1 d.p.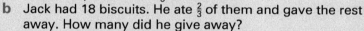

7 Write as recurring decimals.
a $\frac{1}{3}$ **b** $\frac{3}{11}$

8 Which of < or > goes in the box?
a $\frac{1}{3} \square \frac{5}{24}$ **b** $\frac{7}{20} \square \frac{2}{5}$ **c** $\frac{4}{5} \square \frac{3}{4}$ **d** $\frac{7}{9} \square \frac{5}{7}$

9 Jenni worked out that she had spent $\frac{3}{8}$ of last week asleep.
Simon worked out that he had spent $\frac{1}{3}$ of last week asleep.
Who slept more?

10 Write these fractions in order, biggest first.
a $\frac{1}{4}, \frac{2}{5}, \frac{3}{4}, \frac{3}{8}, \frac{1}{6}$ **b** $1\frac{4}{5}, 1\frac{3}{4}, 1\frac{7}{8}, 1\frac{5}{8}$

11 **a** $\frac{3}{7} + \frac{3}{7}$ **b** $\frac{1}{8} + \frac{1}{8} + \frac{1}{8}$ **c** $\frac{3}{5} + \frac{4}{5}$ **d** $\frac{3}{4} - \frac{1}{2}$
e $\frac{7}{8} - \frac{3}{8}$ **f** $\frac{5}{12} + \frac{7}{12} - \frac{3}{12}$ **g** $\frac{12}{13} - \frac{6}{13} + \frac{4}{13}$

12 **a** $\frac{1}{3} + \frac{1}{6}$ **b** $\frac{7}{8} - \frac{1}{2}$ **c** $\frac{2}{3} + \frac{5}{6}$ **d** $\frac{7}{8} - \frac{3}{4}$

13 **a** $\frac{1}{10}$ of £420 **b** $\frac{1}{5}$ of 45 m **c** $\frac{1}{7}$ of 140 ℓ **d** $\frac{1}{4}$ of 60 km
e $\frac{2}{3}$ of £24 **f** $\frac{5}{8}$ of 56 cm **g** $\frac{3}{7}$ of 21 m **h** $\frac{7}{8}$ of 160 g
i $\frac{1}{4}$ of £18 **j** $1\frac{1}{2}$ of 16 m **k** 0.2×25 m

14 Find the answers to these mentally.
a How many minutes is $\frac{5}{6}$ of an hour?
b Jack had 18 biscuits. He ate $\frac{2}{3}$ of them and gave the rest
away. How many did he give away?
c Natalie spends $\frac{5}{8}$ of her allowance and saves the rest.
She gets £32 each month.
How much of this does she spend?
d At a family reunion, there were 150 people.
$\frac{3}{5}$ of them were children.
How many were adults?

15 **a** $\frac{1}{3} \times 24$ **b** $16 \times \frac{1}{4}$ **c** $\frac{3}{8} \times 32$ **d** $70 \times \frac{3}{10}$
e $1\frac{1}{4} \times 24$ **f** $\frac{1}{5} \times 2$ **g** $5 \times \frac{2}{3}$ **h** $\frac{2}{3} \times 10$
i $1\frac{1}{2} \times 7$ **j** 0.4×50 **k** 2.1×30

16 Give the answers as mixed numbers. Cancel if possible.
a $\frac{3}{4}$ of 13 m **b** $\frac{2}{3}$ of 20 km **c** $\frac{7}{8}$ of 12 ℓ **d** $\frac{5}{6}$ of 34 g

17 Jillian's frog can jump 124 cm. Troy's frog can jump $\frac{4}{5}$ of this distance.
How far can Troy's frog jump?

18 Write a division for these. Find the answers.
a How many eighths in 6?
b How many thirds in 5?

19 $84 \times \frac{3}{4} = 63$ so $63 \div \frac{3}{4} = \square$.
What goes in the box?

20 Use this diagram to find the answer to $5 \div \frac{5}{6}$.

You need to know

✓ fractions   page 5

✓ decimals   page 6

✓ percentages   page 6

Key vocabulary

percentage (%)

A bit off

1

The advertisement in the centre is for the internet.
What might the other advertisements be for?

Find some other real-life percentages.
Make a poster or collage of them.

2   Labels on clothes often have the percentages of wool,
polyester, cotton, ...

Find as many labels as you can.

80% LAMBSWOOL
20% NYLON

Percentages to fractions and decimals

Remember

Lucy got 87% in her French test.

'per cent' means out of 100.

$87\% = \frac{87}{100} = 0\cdot87$

87% as a fraction   87% as a decimal

Worked Example

65% of Lucy's class passed the French test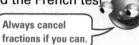

What fraction is this?

Always cancel fractions if you can.

Answer

$65\% = \frac{^{13}65}{_{20}100} = \frac{13}{20}$

We can write percentages greater than 100% as mixed numbers.

Example $142\% = \frac{^{71}142}{_{50}100} = \frac{71}{50} = 1\frac{21}{50}$

It is best to learn these common conversions.

$33\frac{1}{3}\% = \frac{1}{3} = 0\cdot\dot{3}$ $66\frac{2}{3}\% = \frac{2}{3} = 0\cdot\dot{6}$ $12\frac{1}{2}\% = \frac{1}{8} = 0\cdot125$

Exercise 1

1 Write these as fractions in their lowest terms.

    a 30%   b 75%   c 40%   d 5%   e 45%   f 62%

    g $33\frac{1}{3}\%$ h 78%   i 150%   j $66\frac{2}{3}\%$ k 164%   l $12\frac{1}{2}\%$

    m 185%   n 136%   *o 0·5%   *p 3·5%

2 Write the percentages in question **1** as decimals.

3 In a survey it was found that 38% of people think that we will find intelligent life on another planet by 2010. What fraction of people surveyed thought this?

4 There are 100 pupils at Tara's school. 76 of them eat lunch at school.

    a What percentage eat lunch at school?

    b Write your answer to **a** as a decimal.

    c What percentage do not eat lunch at school?

    d Write your answer to **c** as a decimal

    e What do you notice about your answers to **a** and **c**?

    f What do you notice about your answers to **b** and **d**?

5 Write as a decimal.

    a 8·5%   b 15·5%   c 24·5%   *d 0·05%

T **Review 1** Use a copy of this box.

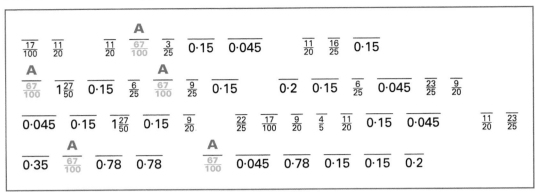

$$\frac{17}{100} \quad \frac{11}{20} \qquad \frac{11}{20} \quad \frac{67}{100} \quad \frac{3}{25} \quad 0{\cdot}15 \quad 0{\cdot}045 \qquad \frac{11}{20} \quad \frac{16}{25} \quad 0{\cdot}15$$

A

$$\frac{67}{100} \quad 1\frac{27}{50} \quad 0{\cdot}15 \quad \frac{6}{25} \quad \frac{67}{100} \quad \frac{9}{25} \quad 0{\cdot}15 \qquad 0{\cdot}2 \quad 0{\cdot}15 \quad \frac{6}{25} \quad 0{\cdot}045 \quad \frac{23}{25} \quad \frac{9}{20}$$

A

$$0{\cdot}045 \quad 0{\cdot}15 \quad 1\frac{27}{50} \quad 0{\cdot}15 \quad \frac{9}{20} \qquad \frac{22}{25} \quad \frac{17}{100} \quad \frac{9}{20} \quad \frac{4}{5} \quad \frac{11}{20} \quad 0{\cdot}15 \quad 0{\cdot}045 \qquad \frac{11}{20} \quad \frac{23}{25}$$

A

$$0{\cdot}35 \quad \frac{67}{100} \quad 0{\cdot}78 \quad 0{\cdot}78 \qquad \frac{67}{100} \quad 0{\cdot}045 \quad 0{\cdot}78 \quad 0{\cdot}15 \quad 0{\cdot}15 \quad 0{\cdot}2$$

Write these as fractions in their lowest terms. Write the letters above the answers.

A 67% = $\frac{67}{100}$ I   17%   U   80%   H   64%   O   92%   N   45%
M   88%   K   12%   T   55%   R   24%   G   36%   V   154%

Write these as decimals. Write the letters above the answers.
P   20%   F   35%   E   15%   L   78%   S   4·5%

Review 2   In a survey it was found that 94% of Americans had been to a burger bar in the last year.
Write 94% as a   **a** decimal   **b** fraction.

Fractions and decimals to percentages

Discussion

1 = 100%

How could you use this to show that these are true? **Discuss**.

10 = 1000%
$\frac{1}{10} = 0{\cdot}1 = 10\%$
$\frac{1}{100} = 0{\cdot}01 = 1\%$
$\frac{1}{8} = 0{\cdot}125 = 12\frac{1}{2}\%$
$\frac{1}{4} = 0{\cdot}25 = 25\%$
$\frac{1}{3} = 0{\cdot}333 \ldots = 33\frac{1}{3}\%$

I ate 100% of the pie. That's a whole pie.

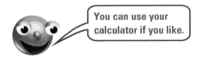

You can use your calculator if you like.

We can **write fractions and decimals as percentages** by writing them as fractions with denominators of 100.

Examples

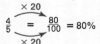
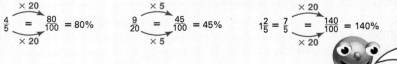

$$\frac{4}{5} \overset{\times 20}{\underset{\times 20}{=}} \frac{80}{100} = 80\% \qquad \frac{9}{20} \overset{\times 5}{\underset{\times 5}{=}} \frac{45}{100} = 45\% \qquad 1\frac{2}{5} = \frac{7}{5} \overset{\times 20}{\underset{\times 20}{=}} \frac{140}{100} = 140\%$$

$$0{\cdot}6 = \frac{6}{10} = \frac{60}{100} = 60\% \qquad 0{\cdot}04 = \frac{4}{100} = 4\%$$

To use this method the denominator must be a factor of 100. See page 48 for factors.

Another way of converting **fractions and decimals to percentages** is to **multiply by 100%**.

Examples $0.86 = 0.86 \times 100\%$ $1.93 = 1.93 \times 100\%$ $\frac{2}{7} = (2 \div 7) \times 100\%$ $2\frac{2}{3} = \frac{8}{3}$
$= 86\%$ $= 193\%$ $= 28.6\%$ (1 d.p.) $= (8 \div 3) \times 100\%$
$= 266.67\%$ (2 d.p.)

Worked Example

Akbar got these marks in four tests.
   71% in Science
   8 out of 11 in Maths
   18 out of 25 in English
   0.77 of the total possible marks in French

a In which test did Akbar do best?

b In which test did Akbar get the lowest percentage?

Answer

a We will change all the marks to percentages.
   8 out of 11 $= \frac{8}{11}$
   $= (8 \div 11) \times 100\%$
   $= 73\%$ to the nearest per cent in Maths.
   18 out of 25 $= \frac{18}{25}$
   $= \frac{72}{100}$
   $= 72\%$ in English
   $0.77 = 77\%$ in French.

   His percentages were Science 71%, Maths 73%, English 72%, French 77%.
   Akbar did best in **French**.

b **Science**

You could use a calculator or written method to do this.

VAT is usually given as $17\frac{1}{2}\%$ rather than 17.5%.

Exercise 2   **Except for question 3, 13 and Review 2.**

1 Do this question mentally.
   Use a copy of this table.
   Fill in the gaps. The first row is done.

Fraction	Decimal	Percentage
$\frac{2}{5}$	0.4	40%
	0.8	
		60%
	0.65	
		55%
$\frac{9}{20}$		

Fraction	Decimal	Percentage
	0.35	
$\frac{4}{25}$		
		3%
	0.06	
$1\frac{3}{4}$		
		136%

2 Convert these to percentages.
   a $\frac{4}{10}$ b $\frac{7}{25}$ c $\frac{9}{20}$ d $\frac{27}{50}$ e $\frac{18}{30}$ f $\frac{42}{60}$
   g $\frac{1}{8}$ h $\frac{1}{3}$ i $\frac{2}{3}$ j $2\frac{3}{4}$ k $3\frac{2}{7}$

3 Convert these to percentages. Give the answer to the nearest per cent.
   a $\frac{4}{11}$ b $\frac{5}{7}$ c $\frac{8}{13}$ d $\frac{6}{17}$ e $\frac{23}{27}$ f $1\frac{7}{34}$

4 Convert these to percentages.
  a 0·56   b 0·565   c 1·23   d 0·875   e 5·68
  f 2·08   g 1·45   h 1·465   i 12·65   j 0·0075
  k 0·0095   l 4·002   m 15·086

5

| $37\frac{1}{2}\%$ | $66\frac{2}{3}\%$ | 30% | $33\frac{1}{3}\%$ | $12\frac{1}{2}\%$ | 40% | 60% | 12% |

Find a percentage from the box for each sentence.
  a One out of three horses in a race lost its rider.
  b Two out of three boys like sport.
  c One out of every eight people at a concert left early.
  d Twelve out of forty people who applied for a job were interviewed.
  e One hundred out of two hundred and fifty pupils at a school had sandwiches for lunch.

6 Estimate the percentage that is red.

  a
  b
  c

7 This pie chart shows what sort of
take-aways Year 7 students like best.
  a What percentage like fish and chips?
  b What fraction like Burger King?
 *c What percentage do *not* like KFC
     best?

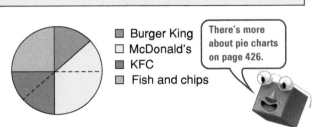

■ Burger King
☐ McDonald's
■ KFC
■ Fish and chips

There's more
about pie charts
on page 426.

8 Are these true or false?
  a $\frac{3}{10} < 29\%$ b $\frac{3}{5} > 38\%$ c $\frac{5}{8} < 65\%$ d $\frac{24}{30} < 80.5\%$
  e $\frac{3}{20} < 12\%$ f $60\% > \frac{13}{20}$ g $\frac{7}{25} < 33\frac{1}{3}\%$ h $90\% > \frac{17}{20}$
  i $66\% < \frac{2}{3}$ j $\frac{28}{40} > 65\%$ k $\frac{18}{25} > 62.5\%$ l $142\% > 1\frac{4}{9}$

9

A   B   C   D

Choose the fraction, decimal or percentage that would be at C.
The answer to **a** is 0·65 since the numbers in order are $\frac{3}{5}$, 0·61, 0·65 68%.
  a 0·65, 68%, $\frac{3}{5}$, 0·61   b 36%, $\frac{2}{5}$, 0·3, $\frac{3}{8}$
  c $\frac{5}{8}$, 62%, 0·65, $\frac{32}{50}$ d $\frac{4}{5}$, 0·87, 85%, $\frac{19}{25}$

10 a Last month Alex had tests in four of her subjects.
       She got 74% for Geography
                18 out of 25 for Maths
                16 out of 20 for Science
                0·72 of the total possible marks in English.
    In which subject did Alex get the highest percentage? What percentage did she get in
    this subject?

*11 What percentage of 2 hours is 12 minutes?

*12 A 600 g rock is analysed and found to contain 24 g of gold.
     What percentage of the rock is gold?

*13 What percentage of the muesli bar is
     a carbohydrate   b protein   c fat?

Muesli Bar 34 g	
carbohydrate	23.6 g
protein	2.2 g
fat	4.9 g

Review 1 Convert these to percentages
a 0·39   b 0·685   c 1·764   d $\frac{3}{10}$ e $\frac{3}{8}$ f $\frac{2}{6}$ g $1\frac{7}{25}$

Review 2 Convert $\frac{9}{13}$ to the nearest per cent.

Review 3 45 cm of a 1 m long shelf was painted.
a What percentage was painted?
b What fraction was *not* painted?

Review 4 Estimate the percentage of the lace that is yellow.

a

b

Review 5 Megan got these marks in four tests.

Number   15 out of 20
Algebra   33 out of 50
Shape, space and measures   21 out of 30
Handling data   69%

In which of these tests did Megan do best? What was her percentage in this subject?

Review 6 Put these in order from smallest to largest. Only use a calculator if you need to.
a $\frac{1}{4}$, $\frac{3}{4}$, 0·6, 70%   b $\frac{1}{8}$, $\frac{2}{5}$, 0·3, 0·45   c $1\frac{1}{2}$, 120%, $1\frac{2}{5}$, $1\frac{3}{8}$
d 0·25, $\frac{25}{40}$, 0·5, $\frac{14}{25}$ e $\frac{5}{8}$, 69%, 0·78, $\frac{7}{11}$

Use a number line to help.

Review 7 Five out of 200 batteries tested, failed the test.
What percentage failed?

---

* **Discussion**

Raylene did a survey on whether people prefer summer or winter. She gave her results as percentages rounded to the nearest per cent.

Like summer best 68%
Like winter best 20%
Like both the same 13%

Link to Handling data.

These add to 101%. Does this mean Raylene's results are wrong? **Discuss**.

Practical

1 Carry out a survey to find one of these percentages.
percentage of female drivers
percentage of cars with just one person inside
percentage of cars with personalised number plates
percentage of people who often eat Indian take-aways
percentage of students who don't have any brothers
percentage of houses that are detached
percentage of students at your school who are the eldest in their family

There is more about surveys on page 403.

2 Each night for a month, keep a note of the times you spend on homework for each subject. At the end of the month, work out what percentage of the total homework time you spent on each subject.

Make a wall chart to illustrate this.

* 3 From the Jobs section of one day's newspaper, find the percentage of jobs that:
   are part time
   are for sales people
   are in factories
   are for outside jobs
   are for people with University degrees
   are suitable for school leavers
Choose a way to show what you found out.

Percentage of – mentally

We sometimes use facts like those in the boxes to find percentages mentally.

Examples
10% of £50 = £5	dividing by 10
10% of 86 m = 8·6 m	dividing by 10
5% of 8 ℓ = 0·4 ℓ	finding 10% of 8 ℓ and halving
15% of 60 g = 9 g	finding 10%, then 5% and adding
25% of 160 cm = 40 cm	dividing by 4

Example Broomfield School raised £5600. 11% of this went to the RSPCA.
11% of £5600 = **£616**

10% of £5600 = £560
1% of £5600 = £56
11% of £5600 = £560 + £56

Examples 135% of 180 m = 180 + 54 + 9
                    = **243 m**

100% of 180 = 180
10% of £180 = 18
30% of 180 = 3 × 18
          = 54
5% of 180 = 9

$3 \times 18 = 3 \times 10 + 3 \times 8$
$= 30 + 24$

28% of 80 ℓ = **22·4 ℓ**

10% of 80 ℓ = 8 ℓ
30% of 80 ℓ = 24 ℓ
1% of 80 ℓ = 0·8 ℓ
2% of 80 ℓ = 1·6 ℓ
28% of 80 ℓ = (30% − 2%) of 80 ℓ
          = 24 − 1·6
          = 22·4 ℓ

Exercise 3 **This exercise is to be done mentally. You may use jottings.**

1 **a** 30% of 80   **b** 60% of 30   **c** 5% of 150   **d** 15% of 120
    e 55% of 300 m   **f** 85% of 30 ℓ **g** 35% of 70 mm   **h** 45% of 150 km
    i 65% of 180 g   **j** $33\frac{1}{3}$ of 96 kg   **k** 125% of £80   **l** 120% of 220 cm
    m 160% of 80 ℓ **n** 175% of 2000 m   **o** 31% of 200 m   **p** 51% of 60 g
    q 19% of 200 km   **r** 52% of 60 sec   **s** 29% of £1500   **t** 12% of 300 m
    u 48% of 60 ℓ **v** 39% of 12 km   **w** 93% of 20 kg

2 How much would Perry save on a shirt in the sale?

3 Jane said that 30% of £45 is the same as 45% of £30.
   Is she correct?

4 How much would be saved on a watch today that usually cost
   a £50   **b** £60   **c** £80?

SHIRTS £50 — *SALE 35% off this price*

WATCHES 45% DISCOUNT TODAY ONLY

5 A file is 120 kilobytes in size.
   How much of it has been downloaded in each of these?

   a 65% downloaded   **b** 85% downloaded   **c** 21% downloaded

6 Find
   a 20% of 1·6 m   **b** 30% of 10·4 ℓ **c** 60% of 8·2 cm   **d** 25% of £9·60
   e 85% of 16·8 ℓ **f** $33\frac{1}{3}$% of 23·4 m   **g** $12\frac{1}{2}$% of £60·40.

7 The femur (thighbone) is a long bone. Its length is about 25% of a person's height. Ann is
   1·5 m tall. About how long is her femur? Give your answer in centimetres.

*8 A jeweller took 10% off the price of all gold rings for one week. At the end of the week she
   added 10% back to the sale prices. These prices were not the same as the original prices.
   Why not?

*9 Carla worked out $17\frac{1}{2}$% of 120 mentally like this.

         10% of 120 = 12
           5% of 120 = 6
         $2\frac{1}{2}$ of 120 = 3
     So $17\frac{1}{2}$ of 120 = 21

   Use her method to work these out.
   a $17\frac{1}{2}$% of 360   **b** $12\frac{1}{2}$% of 620   **c** $62\frac{1}{2}$% of 80   **d** $17\frac{1}{2}$% of £50

Review 1
a 40% of £60   **b** 65% of 80 m   **c** 85% of 120 ℓ **d** 35% of 40 cm
e 125% of 240 g   **f** 165% of 3500 m   **g** 102% of 640 km   **h** 99% of 5000 ml
i 21% of 70 m   **j** 87% of £300

Review 2 There are 60 people in Mrs Redmond's walking group.
45% of them are aged over 70.
How many is this?

Review 3 Jake bought this guitar. How much was the VAT?

£240 + $17\frac{1}{2}$% VAT

143

Percentage of – using a calculator

We can use fractions or decimals **on the calculator** to find 23% of 55.

Using fractions   23% of 55 = $\frac{23}{100}$ × 55

Key (23) (÷) (100) (×) (55) (=) to get 12·65.

Using decimals   23% of 55 = 0·23 × 55

Key (0.23) (×) (55) (=) to get 12·65.

Finding 1% first   23% of 55   1% of 55 = 0·55
                                      23% of 55 = 0·55 × 23

Key (23) (×) (0.55) (=) to get 12·65.

Which way do you think is the quickest?

Key 14·5% as (14.5) (÷) (100) if using fractions or (0.145) if using decimals.

Discussion

Kay found 23% of 55 by finding 1% first.
What would she key on her calculator to find the answer? **Discuss**.

Exercise 4

1 Use fractions to find these.
   a 26% of 8300 g   b 34% of 6500 m   c 18% of 6350 cm   d 87% of £32
   e 44% of 3250 mm   f 66% of 255 kg   g 14% of 5660 kg   h 28% of £3650
   i $33\frac{1}{3}$% of 84 g   j $66\frac{2}{3}$% of 120 km   k $33\frac{1}{3}$% of £114

2 Use decimals to find these.
   a 16% of £26·50   b 38% of £136   c 55% of £72·40   d 84% of £425·50
   e 98% of £24·50   f 46% of £199   g 9% of £44   h 8% of £474·50
   i 14·5% of £52   j 17·5% of £82   k 17·5% of £38   l 181% of £62
   m 167% of 120 km   n 119% of 720 cm

3 Calculate these by finding 1% first.
   a 24% of 724 kg   b 92% of 48 m   c 67% of 36 cm   d 38% of 4·2 km
   e 33% of 25 g   f 185% of 6·4 kg   g 149% of 49 m   h 7% of 1347 cm

4 A TV survey was given to the 550 pupils at Chatfield College.
   a 78% of pupils watched TV every night. How many was this?
   b 46% watched for more than 1 hour every day. How many was this?

5 A jersey with this label has a mass of 650 g.
   What mass of cotton does it have in it?

80%	Wool
14%	Cotton
6%	Polyester

6 An adult's brain weighs about 2% of a person's total weight.
   What is the approximate weight of Sandy's brain if Sandy weighs 56 kg?

7 16% of a 125 g pot of 'Fruit Delight' is fruit. How many grams are *not* fruit?

8 Which of these is true?

   A 85% of £9 < 90% of £8.50   B 85% of £9 > 90% of £8.50   C 85% of £9 = 90% of £8.50

9 Round the answers to these sensibly.

   a A 2p coin is 95% copper, 3·5% tin and 1·5% zinc.
      Five hundred 2p coins weigh 3·56 kg.
      What weight of copper is needed to make five
      hundred 2p coins?

   b It is predicted that by 2021 the population of the
      United Kingdom will be about 63 640 000.
      How many people will live in each of England,
      Wales, Scotland and Northern Ireland then?

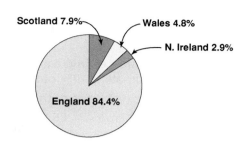

Scotland 7.9% — Wales 4.8% — N. Ireland 2.9% — England 84.4%

10

Blood type	% of population
O	42
A	44
B	10
AB	4

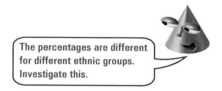

The percentages are different for different ethnic groups. Investigate this.

   a Out of 850 students at a school, how many would you expect to have type **O** blood?
 *b Out of 26 students in a class, how many would you expect to have type **A** blood?
 *c In Matthew's class of 28 pupils, three had type **AB** blood.
      Is this more or less than expected.

*11 Find

   a 25% of 20% of £150   b 75% of 10% of £300.

*12 820 people were surveyed about how much TV they watched. It was found that 60%
watched TV for more than 4 hours each day. Of these, 65% watched for more than 6
hours.

   a How many of those surveyed watched television for more than 4 hours each day?
   b How many watched television for more than 6 hours each day? Round your answer
      sensibly.

*13 44% of the pupils on a technology visit are boys.
25% of the boys and 50% of the girls are from Essex.
What percentage of all the pupils on the visit are from Essex?

*14 A company made £150 million profit.
Next year this is projected to increase by 30% and the following year by a further 20%.
When Marcus read this in the paper he said 'The overall profit increase for the two years is
projected to be 50%'. Explain why Marcus is wrong.

Review 1
a 28% of 3600 kg   b 34% of 8150 m   c 16% of 8420 mm   d 82% of £3140
e 18% of 670 g   f 27% of 920 kg   g 48% of 7580 g   h 16% of 104 m
i $33\frac{1}{3}$% if 132 g   j $66\frac{2}{3}$% of 411 kg   k 185% of 68 m

Review 2 84% of the profit made at a gala day is to go to charity. How much goes to charity
if the total profit is £213·50?

***Review 3** At Burlingford School there are 1240 pupils. 15% wear glasses. Of those who
wear glasses, 63% would like to try contact lenses.
a How many pupils wear glasses?
b How many would like to try contact lenses? Round you answer sensibly.

? Puzzle

Five students estimate the speed of a car as it passes.
Jamie's estimate is 20% out and Susie's estimate is 10% out.
The estimates of the five students are:

    56 km/h   65 km/h   70 km/h   77 km/h   80 km/h

One of the five students correctly estimated the speed.
What was the speed of the car?

★ Practical

1 Design a floor plan for a bungalow or house of total floor area 100 m². Have the bedrooms take up 40% of the space, the lounge and dining room another 40% and the kitchen, bathroom, stairs and hallway 20%.

2 Investigate tipping. You could include some of these: where it began, the countries where you tip, when you tip, the wages of people who get tips.

Percentage increase and decrease

Discussion

Patty sells cakes. She buys them from the bakery for £4. She wants to sell them at a 20% profit.

Patty worked out what price she had to charge like this:
    20% of £4 = £0·80
    Selling price = £4 + £0·80   adding on 20%
                 = £4·80
Patty's friend tells her it is easier to find 120% of £4.

Discuss what Patty's friend means.
How could you find 120% of £4? Does it give the same answer as Patty got? **Discuss**.

To find an **increase** of 30% we can find 30% and add it on or we can find 130%.
130% = 100% + 30%.

To find a **decrease** of 45% we can find 45% and subtract it or we can find 55%.
55% = 100% − 45%.

Worked Example
Marita bought a car for £4800.
When she sold it she made a 15% loss.
How much did she get for the car?

Answer

We can find 15% of £4800 and subtract it.

> 15% of £4800 = £720
> selling price = £4800 − £720
> = **£4080**
>
> **Subtracting 15%**

or

We can find 85% of £4800.

> 85% of £4800 = 0·85 × 4800
> = **£4080**
>
> **Finding 85%**

 100% − 15% = 85%

Exercise 5

1 Mary bought a house for £135 500. She sold it five years later and made a 45% profit. How much did she sell the house for?

2 Sinita bought a bike for £75. When she sold it, she made a 35% loss. How much did she sell the bike for?

3

Was £5   Was £39·50   Was £89·40

SALE 30% OFF

Mark bought these three things in the sale.
How much did he pay altogether?

4 Work out the new price of these.

a   b   c

10% OFF! WAS £210 NOW ?   20% OFF! WAS £6 NOW ?   40% OFF! WAS £180 NOW ?

5 This table shows the pocket money five pupils get each week.
How much would each get after a 10% increase?

Name	Amount
Julieanne	£5
Daniel	£7
Witek	£10
Simon	£3
Cathy	£5.50

6 A stadium seats 3200 people.
Another stand is built so that the number of seats is increased by 15%.
How many people can the stadium seat now?

7 a Increase 80 by 10%.   b Increase 120 by 25%.
   c Increase 120 by 30%.   d Decrease 240 by 35%.
   e Decrease 380 by 65%.   f Decrease 68·5 by 20%.
   g Increase 4·6 by $12\frac{1}{2}$%.

> Young People's
> Savings Bank
> 6% interest
> per year

8 Martha put £1200 in this bank.
   How much will she have after one year?

9 The cost of a holiday for two is £480 in the high season.
   In the low season, the cost is 8% less.

   What is the price of this holiday in the low season?

10 'Betta Biscuits' are sold in packets of 15.
    A special offer has 20% more biscuits free.
    a How many biscuits are in the special offer packet?
    b What fraction of the biscuits in the special offer packet are free?

11 Mohammed bought these at the DIY shop.
    What price did he pay if VAT of $17\frac{1}{2}$% is added on? Give your answers to the nearest
    penny.
    a   b   c

£65   £9·65   £89·50

*12 Lorna bought a horse for £500.
     In January the value of the horse had increased by 12%.
     By June the new value had dropped by 12%.
     Is the horse worth more or less than £500 in June?
     Explain your answer.

Review 1 a Increase 186 by 45%.   b Decrease 5450 by 24%.

Review 2 The weight of a 'Choco-Bar' increased by 20% for no extra cost.
If the original 'Choco-Bar' weighed 120 g, how much does a bar weigh now?

Review 3 The Chocolate Factory reduced the price of its easter eggs by 25%.
a Ruth bought a packet of caramel eggs.
   How much did she pay?
b Tim bought a treat filled egg.
   How much did he pay?
c Lisa bought two packets of cream eggs and a hollow egg.
   How much did she pay?
*d On the day before Easter the eggs were reduced by a further 10%.
    Hamel bought two packets of hollow eggs on this day.
    How much did he pay?

> **EASTER EGGS**
> Caramel Eggs   £2.20 pkt
> Cream Eggs   £2.60 pkt
> Hollow Eggs   £1.80 pkt
> Treat Filled Eggs   £3.40.pkt
> **25% OFF**

Summary of key points

 To write a **percentage as a fraction or decimal**, write the percentage as the number of parts per hundred.

Examples $57\% = \frac{57}{100}$ $85\% = \frac{85}{100}$

Cancel if you can.

$= 0{\cdot}57$ $= \frac{17}{20}$

 $33\frac{1}{3}\% = \frac{1}{3} = 0{\cdot}\dot{3}$ $66\frac{2}{3}\% = \frac{2}{3} = 0{\cdot}\dot{6}$ $12\frac{1}{2}\% = \frac{1}{8} = 0{\cdot}125$

 To write a **fraction or decimal as a percentage**, write it with a denominator of 100 or multiply by 100%.

Examples $0{\cdot}83 = \frac{83}{100} = 83\%$ $\frac{7}{12} = (7 \div 12) \times 100\%$

$= 58{\cdot}3\%$ (1 d.p.)

 $10\% = \frac{1}{10}$ 5% is half of 10% $25\% = \frac{1}{4}$

We sometimes use these facts **to find the percentage of a quantity mentally.**

Examples   20% of 300 m = 60 m   finding 10% and doubling it

15% of 80 g   = 12 g   finding 10%, then 5% and adding

 We sometimes need to use a calculator to find the **percentage of a quantity.**

Using fractions 47% of $8250 = \frac{47}{100} \times 8250$

Key 47 ÷ 100 × 8250 = to get 3877·5

Using Decimals   47% of 8250 = 0·47 × 8250

Key 0.47 × 8250 = to get 3877·5

Finding 1% first   47% of 8250   1% of 8250 = 82·5

47% of 8250 = 82·5 × 47

Key 82.5 × 47 = to get 3877·5

 To find an **increase of 5%** we can find 5% and add it on or we can find 105%.
To find a **decrease of 18%** we can find 18% and subtract it or we can find 82%.
82% = 100% − 18%.

Example   Increase 80 by 45%.

45% of 80 = 36   **or**   145% of 80 = 1·45 × 80

Answer   = 80 + 36   = 116

= 116

Example   Dean bought a stereo for £225. He sold it at a 23% loss.

23% of £225 = £51·75   **or**   77% of £225 = 0·77 × 225

£225 − £51·75 = £173·25   = £173·25

He sold it for £173·25.

Test yourself **Except for questions 9, 11, 12, 13 and 14.**

T

1 Write these as fractions in their lowest terms. ◆A ◆B
   a 65%   **b** $66\frac{2}{3}$%   **c** 128%   **d** 8·5%

2 Fill in the missing entries in these tables. ◆A ◆B ◆C

Decimal	Percentage	Fraction
0·85		
	35%	
		$\frac{13}{20}$
0·08		

Decimal	Percentage	Fraction
		$1\frac{2}{5}$
		$\frac{1}{3}$
	148%	
		$2\frac{5}{8}$

3 Write these as percentages. ◆C
   a 0·63   **b** 0·84   **c** 0·05   **d** 0·11   **e** 1·49   **f** 0·03

4 Estimate how full these jars are. ◆C
   Give your answer as a percentage.
   a   **b**   **c**

5 Put these in order, from smallest to largest. $\frac{1}{4}$, $\frac{3}{8}$, 0·42, $\frac{7}{25}$, 30% ◆C

6 Julia did a survey about gas heating. ◆C
   In Merlin Street 0·62 of the houses used gas.
   In Ruth Street 32 out of 50 of the houses used gas.
   In Brandon Street 7 out of 12 of the houses used gas.
   In which street did the greatest percentage of houses use gas?
   What is this percentage?

7 Find the answers to these mentally. ◆D
   a 45% of 300   **b** 65% of 120   **c** 5% of 480   **d** 15% of 900
   e 31% of 200   **f** 98% of 400   **g** 22% of 40   **h** 51% of 60

8 About 70% of our body weight is water. ◆D
   About how much water is in a person who weighs 55 kg?

9 **a** 19% of £35   **b** 24% of 34 ℓ **c** 36% of 95 m ◆E
   d 83% of £1672   **e** 6·5% of £8

10 Which of these is true?   **A** 40% of £35 < 35% of £40 ◆D
                    B 40% of £35 > 35% of £40
                    C 40% of £35 = 35% of £40

11 a Sam is 160 cm tall. Kim is 85% as tall as this.
How tall is Kim? ◆E

   b Paul is 180 cm tall. Ellie is 94% as tall as Paul.
How tall is Ellie?

12 Joshua got 94%. Mary got 19 out of 20.
Who got the higher mark? ◆E

13 Beth bought a car for £5600. She sold it a year later at a loss of 17%.
For what price did she sell it? ◆F

14 a The same Adidas running shoes were usually £80 in two different shops.
Both shops were offering 'specials' on these shoes.
J&C had them at '£15 off'. B&M had them at '20% off'.
Which shop was cheaper and by how much? ◆F

  *b B&M reduced them a further 15% of the sale price. How much were they then?

7 Ratio and Proportion

You need to know

✓ ratio and proportion   page 6

Key vocabulary

**direct proportion,   equivalent ratio,   proportion,   ratio,
unitary method**

 All mixed up

The chef in this picture is about to mix up
some orange drink from concentrated juice.
She has to mix 3 parts water to 1 part juice.
What does this mean?
What other things do we mix in parts?

Proportion

Worked Example

£1 is worth US$1·62.
How many US$ would Beth get for £20?

Answer

£20 is worth 20 times as much as £1.
£20 is worth 20 × 1·62 = US**$32·40**.

To make purple paint we mix 5 ℓ of blue paint with every 2 ℓ of red paint.

Joanna is making purple paint. She has 8 ℓ of blue paint.
She needs to work out how much red paint to mix with this.

First, she found out how much red paint she would need to mix with **1** ℓ of blue paint.

5 ℓ of blue paint needs **2** ℓ of red paint.
1 ℓ of blue paint needs $\frac{2}{5}$ ℓ of red paint.
So **8** ℓ of blue paint needs $8 \times \frac{2}{5} = 3\frac{1}{5}$ ℓ or **3·2** ℓ of red paint.

$2\ell \div 5 = \frac{2}{5}\ell$

The amount of blue and the amount of red paint are in **direct proportion**.

Worked Example

Six pies cost £9.
How much will seven pies cost?

Find how much **1** pie costs first.

Answer

1 pie costs £$\frac{9}{6}$ or £1·50.
7 pies cost $7 \times £\frac{9}{6}$ = **£10·50**

When we find the cost or amount for **one** first, we call this the **unitary method**.

Exercise 1   **Only use your calculator when you need to.**

1   Three packets of crisps cost £1·20.
    How much will these cost
    a 6 packets   b 12 packets   c 18 packets?

2   A litre of punch contains 300 mℓ of lemonade.
    How much lemonade is there in
    a 2 ℓ punch   b 3 ℓ punch   c 1$\frac{1}{2}$ ℓ punch?

3   £1 is worth 175 Japanese yen.
    How many yen will James get for
    a £30   b £50   c £150?

4 Raphael uses 12 tomatoes to make $\frac{1}{2}\ell$ of sauce.
   a How much sauce can he make with 36 tomatoes?
   b How many tomatoes does he need to make 2 ℓ of sauce?

5 Julian uses this recipe to make pizzas for four people.

Pizza
2 pizza bases
2 onions
250 g cheese
4 tomatoes
120 g bacon

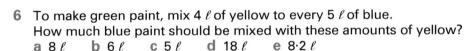

He wants to make a pizza for 10 people.
What quantity of these does he need?
  a tomatoes   **b** bacon   **c** cheese

6 To make green paint, mix 4 ℓ of yellow to every 5 ℓ of blue.
   How much blue paint should be mixed with these amounts of yellow?
   a 8 ℓ **b** 6 ℓ **c** 5 ℓ **d** 18 ℓ **e** 8·2 ℓ

7
  a Eight rolls cost £24.
     What will six rolls cost?
  b Ten filled mushrooms cost £7·50.
     What will six cost?
  c Eight milk shakes cost £9·60.
     What will three cost?
  d Nine tickets to see a band cost £113·40.
     What will five cost?
  e Seven afternoon teas cost £43·75.
     What will 15 cost?
  *f Six fruit and nut bars cost £4·80. Joni has £24.
     How many fruit and nut bars can she buy?

8 5 miles is about 8 km.
   a Manzoor rode his bike 20 miles to the beach.
      About how many kilometres is this?
   b On the way home he got a puncture after 24 km.
      About how many miles from home was he?

*9 A 380 g packet of chocolate coffee beans cost £11.40.
   How much is this per 100 g?

*10 For every £4 Guy earns, he saves £2·20.
   a How much does he save if he earns £85?
   b If he needs to save £100, how much does he need to earn?

*11
  a 400 g of mushrooms cost £2·79.
     How much per kg is this?
  b 120 g of plums cost 84p.
     How much per kg is this?

Review 1 Four chocolate bars cost £1·40.
How much will these cost?
a 8 bars   **b** 12 bars   **c** 20 bars

Review 2 To make rose spray, mix 10 ℓ of water with every 3 ℓ of spray.
How much water should be mixed with these amounts of spray?
a 6 ℓ **b** 12 ℓ **c** 1·5 ℓ

Review 3 Here is a recipe for curry for two people.
How many grams of meat would be needed to make this curry
for three people?

Curry
250g meat
1 onion
1 cup curry sauce
2 tbsp sultanas

Review 4

a Four chocolate frogs cost £5.40.
  What will nine cost?

b Eleven coloured pens cost £20·79.
  What will six cost?

Review 5

a A 650 g packet of salted nuts costs £3·90.
  How much would 100 g cost?

b Ralph has £12. How many grams of salted nuts can he buy?

Writing ratios

2 : 3   4 : 7   1 : 6   5 : 9

All these are **ratios**. We read 2 : 3 as '2 to 3'.

Tammy is making pastry. The ratio of flour to butter is 1 : 2.
This means that for every 1 part of flour, there are 2 parts of butter.

The order in a ratio is important. There is $\frac{1}{2}$ as much flour as butter.
If the ratio of flour to butter is 2 : 1 there would be 2 parts flour to every 1 part butter.
There would be $\frac{2}{1}$ as much flour as butter.

Worked Example

At a camp there are 17 children and 5 adults.
Find the ratio of adults to children.

Answer

We must write the number of adults first.
The ratio of adults to children is **5 : 17**

Worked Example

In water (H_2O) there are twice as many hydrogen atoms as oxygen atoms. Find the ratio of the number of hydrogen to oxygen atoms.

Answer

For every two hydrogen atoms there is one oxygen atom.
The ratio is **2 : 1**.

A ratio can have more than two parts.

Example   At a 'Pet Expo' the ratio of cats to dogs to rabbits was 4 : 6 : 1.

Exercise 2

1 Orange paint has 13 parts yellow and 9 parts red.
  Write the ratio of
  a yellow paint to red paint
  b red paint to yellow paint.

2 The average rainfall in Jakarta is twice that in Manchester.
  What is the ratio of the rainfall in Jakarta to the rainfall in Manchester?

3 Mr Steven took his daughter Pru cross-country skiing.
  Pru is 103 cm tall and it was recommended that she used skis 120 cm long.
  What is the ratio of Pru's height to the length of her skis?

4 a Julian pours 1 bottle of lemonade and 3 bottles of orange into a large bowl.
What is the ratio of orange to lemonade?

b Mary pours 1 bottle of lemonade and 1 bottle of orange into a big jug.
She wants only half as much orange as lemonade in her jug.
What should Mary pour into her jug now?

5 Grace made fruit punch by mixing 2 ℓ of lemonade with 3 ℓ of ginger ale and 1 ℓ of punch mix.

a What is the ratio of lemonade to punch mix to ginger ale?

b What is the ratio of punch mix to lemonade to ginger ale?

Review 1 A drink is made by mixing one part of cordial with three parts of water.
Find the ratio of cordial to water in this drink.

Review 2 Tracy got three times as many marks in a test as Judy.
What is the ratio of Tracy's marks to Judy's marks?

Review 3 There are 9 five-year-olds, 11 six-year-olds and 19 seven-year-olds in a fancy dress competition.
Write the ratio of five-year-olds to six-year-olds to seven-year-olds.

Equivalent ratios

Discussion

Discuss how to show the ratio 4 : 6 on the green line.

Discuss how to show the ratio 4 : 6 on the red line.
Use the red line to write another ratio equal to 4 : 6.
What other ratios are equal to 4 : 6? **Discuss**

● **Discuss** how to shade a circle so that the ratio of the shaded part to the unshaded part is 10 : 2. What other ratio is equal to 10 : 2? **Discuss**

What other ratios could you show by shading parts of a circle? **Discuss**.

To find an **equivalent ratio**
we can divide both parts of a ratio by the same number
or we can multiply both parts of a ratio by the same number.

Examples

$$\div 4 \left(\begin{array}{c} 12 : 8 \\ = 3 : 2 \end{array} \right) \div 4 \qquad \times 6 \left(\begin{array}{c} 2 : 5 \\ = 12 : 30 \end{array} \right) \times 6$$

Worked Example
What number goes in the gap to make these ratios equivalent?
a 1 : 5 = 3 : ____   **b** 12 : 30 = ____ : 5

Answer

a $\times 3 \left(\begin{array}{c} 1 : 5 \\ = 3 : \mathbf{15} \end{array} \right) \times 3$ **b** $\div 6 \left(\begin{array}{c} 12 : 30 \\ = 2 : \mathbf{5} \end{array} \right) \div 6$

We can **simplify a ratio** to its simplest form by dividing both parts by the HCF.

Example $8 : 12 = 2 : 3$

> Both parts of the ratio have been divided by 4.

A ratio is in its **simplest form** when both parts of the ratio have no common factors other than 1.

The numbers in a ratio that is in its simplest form must not be fractions or decimals.

Worked Example
Hamish put $1\frac{1}{2}$ litres of juice and 1 litre of lemonade in a jug.
Write the amount of juice to the amount of lemonade as a ratio in its simplest form.

Answer
The ratio is $1\frac{1}{2} : 1$.

$$\times 2 \left(\begin{array}{c} 1\frac{1}{2} : 1 \\ = 3 : 2 \end{array} \right) \times 2$$

> Multiply both parts of the ratio by 2 to get whole numbers.

The ratio in its simplest form is **3 : 2**.

Exercise 3

1 What number goes in the gap to make these ratios equivalent?
  a $1 : 2 = 3 : \_\_$
  b $3 : 9 = \_\_ : 3$
  c $1 : 4 = 2 : \_\_$
  d $4 : 16 = \_\_ : 4$
  e $1 : 5 = 5 : \_\_$
  f $6 : 30 = 1 : \_\_$
  g $2 : 3 = 4 : \_\_$
  h $2 : 3 = \_\_ : 12$
  i $10 : 4 = 5 : \_\_$
  j $3 : 15 = \_\_ : 5$

2 Give these ratios in their simplest form.
  a $4 : 12$
  b $5 : 20$
  c $3 : 18$
  d $12 : 20$
  e $16 : 24$
  f $20 : 36$
  g $36 : 48$
  h $18 : 32$
  i $6 : 4 : 10$
  j $12 : 9 : 15$
  k $8 : 24 : 32$
  l $20 : 35 : 60$

3 On a school camp there were 15 adults and 40 pupils.
  Write the ratio, in its simplest form, of adults to pupils.

4 This table shows the colour of hats sold.
  Write these as ratios in their simplest form.
  a red hats to green hats
  b green hats to black hats
  c blue hats to black hats
  d black hats to red hats
  e red hats to green hats to black hats

Red	24
Green	16
Blue	21
Black	30

5 A fuel mix is 100 parts petrol to 4 parts oil to 2 parts 'long-life'.
  What is the ratio, in its simplest form, of petrol to oil to 'long-life'?

6 Write these ratios in their simplest form.
  a $2\frac{1}{2} : 3$
  b $1\frac{1}{2} : 2$
  c $3 \cdot 5 : 2$
  d $\frac{1}{2} : 1\frac{1}{2}$
  e $2\frac{1}{5} : 3$
  f $3 \cdot 1 : 5$
  *g $3\frac{1}{3} : 2$
  *h $4 \cdot 5 : 1 \cdot 5$
  *i $1\frac{1}{3} : 4$

*7 Samantha used $2\frac{1}{4}$ m of red and 3 m of green fabric to make a dress.
  Write, in its simplest form, the ratio of red to green fabric.

> Both parts of the ratio must have the same units.

*8 Write these as ratios in their simplest form.
  a 15 mm : 4·5 cm
  b 180 cm : 2·4 m
  c 25 min : $1\frac{1}{4}$ hours
  d 350 g : 2 kg
  e 85 mℓ : 1·7 ℓ
  f 4 cm : 1 km
  g 5 mm : 25 m

*9 Find the ratio $x : y$ if $x = 12z$ and $y = 20z^2$.

Review 1 What number goes in the gap to make these ratios equivalent?
a 1 : 3 = 2 : __
b 1 : 3 = __ : 12
c 8 : 10 = 4 : __
d 2 : 3 = __ : 24
e 6 : 15 = 2 : __

Review 2 René invited 24 friends and 40 relatives to her birthday party.
Write, in its simplest form, the ratio of friends to relatives.

Review 3 Write these as ratios in their simplest form.
a 4 : 6 : 20
b 50 : 20 : 25
c 18 : 24 : 30

Review 4 At a dance, 22 pop, 12 hip hop and 4 techno tracks were played.
Write the ratio of pop to hip hop to techno tracks played.

***Review 5** Write these ratios in their simplest form.
a $\frac{1}{2}$: 4
b $4\frac{1}{4}$: 3
c 2·5 : 10
d 12 min : 2 hours
e 450 g : 3 kg

*Investigation

Triangle Paths

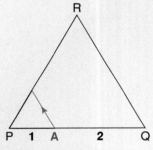
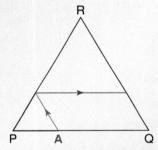
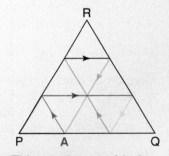

In this triangle, **A** divides the side PQ in the ratio 1 : 2.
From A, a line is drawn parallel to side RQ. This is the red line.

When the red line meets PR, it turns and moves parallel to PQ.
This is the blue line.

This continues with the line turning each time a side is reached, and then moving parallel to another side.
The path finishes at **A**.

Use a copy of these.
What if the triangle had these shapes? Would the path that began at A finish at A?
In what ratio is each side divided? **Investigate**.

a
b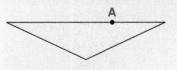
c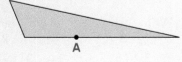

What if A divided the side in the ratio 1 : 3? Investigate.
What if A divided the side in the ratio 2 : 3?
What if ...?

Ratio and proportion

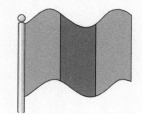

Ratio compares part to part.
Proportion compares part to whole.

Example   This flag has 3 parts, 2 green and 1 red.
The ratio of green to red is 2 parts to 1 part or **2 : 1**.
2 of the 3 parts are green. The proportion of green in the whole flag is $\frac{2}{3}$.

Worked Example
At a camp there are 14 sleeping in tents and 6 in a hall.
a Write the ratio of those in the hall to those in tents in its simplest form.
b What proportion are sleeping in tents? Write this as a percentage.

Answer
a ratio of those in the hall to those in tents = 6 : 14
$\qquad\qquad\qquad\qquad\qquad\qquad\qquad\quad$ = **3 : 7**

b There are 20 people at camp altogether.
14 of them are sleeping in tents.
Proportion sleeping in tents = $\frac{14}{20}$
$\qquad\qquad\qquad\qquad\qquad$ = **70%**

Discussion

One lunch time Lena's class had a stall.
They sold 24 hot dogs, 20 packets of sweets and 36 cans of drink.

How could you work out the proportion of each item sold? **Discuss.**
What ratios can you work out? **Discuss.**

Exercise 4   **Except for question 7 and Review 3.**

1 The RSPCA had 12 kittens to give away. There were 4 grey and 8 ginger ones.
   a What is the ratio of grey to ginger kittens?
   b What proportion of the kittens were   i grey   ii ginger?
   Give your answers as fractions.

2 On a busy road, there were 12 accidents last month.
   Three were injury and nine were non-injury accidents.
   a Give the proportion of injury accidents as a fraction.
   b Give the proportion of non-injury accidents as a percentage.
   c What is the ratio of injury to non-injury accidents?

3 a What is the ratio of sad to happy faces on this banner?
   b What proportion of faces are sad?
   c What proportion of faces are happy?
   Give the answers to b and c as a fraction, decimal and percentage.

4   Rajiv and Rosie each have some stamps in an album.
     The table shows how many stamps they have and how
     many of them are British.

     Who has the greater proportion of British stamps, Rajiv
     or Rosie? You may need to use decimals to compare.

	Number of stamps	Number of British stamps
Rajiv	165	15
Rosie	216	18

5   Bernie planted these 24 plants.
          8 tomato, 10 broccoli, 6 cabbage
     a   Write the ratio of tomato plants to broccoli to cabbage plants.
     b   What proportion of the plants were tomato? Give the answer as a fraction.
     c   What proportion were cabbages? Give the answer as a percentage.
     d   What proportion were either broccoli or cabbage? Give the answer as a decimal.

6   This table shows the points scored in a game.

Name	Nicola	Robyn	Maha
Points	12	24	30

     a   Write these as ratios in their simplest form.
          i   Nicola's to Robyn's to Maha's points
          ii   Robyn's to Maha's to Nicola's points
     b   What proportion of the total points did these score?
          i Nicola   ii Maha   iii Robyn
     c   i   What fraction of Maha's points did Nicola score?
          ii   What is the ratio of Nicola's points to Maha's points?
          iii What do you notice about your answers to i and ii?

*7   Thirty-two dancers, eighty-eight singers and forty backstage crew were on tour.
     a   Write these as ratios.
          i   dancers to singers to backstage crew
          ii   backstage crew to dancers to singers
     b   What proportion of those on tour were
          i dancers   ii singers   iii backstage crew?

Review 1   In a class of 20 drama students, 11 were boys.
a   Give the proportion of boys as a fraction.
b   Give the proportion of girls as a percentage.
c   What is the ratio of boys to girls in the class?

Review 2
a   What is the ratio of blue to yellow hats sold?
b   What is the ratio of blue to yellow to red hats sold?
c   What proportion of hats sold were red? Give the answer as a
     fraction.
d   Give the proportion of yellow hats sold as a percentage.

Colour of hat	Number sold
Blue	20
Red	45
Yellow	15

Review 3   On Thursday and Friday Brett counted the number of cars that stopped in the
'no stopping' area outside the school.
This table shows the number of cars that
stopped outside the school and the number that
stopped in the 'no stopping' area.
On which day did the greater proportion of cars
stop in the 'no stopping' area?

	Number of cars stopping	Number of cars in no stopping area
Thursday	190	21
Friday	219	27

Practical

1 Gather data from your class to make a poster about ratio and proportion.

You could include things like:
    ratio of boys to girls
    proportion of pupils with a computer at home
    proportion of pupils with skateboards
    ratio of those who wear glasses to those who don't
    ratio of left-handed to right-handed pupils
    proportion of pupils who like a particular TV programme
    proportion of pupils who can roll their tongue
    ratio of those who walk to those who come by bus
    proportion of pupils who were not born in Britain.

*2 You will need red, blue and yellow paint and an eye dropper.
You can make other colours using these three colours.
Experiment with different ratios of two colours to see what difference it makes to the final colour.

Example   Have red to blue in the ratio 2 : 1 and
            then try blue to red in the ratio 2 : 1.
            Make a poster of your results.

> This would be good to do at home or in the art room.

Solving ratio and proportion problems

Worked Example
At a party there are 6 fruit bowls and 5 plates of cakes on every table.
There are 48 fruit bowls altogether.
How many plates of cakes are there?

Answer
48 fruit bowls = **8** lots of 6 fruit bowls.
So there must be **8** lots of 5 plates of cakes.
There are **40** plates of cakes.

Worked Example
A nut mix has brazils, cashews and almonds in the ratio 2 : 3 : 5.
How many grams of brazils and cashews are needed to mix with 350 g of almonds?

Answer
5 parts = 350 g
 1 part = 70 g
2 parts = 140 g
3 parts = 210 g
We need **140 g of brazils** and **210 g of cashews** to mix with 350 g of almonds.

Except for question 13.

1 A recipe for a salad has 2 peppers for every 7 tomatoes.
   Ruth makes the salad and uses 6 peppers. How many tomatoes does she use?

2 For every US $16 you get £10.
   How many pounds would you get for US $48?

3 A class voted between going to the cinema and going to the beach. The result was 5 : 3 in favour of the cinema.
   If 20 pupils voted for the cinema, how many voted for the beach?

4 The ratio of non-smokers to smokers at a restaurant was 10 : 2.
   There were 120 non-smokers.
   How many smokers were there?

5 Pizzas are sold in packs of 25.
   There are 2 vegetarian ones in every 5.
   a What fraction are vegetarian?
   b How many of the pack of 25 are vegetarian?

6 One in every 5 families has a single parent.
   Out of 15 million families, how many have a single parent?

7 The scale on a drawing is 1 : 50.
   How far in real life is a distance on the scale drawing of 5·2 cm?

8 Aunt Jannie's pumpkin pie recipe has pumpkin, flour and sugar in the ratio 1 : 3 : 2.
   How many grams of flour and sugar must be added to 500 g of pumpkin?

9 A mixture is made up of blue, black and red jelly beans in the ratio 2 : 5 : 3.
   There are $2\frac{1}{2}$ kg of black jelly beans.
   a How many kg of blue jelly beans are in the mix?
   b How many kg of red jelly beans are in the mix?

10 The ratio of the ages of Selma, Rowan and Mira is 14 : 3 : 1.
   Rowan is 18.
   How old are Selma and Mira?

11 The recipe for punch says to mix juice, ginger ale and tea in the ratio 2 : 5 : 1.
   a How much juice and how much tea is needed to mix with 1 litre of ginger ale?
   b How much punch will be made altogether?

⊞12 Here is a recipe for lasagne for four people.
Ben wants to make lasagne for nine people.
How much of each of these does he need to use?
a mince   **b** cheese sauce
c tomato paste   **d** lasagne sheets

> **Lasagne**
>
> 1 kg mince
> 1 onion
> 600 mℓ cheese sauce
> 30 g tomato paste
> 500 g lasagne sheets

∗13 Linda put some equal sized muffins on the scales. In total, they weighed 3 kg.
She took 4 off the scales. Those left weighed 2 kg.
How many muffins did she put on the scales at the start?

14 The scale on a map is 1 : 50 000.
Selene measured these distances on the map.
How far are they in real life? Give the answers in kilometres.
a 1 cm   **b** 4·6 cm   **c** 15·3 cm   **d** 21·2 cm

> **Antifreeze**
>
> Mix in the ratio
> 3 parts water to
> 2 parts antifreeze

Review 1 How much antifreeze should be added to 6 litres of water?

Review 2 A school voted about changing the uniform. The result was 9 : 1 in favour of changing it.
If 360 voted to change it, how many voted against changing it?

Review 3 John Brown, Susan Smith and Jack Horner got votes in the ratio 3 : 4 : 1.
John Brown got 15 votes.
How many votes did Susan Smith and Jack Horner each get?

Review 4 The lengths of the junior, intermediate and senior cross-country courses are in the ratio 3 : 4 : 9.
a The intermediate course is 2 km long. How long are the other two courses?
b How long are the three courses altogether?

Review 5 This is a recipe for stir fry vegetables for four people.
Tim makes the recipe for five people.
How much of these does he use?
a chopped vegetables   **b** water   **c** black bean paste

> **Stir Fry Vegetables**
>
> 1 kg chopped vegetables
> 2 onions
> 500 ml water
> 30 g black bean paste
> 2 garlic cloves
> 2 tbsp soya sauce

Puzzle

1 a The ratio of Lena's age to her daughter Anthea's age is 4 : 1.
Lena is aged between 20 and 50.
What are all the possible ages, in years, Lena and Anthea could be?
b In four years time the ratios of their ages will be 3 : 1.
How old will Lena and Anthea be then?
c How old will Lena and Anthea be when the ratio of their ages is 2:1?

2

Ratio of games won by Team A to games won by Team B	
Round 1	2 : 1
Round 2	1 : 3
Round 3	2 : 1

Number of games won by Team A	
Round 1	4
Round 2	3
Round 3	2

Which team won the most games?

Dividing in a given ratio

Discussion

Rich Uncle Bob left Aaron and Melanie £42 000 to be shared in the ratio 4 : 3.
Aaron wrote

There are 4 + 3 = 7 shares.
One share is £42 000 ÷ 7 = £6000
4 shares = 4 × £6000   3 shares = 3 × £6000
           = £14 000   = £18 000

Melanie wrote

Out of every £7000, I get £3000 and Aaron gets £4000.
He gets $\frac{4000}{7000}$ or $\frac{4}{7}$ and I get $\frac{3000}{7000}$ or $\frac{3}{7}$.
$\frac{1}{7}$ of £42 000 = £6000
$\frac{4}{7}$ of £42 000 = 4 × £6000 $\frac{3}{7}$ of £42 000 = 3 × £6000
           = £14 000   = £18 000

Discuss Aaron's and Melanie's methods of working out how much each gets.

Worked Example

Three friends, Rachel, Paul and Khalid, paid for a house in the ratio 2 : 3 : 1.
They sold the house for £84 000. How much should each get?

Answer

2 + 3 + 1 = 6
There are six shares altogether.
One share is $\frac{84\,000}{6}$ = £14 000.
Two shares are 2 × £14 000 = £28 000.
Three shares are 3 × £14 000 = £42 000.
Rachel should get £28 000, Paul £42 000 and Khalid £14 000.

> Check that these add up to £84 000.

Exercise 6

1   Melissa had £50. She spent it on CDs and clothes in the ratio 1 : 4.
    How much did she spend on CDs?

2   Thirty-six pupils are going to visit the museum.
    The boys and girls are in the ratio of 4 girls to 5 boys.
    How many boys are there?

3   To make purple paint you mix red and blue paint in the ratio 2 : 7.
    How much of each colour do you need to make
    a 9 ℓ   **b** 27 ℓ   **c** 4·5 ℓ?

4 A drink is made from juice and lemonade in the ratio 2 : 3.
  a How much juice is needed to make 500 mℓ of the drink?
  b How much lemonade is needed to make 1 ℓ of the drink?
  c Answer a and b again if the ratio of juice to lemonade is 3 : 2.
     Do you get the same answers? Explain.

5 To make Joanne's salad dressing you mix oil and vinegar in the ratio 3 : 1.
  How much of each do you need to make these amounts of salad dressing?
  a 400 mℓ b 800 mℓ c 200 mℓ d 1 ℓ
  e Todd made some salad dressing. He mixed the oil and vinegar in the ratio 1 : 3.
     Would it taste the same as Joanne's? Explain.

6 Two friends bought a house for £150 000. Amy paid £50 000. Siobhan paid the rest.
  They later sold this house for £180 000.
  How should the friends share this?

7 a Share £50 in the ratio 3 : 2.   b Share £2400 in the ratio 1 : 5.
  c Share £128 in the ratio 1 : 3 : 4.   d Share £72 000 in the ratio 2 : 3 : 4.
  e Share £850 in the ratio 3 : 1 : 1.   f Share £3600 in the ratio 3 : 4 : 5.
  g Share £720 in the ratio 1 : 2 : 3.   h Share £9600 in the ratio 4 : 3 : 3.
  i Share £85 in the ratio 1 : 5 : 4.   j Share £148·50 in the ratio 3 : 2 : 4.

8 Kate was mixing concrete.
  The mix she used was 1 part cement to 2 parts sand to 4 parts
  aggregate.
  a Write the mix as a ratio.
  b How much sand would Kate need to make 3·5 cubic metres
     of concrete?

9 A fertiliser mixture contains 2 parts nitrogen, 2 parts potash and
  3 parts lime.
  How much of each is in a 70 kg bag?

10 Bob, Rick and Pete paid for a house in the ratio 1 : 3 : 6.
  They sold the house for £180 000.
  How much should each get?

*11 Jake, Jenni and Abbie shared the cost of an evening out in the ratio 2 : 3 : 2. The total cost
  was £164·90.
  How much should each pay? Round the answer sensibly.

Review 1 A 35 cm length of liquorice is shared in the ratio 2 : 3.
What is the length of the longest piece?

Review 2 Pru has 35 tops.
She has four times as many short-sleeved tops as long-sleeved tops.
How many short-sleeved tops does she have?

Review 3
a Share £80 in the ratio 2 : 3 : 5.
b Share £192 in the ratio 3 : 6 : 7.
c Share 5200 g in the ratio 1 : 7 : 8.

Review 4 A farmer planted 40 hectares of crops.
He planted wheat, barley and oats in the ratio 5 : 1 : 2.
How many hectares did this farmer plant in oats?

Review 5 Margot, Richard and Lyn shared a lottery prize of £400 in the ratio 4 : 3 : 1.
How much did each get?

Summary of key points

We can solve problems using **direct proportion**.

Example   Four packets of Easter eggs cost £2·80.

12 packets cost three times as much.

£2·80 × 3 = £8·40. 12 packets cost £8·40.

Example   Nine 'mini-bites' cost £5·22. To find the cost of five mini-bites, first find the cost of one.

One mini-bite costs $\frac{£5·22}{9}$ or £0·58

Five mini-bites cost $\frac{£5·22}{9}$ × 5 = £2·90

5 : 2   4 : 7   12 : 9

These are all **ratios**.

The **order in ratio** is important. 5 : 2 is different from 2 : 5.

Example   There are 4 black rabbits and 3 white rabbits.

The ratio of black rabbits to white rabbits is 4 : 3.

The ratio of white rabbits to black rabbits is 3 : 4.

We can make an **equivalent ratio** by multiplying or dividing both parts of the ratio by the same number

Examples

×5 (2 : 5) ×5   ÷4 (8 : 12) ÷4
= 10 : 25   = 2 : 3

A ratio is in its **simplest form** when both numbers in the ratio have no common factors other than 1.

Example   In its simplest form, the ratio 8 : 12 is 2 : 3.

The numbers in a ratio in its simplest form must not be fractions or decimals.

Examples $3\frac{1}{2}$: 5 = 7 : 10   **Both parts have been multiplied by 2 to get whole numbers.**

4·2 : 3 = 21 : 15   **Both parts have been multiplied by 5 to get whole numbers.**

= 7 : 5

Proportion compares part to whole.

Ratio compares part to part.

Example

The ratio of red to blue is 1 part to 5 parts or 1 : 5.

The proportion that is red is 1 part out of 6 parts or $\frac{1}{6}$.

We can solve **ratio and proportion** problems.

Example   The ratio of boys to girls in a cricket club is 5 : 3.

If there are 20 boys then we can work out the number of girls.

20 is **4** lots of 5.

There must be **4** lots of 3 or **12** girls.

G We can **divide in a given ratio**.

Example There are 45 pupils in the badminton club.

The ratio of boys to girls is 4 : 5.

Out of every 9 pupils, 4 are boys.

$\frac{4}{9}$ are boys.

$\frac{1}{9}$ of 45 = 5

$\frac{4}{9}$ of 45 = 4 × 5

= 20

There are **20** boys in the badminton club.

Example Divide £5650 in the ratio 2 : 3 : 5.

There are 2 + 3 + 5 = 10 shares altogether.

One share is $\frac{5650}{10}$ = £565

Two shares is 2 × £565 = £1130

Three shares is 3 × £565 = £1695

Five shares is 5 × £565 = £2825

£5650 divided in the ratio 2 : 3 : 5 is **£1130 : £1695 : £2825**.

Test yourself

1 5 plants cost £5·20.
How much will these cost?
a 10 plants   b 15 plants   c 20 plants

2 a Seven sandwiches cost £9·45.   b Nine filled rolls cost £25·65.
What will 15 cost?   What will five cost?

3 At a school fête, 14 stalls were outside and eight were inside.
Find the ratio of
a inside to outside stalls
b outside to inside stalls.

4 Charlotte has twice as many pairs of shoes as Katie.
Find the ratio of the number of shoes Charlotte has to the number Katie has.

5 What goes in the gap to make these ratios equivalent?
a 2 : 3 = 8 : __   b 28 : 12 = 7 : __   c 4 : 9 = __ : 45

6 Write each ratio in its simplest form.
a 16 : 20 : 24   b 30 : 120 : 75   c $2\frac{1}{2}$: 3   d 3·6 : 1

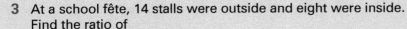
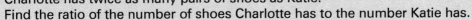

7 20 pupils ordered pizza. Seven ordered barbeque chicken and 13 ordered meat lover's.   **(E)**
  a Give the proportion that ordered meat lover's as a fraction.
  b Give the proportion that ordered barbeque chicken as a percentage.
  c What is the ratio of those who ordered barbeque chicken to those who ordered meat lover's?

8 A recipe for fertiliser says to add 5 litres of water to every 2 litres of fertiliser.
How much water should be added to 6 litres of fertiliser?   **(F)**

> **Fertiliser**
> Mix 5 litres
> of water to
> every 2 litres
> of fertiliser

9 At an exhibition there were water colours, oil paintings and screen prints in the ratio 6 : 9 : 4. There were 18 oil paintings.   **(F)**
  a How many water colours were there?
  b How many screen prints were there?

10 A salad is made up of rice, beans and carrots in the ratio 7 : 3 : 5.   **(F)**
Marian used 1 kg of carrots to make the salad.
What would be the total weight of Marian's salad?

11 A recipe for 5 people used 250 g of mushrooms.   **(A) (F)**
What weight of mushrooms is needed for 11 people?

12 Natasha was given a 700 g bag of sweets   **(G)**
She shared this with her sister in the ratio 3 : 4.
Her sister got the bigger share.
How many grams of sweets did Natasha and her sister each get?

13 Bett and Fred share a 56 hour job in the ratio 5 : 3.   **(G)**
How many hours does Bett work?

14 a Divide £850 in the ratio 1 : 2 : 2.   **(G)**
  b Share 1620 g in the ratio 5 : 4 : 3.

*15 A concrete mix consists of cement, sand and aggregate in the ratio of 1 : 2 : 5.   **(G)**
How much cement is needed to make 2 cubic metres of concrete?

Algebra Support

Unknowns

In each of these, ■ stands for an **unknown** number.

8 + ■ = 12   3 × ■ = 21   24 + 36 = ■

In 8 + ■ = 12, ■ stands for 4 because 8 + 4 = 12

Practice Questions 3, 10, 13, 15, 16

Expressions

Paul has x coins in his hand. He is given three more coins.
An expression for the number of coins Paul has now is $x + 3$.

Here are some more examples of expressions.

add 5 to a number	$n + 5$
subtract 3 from a number	$n - 3$
subtract a number from 6	$6 - n$
divide a number by 3	$\frac{n}{3}$
multiply a number by 5	$n \times 5$ or $5n$
multiply a number by 3 then add 4	$n \times 3 + 4$ or $3n + 4$
add 4 to a number then multiply by 3	$(n + 4) \times 3$ or $3(n + 4)$

We write $n \times 5$ without a multiplication sign as $5n$.

Practice Questions 5, 6, 11

Formulae

age in years = age in months divided by 12   is a **formula**.
If we know the age in months we can find the age in years.

Formulae is the plural of formula.

A **formula** is a rule for working something out.

Example   Emily does jobs for her parents.
She uses this formula to work out how much she gets paid.
amount paid = hours worked × £2 + £3
We can use the formula to work out how much she gets paid for working 6 hours.
Amount paid = 6 × £2 + £3
= £12 + £3
= £15

Practice Questions 7, 8, 12

169

Counting on and back

Counting on from 4 in steps of 5 we get

4   9   14   19   24   •••
   +5   +5   +5   +5   +5

Counting back from 72 in steps of 4 we get

72   68   64   60   56   52   •••
  −4   −4   −4   −4   −4   −4

We usually write the numbers with commas in-between.

72, 68, 64, 60, 56, 52, ...

Counting back from 40 in steps of 10 we get

40, 30, 20, 10, 0, −10, −20, ...

Sometimes we count on to find the next shape in a **pattern**.

Example   Wesley makes bags. He puts patterns on the bags using strips of tape.

1 bag
4 strips

2 bags
8 strips

3 bags
12 strips

The pattern of the number of strips is 4, 8, 12, ...
Counting on in steps of 4 we get 4, 8, 12, 16, 20, 24, 28, ...
Four bags will have 16 strips, five bags 20 strips and so on.

Practice Questions   **1, 2, 4, 9, 14**

Practice Questions

1   Write down the first six numbers you get when you
   a   start at 0 and count on in threes   **b**   start at 0 and count on in nines
   c   start at 60 and count back in fives   **d**   start at 200 and count back in tens
   e   start at 200 and count back in steps of twenty-five
   f   start at 72 and count on in steps of six
   g   start at 4 and count back in steps of two.

2   What is the next number?
   a   4, 8, 12, ...   **b**   3, 6, 9, ...   **c**   5, 10, 15, ...   **d**   8, 16, 24, ...
   e   20, 40, 60, ...   **f**   100, 90, 80, ...   **g**   40, 35, 30, ...   **h**   12, 19, 26, ...
   i   17, 26, 35, ...   **j**   81, 72, 63, ...   **k**   100, 93, 86, ...

3   What number does ■ stand for in these?
   a   ■ + 4 = 6   **b**   ■ + 5 = 10   **c**   8 + ■ = 14   **d**   ■ + 9 = 16
   e   9 + ■ = 20   **f**   ■ + 45 = 80   **g**   33 + ■ = 50   **h**   4 × ■ = 12
   i   5 × ■ = 20   **j**   7 × ■ = 21   **k**   4 × ■ = 36   **l**   4 × ■ − 5 = 15
   m   3 × ■ + 6 = 12   **n** $\frac{■}{4} = 10$ **o** $\frac{■}{3} = 6$

4 Find the missing numbers in these sequences. Explain the rule for the sequence.
For **a** the rule is 'start at 1 and count on in steps of 2'.
a 1, 3, ☐, ☐, 9, ...   **b** 10, 20, ☐, ☐, 50, ...
c 6, 10, ☐, ☐, 22, ...   **d** 50, 44, ☐, ☐, 26, 20, ☐, ☐, ...
e ☐, 8, 11, 14, ☐, ☐, ...   **f** ☐, ☐, 19, 16, 13, ☐, ☐, ...
g ☐, ☐, 45, 49, ☐, 57, 61, ☐, ...

T

5 Use a copy of this.
Write the expression on the left without a multiplication sign.
Join it to the answer on the right with a line.

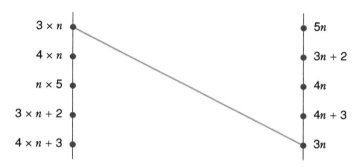

6 Write an expression for these. Let the unknown number be n.
a subtract 5 from a number   **b** add 6 to a number
c multiply a number by 9   **d** divide a number by 8
e multiply a number by 5 then add 4
f multiply a number by 6 then subtract 3
g add 4 to a number then multiply by 3
h subtract 3 from a number then multiply by 4
∗**i** multiply a number by itself
∗**j** multiply a number by itself then multiply by 5
∗**k** multiply a number by 4 then multiply the answer by itself.

Write your expressions without a multiplication sign.

7 A formula for finding the cost of sweets is
cost = cost of one packet × number of packets.
Use this formula to find the cost if
a cost of one packet = £2, number of packets = 5
b cost of one packet = £1·50, number of packets = 4.

8

Microwave
Time = weight in pounds × 10 + 20

Electric Oven
Time = weight in pounds × 25 + 40

These formulae show the time, in minutes, to cook meat in the microwave and in an electric oven.
a How long will it take to cook 3 lb of meat in the microwave?
b How long will it take to cook 5 lb of meat in an electric oven?
c How much quicker is it to cook 4 lb of meat in the microwave than in the oven?

9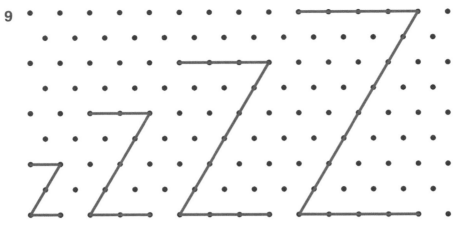

    a 5 dots are needed to draw the first shape.
       How many dots are needed to draw each of the other Z shapes?
    b Write down the sequence of numbers made by the number of dots in these four shapes.
    c Continue the sequence, by counting on.
    d How many dots will be needed to draw the 10th Z?

10 What number does ▲ stand for in these?
    a $10 - ▲ = 4$ **b** $20 - ▲ = 10$ **c** $16 - ▲ = 11$ **d** $25 - ▲ = 16$
    e $100 - ▲ = 80$ **f** $37 - ▲ = 29$ **g** $50 - ▲ = 21$

11 Write an expression for the number of marbles Kishan has. The answer to **a** is *m* – **5**.
    a Kishan had *m* marbles. He gave 5 away.
    b Kishan had *p* marbles. He was given 8 more.
    c Kishan had *n* marbles. He lost 6.
    d Kishan had *x* marbles. He gave 9 to a friend.

12 You will need to draw this diagram.

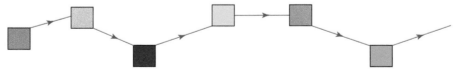

    Write down a formula such as 'double each number and subtract five'.
    Choose a starting number. Use your formula to find the next five or six numbers to make a chain.
    Give your number chain to someone else to see if they can work out your formula.

13 ■ + ■ = 10   ■ – ■ = 8
    What numbers might ■ stand for in these?

14

+	1	2	3	4	5
1	2	3	4	5	6
2	3	4	5	6	7
3	4	5	6	7	8
4	5	6	7	8	9
5	6	7	8	9	10

There are many number patterns in this addition table.
In the red square 'box', the numbers on each diagonal (8 and 10, 9 and 9) add to the same number.
Is this true for all square 'boxes'?

In the blue rectangular 'box', the numbers read clockwise, beginning at the top left, are 3, 4, 5, 6, 5, 4, 3. Do all rectangular 'boxes' have a similar number pattern?

What other number patterns can you find in this addition table?

15 Each letter from A to G is a code for one of these digits: 1, 3, 4, 5, 6, 8 or 9. Crack the code.

D + D = A	D × D = BC	D + E = BF	G × C = 30
E + E = BA	E × E = AB	D × E = FC	

16 Replace each of the letters with one of the digits 0, 1, 2, 3, 4, 5, 6, 7, 8, 9 so that the additions are correct.

```
1   S O M E        2      S U N        3      F O U R
  + C A N               + B U R N           + F I V E
  ─────────               ─────────           ─────────
    H E A R               F E V E R             N I N E
```

Is there more than one possible answer?

8 Expressions, Formulae and Equations

You need to know

✓   unknowns   page 169

✓   expressions   page 169

✓   formulae   page 169

Key vocabulary

algebra, algebraic expression, brackets, collect like terms, commutative, denominator, equation, evaluate, expression, formula, formulae, multiply out, numerator, relationship, simplest form, simplify, solution, solve, substitute, symbol, therefore ($\therefore$), transform, unknown, value, variable

▶▶ Get the message

This is the **Semaphore Alphabet**. Symbols stand for letters.

Decode this message.

In Algebra, letters stand for numbers.

Make up a message of your own using the Semaphore Alphabet.
You could draw stick people like this.

I   Z

Give your message to a friend to decode.

Understanding algebra

Brad has a number of borrowed videos plus 8 videos of his own.
We don't know how many borrowed ones he has.
We could say he has $n + 8$ videos.

$n + 8$ is an **algebraic expression**.
n is a letter **symbol** standing for an **unknown** number.
In an expression, n can have any **value**.

If we know Brad has 11 videos altogether we can write
$n + 8 = 11$.

$n + 8 = 11$ is an **equation**. n has a particular value.

> Sometimes n is called a variable because its value can vary.

$a + b = 8$ is also an equation. a and b can have many different values but they must always add to 8. For example, $a = 3$, $b = 5$ **or** $a = 2 \cdot 5$, $b = 5 \cdot 5$.

An equation always has an equals sign.

Exercise 1

1 Which of these are expressions and which are equations?
Explain how you can tell.
$3n - 4$ $3n - 4 = 7$ $x + y = 8$ $\frac{5x}{2}$ $3(x + 7)$

2 a Is $2p + 3 = 4$ an expression or an equation?
   b Can p have *any* value or does it have a *particular* value?

Review
a Is $4x - 2$ an expression or an equation? How can you tell? Can x have any value?
b Is $a - 6 = 9$ an expression or an equation? Can a have *any* value or does it have a *particular* value?

Brad's local video store delivers hire videos. The cost to hire a video is given by $C = 4v + 5$.
C is the cost, in pounds and v is the number of videos hired.
The 5 on the end is the delivery cost, £5.
$C = 4v + 5$ is called a **formula**.
C and v are unknowns related by the formula. Once v is known we can work out C.

Formulae and equations both have an equals sign. The letters in a formula always stand for something specific like cost, speed or number of books, ... The formula is the rule which gives the **relationship** between these things. There is always more than one unknown in a formula.

Exercise 2

1 Bobbie was asked if $a = 5b - 3$ was a formula or an equation. Can he tell just by looking at it? Explain.

2 The relationship between the number of coloured circles, C, and the diagram number, n, is given by $C = 3n + 2$.
Is $C = 3n + 2$ an equation or a formula? Explain.

diagram 1   **diagram 2**   **diagram 3**

Review   One of these is an equation, one is an expression and one is a formula.
Which is which?

$$D = 5p - 2 \qquad 5b - 2 \qquad 5m - 2 = 18$$

Expressions are usually written without a multiplication or division sign.

$2 \times n$ is written as $2n$.
$1 \times n$ or $n \times 1$ is written as n.
$p \times q$ or $q \times p$ is written as pq.
$(a + b) \div c$ is written as $\frac{a+b}{c}$.
$x \times (y + 4)$ is written as $x(y + 4)$
$3 \times 2a$ is written as $6a$
$n \times n$ is written as n^2 and is said 'n squared'.
$n \times n \times n$ is written as n^3 and said as 'n cubed'.
$n \times n \times n \times n$ is written as n^4 and said as 'n to the power of 4'.
$n \times n \times n \times n \times n$ is written as n^5 and said as 'n to the power of 5'.
$*2n \times 2n$ is written as $(2n)^2$ or $4n^2$.

> We write the number first.

> We usually write the letters in alphabetical order.

> 3 lots of $2a$ is $6a$.

$$*n \times n^2 = n \times n \times n \qquad *2a \times a^2 = 2 \times a \times a \times a$$
$$= n^3 \qquad\qquad\qquad\quad = 2 \times a^3$$
$$= 2a^3$$

Exercise 3

1 Write these without multiplication or division signs.
   a $3 \times x$ b $4 \times b$ c $9 \times (x + 3)$ d $a \times b$ e $s \times t$ f $(n - 4) \times 3$
   g $p \times p$ h $m \times m$ i $n \times n \times n$ j $5 \times b \times b$ k $a \times (b + c)$ l $3 \times n \times m$
   m $(n + m) \div p$ n $(p + 4) \div q$ o $3 \times 4a$ p $5 \times 2b$ q $3 \times 4n$
   *r $3a \times 3a$ *s $5x \times 5x$ *t $a \times a^2$ *u $4p \times p^2$

2 Write these with multiplication signs.
   a $4p$ b $4(x + 3)$ c $8(n - 2)$ d x^2 e $5b^2$ f $4ab$
   g $7q^2$ h $8pq$ i $(4n)^2$ j m^3 k $3p^3$

3 Write an expression for these.
   a three times the value of d
   b one third of n
   c m multiplied by n
   d twice m, plus n
   e two times the product of m and n
   f half the sum of a and b
   g three times the sum of f and g.

> Write the expressions without $\times$ or $\div$ signs.

4 Explain the difference between
   a $3p$ and $p + 3$ **b** n^2 and $2n$ **c** n^3 and $3n$ **d** $5(n + 2)$ and $5n + 2$
   e $5m^2$ and $(5m)^2$ *f $2n \times 3n$ and $6n$.

5 Write an expression for the area of these.

   a   4 metres   **b**   3n centimetres   **c**   4p kilometres

               x metres   4 centimetres   3p kilometres

Review 1   Write these without multiplication or division signs.
  a $4 \times p$ **b** $(b - 3) \times 4$ **c** $a \times a \times a$ **d** $4 \times m \times m$ **e** $(x + y) \div a$ **f** $5 \times 7a$
*g $4a \times 4a$ *h $m^2 \times m$ *i $5x \times x^2$

Review 2   Write an expression for these.
  a p multiplied by q **b** three times a, plus b
  c three times the sum of x and y.

Operations used in algebra follow the same rules as those used in arithmetic.

Arithmetic
$4 + 5 = 5 + 4$
$4 \times 5 = 5 \times 4$
$2 + (4 + 5) = (2 + 4) + 5$
$2 \times (4 \times 5) = (2 \times 4) \times 5$

Algebra
$a + b = b + a$
$a \times b = b \times a$ or $ab = ba$
$a + (b + c) = (a + b) + c$
$a \times (b \times c) = (a \times b) \times c$ or $a(bc) = (ab)c$

 This is the *commutative* rule.

Exercise 4

1 Write true or false for these. Explain.
  a $3 + n = n + 3$ **b** $n + m = m + n$ **c** $x + y = xy$ **d** $3b = 3 + b$
  e $pq = qp$ **f** $(a + b) \times 3 = 3(a + b)$ **g** $3ab = \frac{ab}{3}$ **h** $4nm = 4(n + m)$
  i $ab \times 3 = 3ab$ **j** $st \times r = str$ **k** $3 + (x + y) = (3 + x) + y$ **l** $5 + c + d = 5cd$
  m $a \times b \times 4 = 4ab$ **n** $(x + y) \times 7 = 7xy$ **o** $a + b - c = a - c + b$ **p** $a - b - c + d = a + b + c - d$

Review   Write true or false for these.
  a $s + t = t + s$ **b** $xy = yx$ **c** $abc = acb$ **d** $(4 + n) \times 5 = 5(4 + n)$
  e $5 + m + n = mn + 5$ **f** $5(n + m) = 5nm$ **g** $x + y \times 7 = 7xy$

Collecting like terms

We can simplify expressions by adding and subtracting **like terms**. This is called **collecting like terms**.

Arithmetic
$4 + 4 + 4 = 3 \times 4$
$8 + 8 + 8 + 8 = 4 \times 8$
$3 \times 6 + 4 \times 6 = 7 \times 6$
$5 \times 4 - 2 \times 4 = 3 \times 4$
$2 + 3 \times 2 + 2 = 5 \times 2$

Algebra
$a + a + a = 3a$
$n + n + n + n = 4n$
$3c + 4c = 7c$
$5d - 2d = 3d$
$b + 3b + b = 5b$

Algebra

Examples $b + b + b + b = 4b$ $2n + 4n = 6n$

$2p + 4p + 8 = 6p + 8$ $8x - 6x + 3 = 2x + 3$

$a + b + b + c = a + 2b + c$ $8y - 3y - y = 5y - y$

$= 4y$

Work from left to right.

$5a + 2a - 6a = 7a - 6a$

$= a$

$1a = a$

$5p + q - 3p + 4q = 5p - 3p + q + 4q$

$= 2p + 5q$

$7a + 6 - 4a - 2 = 7a - 4a + 6 - 2$

$= 3a + 4$

It is best to write the 'like terms' next to each other. Remember to include the sign in front when you move terms.

Worked Example

Write an expression for the perimeter of this rectangular pool.
Simplify your expression.

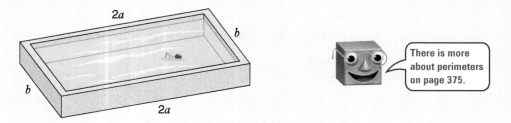

$2a$

b

b

$2a$

There is more about perimeters on page 375.

Answer

The perimeter is the distance around the outside of the rectangle.
An expression for the perimeter is $2a + b + 2a + b = 2a + 2a + b + b$
$= \mathbf{4a + 2b}$

Exercise 5

Simplify the expressions in questions 1–4.

1 **a** $a + a$ **b** $x + x + x$ **c** $y + y + y + y + y$ **d** $5n + 3n$ **e** $4x + 3x$

f $2y + 7y$ **g** $6a - 3a$ **h** $a + a - a$ **i** $10b - 3b$ **j** $9x - 8x$

k $20t - 6t$ **l** $45b - 16b$ **m** $27a + 36a$ **n** $42x - 23x$

2 **a** $2b + b + 3b$ **b** $3y + 2y + 2y$ **c** $4x + 3x + 5x$ **d** $7b + 2b + 4b$

e $7y + 2y - 3y$ **f** $10x - 8x + 2x$ **g** $5x - 3x + 6x$ **h** $7m - m - 3m$

i $15j - 2j - 13j$ **j** $14a - 7a - 3a$ **k** $28p - 16p - 7p$ **l** $20m - 17m - m$

＊**m** $12x - 15x + 8x$ ＊**n** $20a - 32a + 16a$ ＊**o** $16y - 24y + 10y$

3 **a** $8x + x + 2$ **b** $2m + 3m + 4$ **c** $9a + 5 + a$ **d** $4p + 7 + 3p$

e $8m + 7 - 2m$ **f** $6b + 10 + 5 + 3b$ **g** $5 + 8x + 2 - 2x$ **h** $12m + 3 + 6m + 2$

i $8 + 7y - 4 - 3y$ **j** $5p + 8 - 3 - 3p - 2$ **k** $10m + 8 - 3m - 4 - m$ **l** $11x + 13 - 10x - 5 - x$

m $4n - 3 + 8n + 6$ **n** $6a - 4 - 2a + 8$ ＊**o** $7x - 5 - 3x - 2$ ＊**p** $5y - 3 + 2y - 6$

＊**q** $8y - 4 - 7y - 3$ ＊**r** $9a - 8 - 8a - 4 - a$ ＊**s** $4y + 10 - 5y - 3$

4 **a** $b + b + a + a$ **b** $2c + 3c + d$ **c** $s + 2s + t$ **d** $5p + 4q + 2p + q$

e $3x + 8y - x - y$ **f** $3a + 2b + 4a - b$ **g** $4p + 3q - 2p + 8q - p$ **h** $12x + 10y - 7x - 5y + 2x$

＊**i** $10x - 3y - 9x + 5y$ ＊**j** $14p - 5q - 7p - 3q$ ＊**k** $12a - 7b - 6a - 9b$

＊**l** $15x - 3y - 7x - 4y$ ＊**m** $5n + 3m - 6n - 2m$ ＊**n** $12a - 4b - 15a + 6b$

5 Write an expression for the perimeter of these gardens. Simplify your expression.

a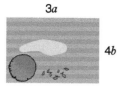
3a
4b

b
4y
3y + 1

* c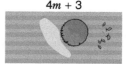
4m + 3
7n + 2

6 The number in each box is found by adding the numbers in the two boxes below it. Find the missing expressions. Write the expressions as simply as possible.

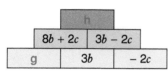
a	
a + b	b + c
a	b	c

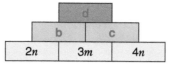
d	
b	c
2n	3m	4n

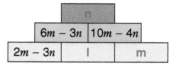
f	
3p + 5q	3p − q
e	3p	− q

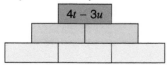
h	
8b + 2c	3b − 2c
g	3b	− 2c

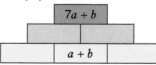
k	
5a + 3b	10a + 4b
i	5a	j

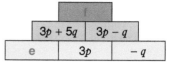
n	
6m − 3n	10m − 4n
2m − 3n	l	m

7 Here are some algebra cards.

a+2	a÷2	2a−1	2a+a	2a

a−2	a²	a+a	2a+2	3a+3

a Which card will always give the same answer as
i $\frac{a}{2}$ **ii** $2 + a$ **iii** $3a − a + 2$ **iv** $5a − 2a + 3$ **v** $a \times a$?
b Which two cards will always give the same answer as $2 \times a$?
c When the expressions on two cards are added they can be simplified to give $5a + 2$. Which two cards are they? Is there more than one answer?
d Write a new card that will always give the same answer as
i $4a + 2a$ **ii** $3a + 5a$ **iii** $8a − 3a$ **iv** $5a + 3 − 2a + 4$.

8 The answer is $3p + 4q$. What could the question have been?

9 Draw some shapes that have a perimeter of $8x + 12$.

* 10 Use a copy of these.
Write possible expressions for the empty boxes.

a
4t − 3u

b
7a + b
a + b

* 11 Show this is a magic square by adding the expressions in each row, column and diagonal.

a−b	a+b−c	a+c
a+b+c	a	a−b−c
a−c	a−b+c	a+b

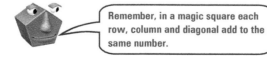
Remember, in a magic square each row, column and diagonal add to the same number.

T | **Review 1**

					E	
$\overline{11a+b}$	$\overline{10a+7b}$	$\overline{a}$ $\overline{8a+2b}$	$\overline{8a+2b}$	$\overline{12a}$	$\overline{2a+2b}$	

$\overline{6a+4}$ $\overline{11a+b}$ $\overline{5a+4b}$ $\overline{5a+4b}$ $\overline{6a+5}$ $\overline{11a+b}$ $\overline{10a+7b}$

$\overline{3a+b}$ $\overline{6a+5}$ $\overline{5a+5}$ $\overline{2a+2b}$ $\overline{3a+b}$ $\overline{6a+1}$ $\overline{3a+b}$ $\overline{3a+1}$ $\overline{a}$

$\overline{6a}$ $\overline{11a+b}$ $\overline{5a}$ $\overline{\text{E } 12a}$ $\overline{5a+4b}$ $\overline{10a+7b}$

Use a copy of the box.
Simplify these expressions. Write the letter that is beside each expression above the answer in the box.

E $9a + 3a = 12a$	**O** $7a - 6a$	**N** $8a - 3a$	**P** $11a + 2a - 7a$
H $2a + 3 + 4a + 2$	**B** $3a + 2a + 4 + a$	**I** $2a + 3a + 5$	**Y** $2a + 3 + 4a - 2$
T $6a + 2b - 3a - b$	**W** $10a + 3 - 7a - 2$	**A** $8a + 2b + 3a - b$	**R** $7a + 3b - 5a - b$
S $13b - 2a - 6b + 12a$	**C** $16a - 3b - 8a + 5b$	**L** $9a - 2b - 4a + 6b$	

Review 2   This diagram shows a park.
Write an expression for the perimeter of the park.
Simplify the expression.

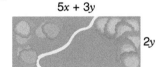

$5x + 3y$

$2y$

Review 3   The number in each box is found by adding the numbers in the two boxes below it.
Find the missing expressions.
Simplify them.

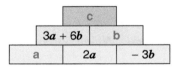

	c	
$3a + 6b$		b
a	$2a$	$-3b$

Review 4   Draw some shapes with perimeter   **a** $4a + 6$ **b** $8b + 6y$

Simplifying expressions by cancelling

We can **simplify expressions** by **cancelling**.

Arithmetic

$$\frac{4}{4} = 1$$

$$\frac{\cancel{3}^1 \times 4}{\cancel{3}^1} = 4$$

$$\frac{\cancel{8}^2 \times 5}{\cancel{4}^1} = 2 \times 5$$

$$\frac{5 \times \cancel{6}^1}{\cancel{6}^1} = 5$$

$$\frac{\cancel{10}^5 \times 7}{\cancel{6}^3} = \frac{5 \times 7}{3}$$

$$\frac{5^2}{5} = \frac{\cancel{5}^1 \times 5}{\cancel{5}^1} = 5$$

Algebra

There is more about cancelling fractions on pages 5 and 116.

$$\frac{n}{n} = 1$$

$$\frac{\cancel{3}^1 n}{\cancel{3}^1} = n$$ Divide numerator and denominator by the common factor 3.

$$\frac{\cancel{8}^2 n}{\cancel{4}^1} = 2n$$ Divide numerator and denominator by the common factor 4.

$$\frac{5\cancel{a}^1}{\cancel{a}^1} = 5$$ Divide numerator and denominator by the common factor a (6 in the arithmetic example).

$$\frac{\cancel{10}^5 a}{\cancel{6}^3} = \frac{5a}{3}$$ Divide numerator and denominator by the common factor 2.

$$\frac{a^2}{a} = \frac{\cancel{a}^1 \times a}{\cancel{a}^1} = a$$ Divide numerator and denominator by the common factor a (5 in the arithmetic example).

Exercise 6

1 Simplify these expressions by cancelling.

a $\frac{m}{m}$ b $\frac{c}{c}$ c $\frac{y}{y}$ d $\frac{n}{n}$ e $\frac{3a}{3}$ f $\frac{6b}{6}$

g $\frac{7m}{7}$ h $\frac{9y}{y}$ i $\frac{5x}{x}$ j $\frac{7x}{x}$ k $\frac{8p}{p}$ l $\frac{5x}{x}$

m $\frac{8n}{4}$ n $\frac{10p}{2}$ o $\frac{14n}{7}$ p $\frac{21y}{7}$ q $\frac{24x}{8}$ r $\frac{54b}{9}$

s $\frac{42m}{7}$ t $\frac{35a}{7}$ u $\frac{21b}{b}$ v $\frac{64y}{8}$ w $\frac{72x}{9}$

2 Simplify these expressions by cancelling.

a $\frac{12a}{8}$ b $\frac{20x}{8}$ c $\frac{15y}{10}$ d $\frac{45m}{27}$ e $\frac{18b}{12}$ f $\frac{54y}{42}$

g $\frac{72x}{48}$ h $\frac{64m}{40}$ i $\frac{54p}{72}$ j $\frac{45n}{75}$ k $\frac{48n}{60}$ l $\frac{56p}{32}$

m $\frac{125a}{75}$ n $\frac{121b}{44}$ o $\frac{81x}{72}$ p $\frac{112y}{80}$ q $\frac{117p}{65}$

3 Simplify these expressions.

a $\frac{b^2}{b}$ b $\frac{m^2}{m}$ c $\frac{2a^2}{2}$ *d $\frac{4n^2}{4}$ e $\frac{3n^2}{3}$ f $\frac{a^3}{a}$

g $\frac{n^3}{n^2}$ *h $\frac{12m^2}{4m}$ *i $\frac{15b^2}{3b}$ *j $\frac{4a^2}{2a}$ *k $\frac{8x^3}{12x}$

*4 What is the length of these rectangles?

a [20x² | 5x]   b [48n² | 6n]   c [72y² | 8y]

*5 $\frac{\square}{\square} = 2x$ Jackie wrote $\frac{8x}{4} = 2x$ to make this true.

What else could Jackie have put in the two boxes to make this true?

T **Review**

2 10 12	$\frac{3x}{2}$ $\frac{9x}{2}$ $\frac{3x}{2}$ 6x	$\overset{A}{1}$ 12 $\frac{3x}{2}$	8x 10x $\frac{3x}{2}$

$\overline{6x}$ $\overset{A}{1}$ $\overline{3x}$ $\overline{\frac{3x}{2}}$ $\overline{6x}$ $\overline{7x}$ $\overline{x}$ $\overline{\frac{3x}{2}}$ $\overline{6}$ $\overline{12}$ $\overline{2}$ $\overline{3x}$ $\overline{\frac{5x}{2}}$ $\overline{7x}$ $\overline{12}$ $\overline{8x}$ $\overline{10x}$

Use a copy of this box.
Simplify these expressions by cancelling.
Write the letter that is beside each expression above the answer in the box.

A $\frac{x}{x} = 1$ O $\frac{2w}{w}$ F $\frac{6x}{x}$ U $\frac{10x}{x}$ R $\frac{12x}{x}$

M $\frac{6x}{2}$ Z $\frac{x^2}{x}$ H $\frac{20x}{2}$ T $\frac{24x^2}{3x}$ S $\frac{30x}{5}$

I $\frac{42x}{6}$ B $\frac{20x}{8}$ E $\frac{24x}{16}$ *Y $\frac{63x^2}{14x}$

Brackets

We can write an expression **without brackets** by following the rules of arithmetic.

Arithmetic

$6 \times 52 = 6 \times (50 + 2)$

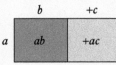

$6 \times (50 + 2) = (60 \times 50) + (6 \times 2)$
$= 300 + 12$
$= \textbf{312}$

$5 \times 39 = 5 \times (40 - 1)$

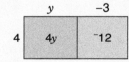

$5(40 - 1) = 5 \times 40 + 5 \times {}^-1$
$= 200 - 5$
$= \textbf{195}$

Algebra

$a(b + c) = a \times (b + c)$

$a \times (b + c) = (a \times b) + (a \times c)$
$= \textbf{\textit{ab}} + \textbf{\textit{ac}}$

$p(q - r) = p \times (q - r)$

$p(q - r) = p \times q + p \times {}^-r$
$= \textbf{\textit{pq}} - \textbf{\textit{pr}}$

There is more about multiplying positives and negatives on page 40.

Examples

$3(x + 7)$

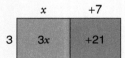

$\textbf{3(x + 7) = 3x + 21}$

$4(y - 3)$

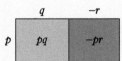

$4(y - 3) = \textbf{4y - 12}$

$6(2b + 3c)$

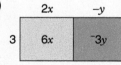

$6(2b + 3c) = \textbf{12b + 18c}$

$3(2x - y)$

$3(2x - y) = \textbf{6x - 3y}$

Exercise 7

1   Write these expressions without brackets.

Use the grids to help.

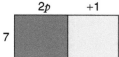

   a $3(a + 2)$

   b $4(x + 2)$

   c $5(n + 4)$

   d $7(2p + 1)$

e $3(m - 6)$

f $4(y - 3)$

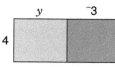

g $2(x - 5)$

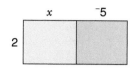

h $3(2m - 4)$

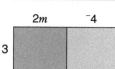

2 Write without brackets.
- **a** $4(a + 2)$
- **b** $3(x + 5)$
- **c** $5(y + 3)$
- **d** $7(b - 2)$
- **e** $9(a + 4)$
- **f** $8(m - 6)$
- **g** $7(n + 1)$
- **h** $4(p - 2)$
- **i** $6(b - 5)$
- **j** $17(c + 1)$
- **k** $14(p - 2)$
- **l** $16(m + 2)$
- **m** $24(n - 1)$
- **n** $2(a - 13)$
- **o** $3(n + 12)$

3 Write without brackets.
- **a** $3(x + y)$
- **b** $4(a - b)$
- **c** $5(2p + 3)$
- **d** $4(3 + 2r)$
- **e** $2(2a - 5)$
- **f** $3(5a - 6)$
- **g** $3(4x + 1)$
- **h** $4(45x - 3)$
- **i** $6(5d - 29)$
- **j** $2(x + 3a)$
- **k** $3(a - 2b)$
- **l** $3(2x - 5a)$
- **m** $4(2a + x)$
- **n** $3(x - 3a)$
- **o** $3(2y - x)$
- **p** $4(6a + 5b)$
- **q** $8(19a - 3b)$
- **r** $7(29x - 5y)$
- **s** $12(25p - 16q)$

*4 Write without brackets and then simplify.
- **a** $3(x + 4) + 2(x + 1)$
- **b** $4(a + 3) + 3(a + 2)$
- **c** $2(2x + 3) + 3(x + 4)$
- **d** $3(a + 5) + 4(2a + 1)$
- **e** $3(a + 4) + 2(3a - 1)$
- **f** $4(b + 5) + 3(1 + 5b)$
- **g** $2(3x - 1) + 3(x + 4)$
- **h** $2(2a - 3) + 3(3 + 4a)$
- **i** $4(5x + 1) + 3(5 - 4x)$
- **j** $5a + 2(3 + a)$
- **k** $3(5y + 2) - 5$
- **l** $6a + 3 + 2(3a + 2)$
- **m** $4(y - 1) + 2y$
- ***n** $15a - (5a - 6)$
- ***o** $5(2a - b) - 3(2a - 5b)$
- ***p** $7(39x - 8) - 5(18x + 12)$

Review 1 Use a copy of the box below.
Write each of the expressions without brackets. Find your answer in the box. Cross out the letter above it.
What three-word sentence do the letters that are left make?

| $4(n + 3)$ | $7(n - 5)$ | $9(n + 3)$ | $36(n - 1)$ | $2(n + 4)$ |
| $3(2n + 6)$ | $5(8 - n)$ | $10(2n + 3m)$ | $6(2n - m)$ | $4(n - 2m)$ |

C	R	A	A	T	Y	S	F	I	C	S	A	N	H	T	E	V	O	A	M	T	I	T
$6n+18$	$9n+12$	$40-5n$	$5n+9$	$4n+3$	$4n+12$	$6n+6$	$4n-8m$	$7n-35$	$8n+1$	$20n+30m$	$9n+3$	$7n-5$	$9n+27$	$40-n$	$36n-36$	$12n+13m$	$6nm$	$12n-6m$	$36n-1$	$2n+8$	$2n+4$	$4n+7$

Review 2 Write these without brackets.
- **a** $4(x + 5)$
- **b** $7(b - 3)$
- **c** $9(n + 8)$
- **d** $14(a - 1)$
- **e** $2(x + y)$
- **f** $3(a - b)$
- **g** $5(m + n)$
- **h** $16(p - q)$
- **i** $4(2x + 3)$
- **j** $4(3y - 7)$
- **k** $5(6 - 2a)$
- **l** $3(2b + 3c)$
- **m** $7(8y + 3w)$
- **n** $4(2x - 3b)$
- **o** $4(3y - x)$

*Review 3 Write these without brackets and simplify.
- **a** $3(x + 7) + 4(x + 2)$
- **b** $3(4a + 2) + 2(1 - 2a)$
- ***c** $15(21p - 3) - 19(7p - 9)$

Brackets – a game for a group

You will need three sets of these cards.

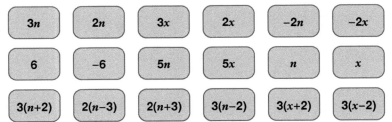

| 3n | 2n | 3x | 2x | –2n | –2x |

| 6 | –6 | 5n | 5x | n | x |

| 3(n+2) | 2(n–3) | 2(n+3) | 3(n–2) | 3(x+2) | 3(x–2) |

To play
- Choose a dealer.
- The dealer deals three cards face down to each player.
- The dealer then places the next card face up on the table.
- The rest of the pack is placed face down in a pile.
- Take turns, starting with the dealer and then the player on the dealer's left and so on.
- The object of the game is to get a set of three cards that make a correct statement. For example,

is a winning set of three cards since $3(n - 2) = 3n - 6$.
- At your turn you may either
  — use the three cards in your hand to make a winning set
  — pick up the card on the table and discard one card from your hand
  — pick up a card from the pile and discard one card from your hand.
  Discarded cards are put on the bottom of the pile.
  If the card on the table is picked up, it is replaced with the top card from the pile.
- The first person to get a winning set of three cards gets 1 point and is the dealer for the next round.
- If a player puts out a winning set which is not in fact a winning set, this player loses 1 point.
  You could keep a score sheet like this.

Points scored	✓✓✓✓✓
Points lost	✓✓
Total	3

- The winner is the person with the most points after a set time.

Substituting

Lauren made some sandwiches. She made some ham sandwiches and 10 egg ones. We could say she made $n + 10$ sandwiches.

Lauren said 'I made 12 ham sandwiches'.
We can now work out the value of $n + 10$.

$n + 10 = 12 + 10$ 12 has been put in place of n.
$\quad\quad = 22$ We call this **substitution**.

We **evaluate** an expression by **substituting** values for the unknown into the expression.

Evaluate means 'find the value of'.

When we evaluate the expression, the **order of operations** is the same as for arithmetic.

Brackets
Indices
Division and **M**ultiplication
Addition and **S**ubtraction

Remember, in 4^2 and 5^3 the 2 and 3 are indices.

Remember this by remembering the word **BIDMAS**.

Example   In $4 + 3a$, the multiplication is worked out before the addition.
In $4 - a^2$, the square is worked out before the subtraction.
In $4(x - 8)$, the expression in the brackets is worked out first.

Worked Examples
If $a = 4$, $b = 1\cdot5$ and $c = 2$, evaluate
a $3a + 1$ b $9 - 3c$ c $3(a + b)$ d $\frac{2a + 1}{2b}$ e $3a^2 - 5$

Answers

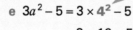

a $3a + 1 = 3 \times 4 + 1$ b $9 - 3c = 9 - 3 \times 2$ c $3(a + b) = 3(4 + 1\cdot5)$
$\quad\quad\quad = 12 + 1$ $\quad\quad\quad = 9 - 6$ $\quad\quad\quad\quad = 3 \times 5\cdot5$
$\quad\quad\quad = \mathbf{13}$ $\quad\quad\quad = \mathbf{3}$ $\quad\quad\quad\quad = \mathbf{16\cdot5}$

Work out brackets first.

d $\frac{2a + 1}{2b} = \frac{2 \times 4 + 1}{2 \times 1\cdot5}$ e $3a^2 - 5 = 3 \times 4^2 - 5$
$\quad\quad = \frac{8 + 1}{3}$ $\quad\quad\quad = 3 \times 16 - 5$
$\quad\quad = \frac{9}{3}$ $\quad\quad\quad = 48 - 5$
$\quad\quad = \mathbf{3}$ $\quad\quad\quad = \mathbf{43}$

Work out the square first.

Exercise 8

Except for questions 7, 8 and Review 3.

1   If $y = 4$ find the value of
   a $2y$ b $2y - 2$ c $3y + 4$ d $18 - 3y$ e $20 - 2y$
   f $3(y + 2)$ g $\frac{y + 2}{2}$ h $4(y - 2)$ i $7(8 - y)$ j $\frac{y}{2} + 3$
   k $2y^2$ l $24 - y^2$ m $2y^2 - 5$ n y^3 o $\frac{3y}{y - 1}$.

2   If $p = 2$ and $q = 4$ evaluate
   a $p + q$ b $p + q - 1$ c $2p + q$ d $3q - p$ e $2p + 3q$
   f $20 - 2p - q$ g $2q - 3p$ h $3(p + q)$ i $4(p - q)$.

3   Repeat question **2** for $p = 8$ and $q = 49$.

Algebra

4 If $a = 2$, $b = 3$, $c = 6$ and $d = 4$, evaluate

   a $\frac{d}{2} + 1$,   **b** $5(c - 2)$ **c** $3b + c$ **d** $20 - 2da$ **e** $ab - c$ **f** $2a + b - c$

   g $3b^2$ **h** $4a^2 + 4a$ **i** $2c^2 - 2d$ **j** $\frac{2d^2}{a}$ **k** $4 - 3b$ **l** $2(b - c)$

   m $\frac{b + c}{a}$ **n** $\frac{2b^2 - a}{d}$ **o** $2a^2 - 2b^2$ **p** $3(a^2 - c)$.

5 Find the value of these when $n = 1 \cdot 5$.

   a $2n + 3$ **b** $4n - 1$ **c** $3(n + 1 \cdot 5)$ **d** $5(8 - 2n)$ **e** $10 - 4n$

6 Find the value of these when $x = 3 \cdot 6$.

   a $5x$ **b** $5x - 3$ **c** $10x - 4$ **d** $24 - 5x$ **e** $4(x - 3)$.

7 Repeat question **6** for $x = 1 \cdot 9$.

8 $2n + 1$, $2n + 3$, $2n + 5$, $2n + 7$, $2n + 9$

   a Substitute $n = 1$ into each of the above expressions.
       You should get five consecutive odd numbers.
       Explain why.
   b What do you get if $n = 2$?
   c Which value of n would give the numbers 13, 15, 17, 19, 21?

*9 The number of rolls in a stack with n rolls in the bottom row is $\frac{n(n + 1)}{2}$.

Find the number of rolls in a stack with these numbers of rolls in the bottom row.

   a 5   **b** 10   **c** 16

* 10 The number of crosses in design n is given by $\frac{n^2 + 3n + 2}{2}$.

design 1   **design 2**   **design 3**

Find the number of crosses in design number

   a 8   **b** 12   **c** 25.

Review 1

7	5	98	22	12	8	5	8	98		18	9	0	13	8		12

13	12	4	5	10	0		11	22	0	4	0		5	12

5	13		4	A 6	5	8	5	8	98

Use a copy of this box.
If $p = 2$, $q = 5$ and $r = 7$ evaluate these. Write the letter that is beside each question above its answer.

A $3p = 6$ **I** $r - p$ **E** $p + q - r$ **W** $2p + r$ **H** $3q + r$

O $2r - q$ **D** $2(p + r)$ **S** $3q - p$ **N** $\frac{16}{p}$ **T** $2(p + 4)$

K $2(r - p)$ **G** $2r^2$ **L** $\frac{r}{7} + 6$ **R** $\frac{3p^2}{2} - 2$

Review 2   If $x = 3$, $y = 5$ and $z = 1$ find the value of

a $x + y$ b $y - x + z$ c $3(x + y)$ d $9y - z$ e $\frac{21}{x}$ f $\frac{y}{5} + 3$

g $2y^2$ h $y^2 - x^2$ i $\frac{x + 2}{y}$ j $\frac{y^2 - z}{x}$ k $20 - (y^2 - x^2)$.

Review 3   The number of diagonals in a polygon with n sides is given
by the expression $\frac{n(n - 3)}{2}$.
How many diagonals are there in a polygon with

a   6 sides   b   15 sides   c   20 sides   d   18 sides?

Investigation

[T]

Substituting

You will need a spreadsheet for question 1. You could also use it for questions 2 and 3.

1   Marcia used a spreadsheet to see if this is
true for different values of x and y.

$$4(x + y) = 4x + 4y$$

She did this by checking if columns C and
D on her spreadsheet were the same.
Use a spreadsheet to check if these
statements are true for different values of n.

	A	B	C	D
1	8	4	= 4*(A1+B1)	= 4*A1+4*B1
2	-6	2	= 4*(A2+B2)	= 4*A2+4*B2
3	2.5	-7	= 4*(A3+B3)	= 4*A3+4*B3

a $\frac{8n}{2} = 4n$ b $3(n + 4) = 3n + 12$ c $6(2n - 3) = 12n - 18$

> Check if columns C
> and D are the same.

2   In arithmetic, if we are told that $56 + 83 = 139$,
then we know that $83 + 56 = 139$
$139 - 83 = 56$
$139 - 56 = 83$.

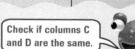

In algebra, if $m + n = 7$, does this mean these are true? **Investigate**.

$n + m = 7$ $7 - n = m$ $7 - m = n$

What if $m + n = 7$ is replaced with $m + n = 100$ or $m + n = 5.4$ or $m + n = p$?

3   If $ab = 48$ does this mean these are true? **Investigate**.

$ba = 48$ $\frac{48}{b} = a$ $\frac{48}{a} = b$

What if $ab = 48$ is replaced with $ab = 64$ or $ab = 4.8$ or $ab = -8$ or $ab = c$?

***4**   a   Show that this is a magic square if
$x = 7$, $y = 2$ and $z = 3$.
b   Find some other whole number
values for x, y and z to make this a
magic square.

$x-y$	$x+y-z$	$x+z$
$x+y+z$	x	$x-y-z$
$x-z$	$x-y+z$	$x+y$

> Show by collecting like
> terms that the square
> is always magic.

c   What values of x, y and z give this magic square?
d   Are there other values of x, y and z that give a magic square with
the numbers 1 to 9?
e   What decimal values of x, y and z give a magic square?

2	7	6
9	5	1
4	3	8

Algebra

Practical

You will need a graphical calculator.
Jodie used a graphical calculator to find the value of $4x^2 - 2$ when $x = 3$.
She keyed ⟨3⟩ ⟨→⟩ ⟨ALPHA⟩ ⟨x⟩ ⟨EXE⟩ to give x the value 3.
then ⟨4⟩ ⟨ALPHA⟩ ⟨x⟩ ⟨x²⟩ ⟨−⟩ ⟨2⟩ ⟨EXE⟩ to find the value of $4x^2 - 2$ when $x = 3$.
The ⟨ALPHA⟩ key is used to get letters on the screen.
The keying sequence on *your* calculator may be a little different.
Find out how to do this on *your* calculator.

Use your calculator to find the value of these when $x = 1$, $x = 2$, $x = 3$, $x = 4$ and $x = 5$.

a $x^2 + 3$	**b** $x^2 - 1$	**c** $2x^2 + 3$	**d** $2x^2 - 4$
e $3x^2 + 4$	**f** $3x^2 - 7$	**g** $4x^2 - 5$	

Formulae

Formulae are used by many people in their jobs.
Usually the **formulae are written in symbols**.

Example Max makes picture frames.
He uses the formula $P = 2l + 2w$ to work out the length of wood
he needs to make a frame.
If $l = 12$ cm and $w = 10$ cm
$P = 2 \times 12 + 2 \times 10$
$\quad = 24 + 20$
$\quad = \mathbf{44\ cm}$

> Remember to do multiplication before addition.

Exercise 9   Only use a calculator if you need to.

1 A formula for finding force from mass, m, and acceleration, a, is $F = ma$.
Find F if **a** $m = 5$ and $a = 8$ **b** $m = 12$ and $a = 4$.

2 A formula for finding speed, S, from distance, D, and time, T, is $S = \frac{D}{T}$.
Find S if **a** $D = 150$ and $T = 5$ **b** $D = 320$ and $T = 4$.

3 $g = \frac{p}{8}$ is a formula for changing pints, p, into gallons, g.
Find g if **a** $p = 40$ **b** $p = 160$ **c** $p = 192$.

4 $P = 2l + 2w$ gives the perimeter of a rectangle.
Find P if
a $l = 5$ cm, $w = 2$ cm **b** $l = 6 \cdot 4$ cm, $w = 3$ cm **c** $l = 8 \cdot 5$ cm, $w = 4 \cdot 5$ cm.

5 $V = IR$ gives the voltage, V, in an electrical circuit with current I and resistance R.
Find V if **a** $I = 2 \cdot 5$, $R = 12$ **b** $I = 0 \cdot 6$, $R = 3$ **c** $I = 1 \cdot 2$, $R = 8$.

6 The cost in pounds of hiring a car from 'Have Wheels' is given by the formula $C = 30d + 20$.
d is the number of days hired.
Find the cost of hiring a car for
  a 3 days   **b** a week   **c** a fortnight   **d** 16 days.

7 The mean length of 4 guinea pigs is given by $M = \dfrac{a + b + c + d}{4}$,
where M is the mean length and a, b, c and d are the lengths of the guinea pigs.
Find M if $a = 17$ cm, $b = 16$ cm, $c = 16.5$ cm, $d = 18.5$ cm.

> There is more about the mean on page 413.

8 The weekly wage of a car salesperson is worked out using $W = 200 + 100n$.
n is the number of cars sold and W is the weekly wage, in pounds.
  i How much is Ivan's weekly wage if the sells
    a 4 cars   **b** 3 cars   **c** no cars   **d** 1 car   **e** 6 cars?
  ii How many cars would Ivan need to sell in one week if he wanted to earn
    a £400   **b** £1200?

9 The distance travelled, S metres, by a car is given by the formula $S = 20 + 0.5t^2$.
t is the time in seconds.
How far has the car travelled after
  a 1 sec   **b** 2 sec   **c** 4 sec   **d** $\frac{1}{2}$ sec?

* 10 'Bank Simple' use the formula $I = \frac{PRT}{100}$ to calculate how much interest, I, you earn when you
invest P pounds at $R\%$ interest for T years.
How much interest would you earn if you invested
  a £100 at 5% interest for 1 year   **b** £80 at 8% interest for 2 years
  c £150 at 6% interest for 6 months   **d** £3500 at $7\frac{1}{2}\%$ interest for 3 years
  e £2750 at 4.25% interest for $3\frac{1}{2}$ years?

* 11 The child's dose for Dr Medicine's Cough Mixture is worked out using the formula
$C = \dfrac{a}{a + 10} \times d$, where C is the child's dose, a is the age of the child and d is the normal adult
dose.
  i The normal adult dose is 24 mℓ. What is the child's dose for a child aged
    a 10   **b** 5   **c** 3   **d** 1   **e** 13?
  ii Dr Medicine's Painkiller uses the same formula. The normal
    adult dose is 10 mℓ. What is the child's dose for a child aged
    a 15   **b** 2   **c** 5   **d** 11?
  iii Robert is 30. What happens if you substitute his age for **a** in the formula?

> Round your answers to the nearest mℓ.

Review 1 $m = \frac{c}{100}$ where m is metres and c is centimetres.
Find m if   **a** $c = 200$ **b** $c = 2600$ **c** $c = 6500$ **d** $c = 18\,000$

Review 2 $S = 20 + 4t$ is a formula which gives the distance, S metres, travelled in t seconds
by a car going downhill.
What distance is travelled in   **a** 10 seconds   **b** 1 minute?

Review 3   The distance, d, in metres, an object has fallen after being dropped t seconds ago
is given by the formula $d = 4.9t^2$.
  a Mrs Woods dropped her purse from the top of a spiral staircase.
    How far would it have fallen in 3 seconds?
  b A skydiver opens his parachute 10 seconds after he jumps.
    How far does he fall before he opens it?

Practical

1 **You will need** a tape measure.
The formula for finding the area, in cm^2, of skin on your body is $S = 2ht$ where h is your height, in cm, and t is the distance around your thigh, in cm.
Use the formula to work out the area of your skin.

*2 Research the formulae used by someone in a job of your choice.
You could use the Internet or e-mail to help.

> You could choose another subject like geography or science to research instead.

Puzzle

Replace each of the letters by one of the digits 0, 1, 2, 3, 4, 5, 6, 7, 8, 9 so that the subtractions are correct.

a
```
  F I V E
- F O U R
  ───────
    O N E
```

b
```
  S E V E N
-   F O U R
  ─────────
  T H R E E
```

Is there more than one possible answer?

Writing expressions and formulae

Writing expressions

April has n pairs of shorts and some t-shirts.
She has three times as many t-shirts as shorts.
An expression for the number of t-shirts, she has is $3 \times n = 3n$.
April threw out half of her pairs of shorts.
An expression for the number she threw out is $\frac{n}{2}$.

Exercise 10

1 You have d sweets.
  a Molly has 4 more than you.
    How many does Molly have?
  b You give away 3 sweets.
    How many do you have left?
  c Shabir has twice as many as you had at the start.
    How many does Shabir have?
  d Shabir shares her sweets equally between herself and 3 friends.
    How many do they each get?

2 Katie makes cheese, egg, ham and tomato sandwiches. S stands for the number of cheese sandwiches Katie makes.

a Katie makes three times as many egg sandwiches as cheese ones.
How many egg ones does she make?

b Katie makes the same number of ham as cheese sandwiches.
How many ham sandwiches does she make?

c Katie makes 10 more tomato than cheese sandwiches.
How many tomato sandwiches does she make?

d How many sandwiches does she make altogether?
Write you answer as simply as possible.

3 In a sale, a comic costs c pence. Hadley gives £3 to pay for 4 comics. Which of these expressions gives his change in pence?

A $4c - 3$ B $4c - 300$ C $2 - 4c$ D $300 - 4c$

4 Ellen has 3 bags of sweets.
Each bag has p sweets inside.
Ellen takes some sweets out.
Now the total number of sweets in Ellen's three bags is $3p - 6$.
Some of these statements could be true.
Which ones?

A Ellen took one sweet out of each bag.
B Ellen took two sweets out of each bag.
C Ellen took three sweets out of one bag and none out of the others.
D Ellen took six sweets out of one bag and none out of the others.
E Ellen took six sweets out of each bag.

5 Catherine takes n minutes to eat her breakfast each day.
She takes twice as long to eat her lunch.

a Write an expression for the time it takes Catherine to eat her lunch.
b Write an expression for the time it takes Catherine to eat her lunch over 5 days.

6 The picture shows some packets of rice and a 5 kg weight.
Each packet of rice weighs n kg.

a Write an expression for the total mass in the weighing pan.
b Write an expression for the total mass in each of these.

Pan 1   **Pan 2**   **Pan 3**

7 Menzier bought p pears and some bananas.

a He had three times as many bananas as pears.
How many bananas did he have?

b He ate all the bananas except 4. How many bananas did he eat?

*c He ate a quarter of the pears then gave two to his sister. How many pears did he have left?

*8 Rob had £$2x$. Hitesh had £3.
Rob and Hitesh spent all their money on a raffle ticket.
It won three times as much as the ticket had cost.
Write an expression, using brackets, for the amount they won.

*9   Ella, Todd and Debbie each have a number of toffees in a bag.
Ella has 3 more than Todd.
Debbie has 5 times as many as Todd.
Debbie calls the number of toffees she has t.
Write an expression using t for the number of toffees in Ella's bag.

Review 1   Sam had p trees in his garden.
Write an expression for the number of trees someone has if they have
a   three more than Sam   **b**   ten fewer than Sam
c   four times as many as Sam   **d**   half as many as Sam.

Review 2   Emlyn has blue, red, green and yellow model planes.
p stands for the number of blue planes he has.
a   Emlyn has twice as many red planes as blue planes.
How many red planes does he have?
b   Emlyn has 4 more green planes than blue planes.
How many green planes does he have?
c   Emlyn has the same number of yellow planes as blue planes.
How many yellow planes does he have?
d   How many planes does he have altogether?
Write your answer as simply as possible.

Review 3   A ride at a fair costs r pence. Matthew gives £8 to pay
for 6 rides. Which of these expressions gives his change in pence?
A $6r - 8$ **B** $800 - 6r$ **C** $8 - 6r$ **D** $6r - 800$

Review 4   There are 8 chocolate eggs in a packet.
a   Jan bought x packets of these eggs plus 5 loose eggs.
Write an expression for the number of eggs Jan bought.
b   Chelsea bought n packets of eggs. She ate 3 eggs.
Write an expression for the number of eggs Chelsea had then.
*c   Marios bought 3 times as many eggs as Jan. Write an expression, using brackets, for the
number of eggs Marios bought.

Writing formulae

The area of a rectangle is found by multiplying the length by the width.
We can write a formula for this using A for area l for length and w width.

$$A = lw$$

Exercise 11

1   Write a formula for these.
a   the cost, c, of n chocolate bars at 35p each
b   the cost, c, of h hair ties at £2 each
c   the total money raised, r, in a sponsored walk at 20p per kilometre, k.
d   the total weight, w, of a cat weighing c kg and a dog weighing 20 kg
e   the total cost, c, of 3 apples at a pence each and a pear for 50p
f   the final length, l, of a 4 m piece of wood with c cm cut off
g   the final length, l of an 8 m piece of wood with c cm cut off

> The answer to **a**
> is $c = 35n$.

> Be careful with the
> units. 1 m = 100 cm.

2 Here is a way of converting degrees Celsius, C, to *approximate* degrees Fahrenheit, F.

Multiply the degrees Celsius by two and add thirty

Write a formula for this.

3 Three equilateral triangles are drawn in the corners of
another equilateral triangle.
Write a formula which has p, q and r as variables.

$r = $ _____

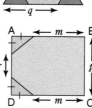

*4 ABCD is a square of side p cm.
Two identical isosceles triangles are drawn in two of the
corners.
Express r in terms of m and p. Simplify your answer.

Review 1 Write a formula for these.

a the total height, h, of a man of height m cm and his hat of height 12 cm
b the cost, c, of 5 packets of biscuits at £n each plus a cake at £5
c the final length, l, of a 2 m piece of rope with r cm cut off

Review 2 This is an isosceles trapezium.
Write a formula for the perimeter, P, in terms of a,
b and c.

$P = $ _____

Remember an isosceles
trapezium has two sides
of equal length.

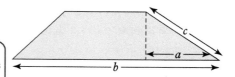

⭐ **Practical**

1 You will need a spreadsheet.
Choose one of these formulae or a formula of your own.
Use a spreadsheet to print a table of values.

$$\text{petrol used} = \frac{\text{distance travelled} \times 8}{100}$$

$$\text{speed} = \frac{\text{distance}}{\text{time}}$$

perimeter of rectangle $= 2\,l + 2\,w$
sale price $= 0{\cdot}95 \times$ original price

Example $\text{petrol used} = \dfrac{\text{distance travelled} \times 8}{100}$

	A	B	C	D	E	F
1	distance	50	= B1+50			
2	petrol	= B1*8/100	= C1*8/100			
3						

2 You will need a spreadsheet or graphical calculator.
You are going on holiday overseas.
Choose a country you would like to visit.
Find out the conversion rate for the currency of that country.
Use a spreadsheet to find out how much of that currency you would get for
£10, £50, £100, £200, £500, £1000, ...

You could change this row
to the distances you want.

Algebra

Investigation

Finding formulae

1

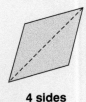 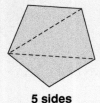 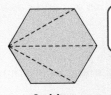

4 sides   **5 sides**   **6 sides**

> These diagonals do not intersect. They all begin at one vertex.

Polly wanted to know if there was a formula for the number of diagonals from *one* vertex of a shape with n sides.

Draw some shapes with 7, 8, 9, ... sides.

Draw the diagonals. Begin all the diagonals at one vertex.

Copy and fill in this table.

Try and find a formula which gives the number of diagonals, d, in a shape with s sides.

Number of Sides, s	4	5	6	7	8	9	10
Number of Diagonals, d	1	2	3				

$d = $ _____

2 a

 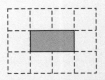 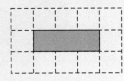

1 m × 1 m pond   **1 m × 2 m pond**   **1 m × 3 m pond**

Gerry made ponds. They were all 1 m wide but of different lengths. He put paving stones around the edge.

He wanted to know if there was a formula for the number of paving stones.

Draw some more ponds that are 1 m wide.

Copy and fill in this table.

Try and find a formula which gives the number of paving stones, N, for a pond of length, l.

Length of Pond, l	1	2	3	4	5	6	7	8
Number of Paving Stones, N	8	10	12					

$N = $ _____

What if the pond was 2 metres wide?

What if the pond was w metres wide and l metres long?

> Hint: a 3 m long pond has 2 × 3 paving stones along each side and 3 paving stones at each end.

*** b** Chris made fish tanks with tiles.

For each side of the tank, he started with a row of 1, 2, 3, 4, 5, ... clear tiles and then surrounded these with green tiles.

The tiles were all 10 cm square. The base of the tank was made up entirely of green tiles.

Write a formula for finding the number of green tiles, G, needed for a fish tank with n clear tiles on each side.

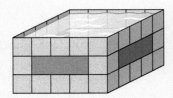

3

| 1 table | 2 tables | 3 tables |

'Connie's Conference Dinners' arrange tables in rows.
Connie wanted to know if there was a formula for calculating the number of people, p, that could be seated at t tables arranged in a row.
Draw the diagrams for 4, 5, 6, ... tables arranged in a row.
Copy and fill in this table.

Number of tables, t	1	2	3	4	5	6	7	...
Number of people, p	6	10						

Find a formula which gives the number of people, p, that can be seated at t tables arranged in a row.

$p =$ _____

What if each table seated 3 along each side?
What if each table seated 3 along each side and 2 at each end?
What if each table seated a along each side and b at each end?

Writing equations

Last week Vicki delivered n bundles of pamphlets. She got £5 per bundle plus £4 extra.
An expression for the amount she earned is $5 \times n + 4$. We write this as $5n + 4$.

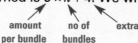

amount   no of   extra
per bundle bundles

If we are told she earned £54 last week, we can **write an equation**. $\qquad 5n + 4 = 54$

Worked Example
Tim bought C tickets to a concert at £15 each. Altogether they cost £75.
Write an equation for this.

Answer
$\qquad C \times 15 = 75 \qquad$ We write this as **$15C = 75$**.
$\qquad \uparrow \qquad \uparrow \qquad \uparrow$
no. of   cost of total
tickets a ticket cost

Worked Example
'Fat Free' pies cost £3 each. Muffins cost £2.
Angie bought p pies and a muffin. Altogether she paid £20.
Write an equation for this.

Answer
$\qquad p \times 3 + 2 = 20. \qquad$ We write this as **$3p + 2 = 20$**.
$\qquad \uparrow \quad \uparrow \quad \uparrow \qquad \uparrow$
no. of cost of cost of total
pies each pie muffin cost

Algebra

Timberlands planted 150 rows with n trees in each row.
The next year they planted 10 more trees in each row.
Altogether they had planted 4500 trees. Write an equation for this.

Answer

Each row now has $n + 10$ trees. There are 150 rows.
An expression for the number of trees is $150 \times (n + 10)$.
We can write the equation **$150(n + 10) = 4500$**.

Exercise 12

1 Write an equation for each of these. Use n for the unknown.
 a I think of a number. I subtract 7. The answer is 16.
 b I think of a number. I divide it by 9. The answer is 4.
 c I think of a number. I multiply it by 3. The answer is 18.
 d I think of a number. I add 12. The answer is 20.
 e I think of a number, multiply it by 3 and then add 7. The answer is 19.
 f I think of a number, add 2 and then multiply the result by 4. The answer is 36.
 g Three more than double a number is 15.
 h Twice the length of a hedge is 14 metres. (n is the length of the hedge.)
 i Simon bought some peanut bars. They cost £2 each.
    The total cost was £16. n is the number of peanut bars.
 j Russell ran round a track 24 times.
    He ran 4800 m in total. n is the length of the track.
 k Bianca bought n bags of sweets at £1 each and a drink for £2. The total cost was £6.
 l 25 paving stones are stacked on top of one another. The total height of the stack is
    88 cm. n is the thickness, in cm, of one paving stone.
 m When Allanah doubles her lucky number and then adds 4, she gets an answer of 30.
    n is Allanah's lucky number.
 n Sian bought some CDs for £11 each and a CD case for £4.
    The total cost was £37. n is the number of CDs bought.
 o In a sale £3 was taken off the price of a t-shirt. William bought 5 of these t-shirts.
    Altogether he paid £55. n is the price of one t-shirt.
 p A trailer unit of 2 tonnes was added to each of the lorries at 'Harry's Haulage'.
    The total weight of 5 lorries was then 35 tonnes. n is the weight of each lorry
    without the trailer unit.

2 Write an equation for each of these.
 a Two numbers, a and b, add to 11.
 b The difference between two numbers, x and y, is 12.
 c Janna and Eugene have a total of £20. Use j and e as the unknowns.
 d Victoria and Edward have a difference of £120 in their bank accounts.
    Use v and e as the unknowns.

*3 Write an equation for each of these.
 a Dylan has scored 4 times as many goals this season as David. Together they have
    scored 20 goals. Use g as the unknown.
 b If Dean gave Mira 4 of his goldfish, they would both have the same number of goldfish.
    Use d and m as the unknowns.
 c Three times Isla's pocket money is the same as twice Chelsea's pocket money.
    Use i and c as the unknowns.

Review 1 Write an equation for each of these. Use n for the unknown.

a   I think of a number. I multiply it by 6. The answer is 54.
b   I think of a number. I multiply it by 5 and then add 4. The answer is 34.
c   When Fred divided his height by 4 the answer was 42 cm. n is Fred's height.
d   Rupert borrowed some CDs from a friend. He lost 2 of them.
     He then had 8 borrowed CDs left. n is the number he originally borrowed.
e   Rosalind bought some bags of crisps for £2 each and a pizza for £4.
     Altogether she spent £14. n is the number of bags of crisps.
f   Three lorries, each n tonnes, carry soil to a building site. In total they dump 2 tonnes of soil.
     The three lorries then have a total mass of 17 tonnes.

Review 2 Write an equation for each of these.

a   Dale and Shelley have a total of 7 pets. Use d and s as the unknowns.
b   Suzy paid £3 less then Stacey for the same top. Together they paid £36.
     Use s as the amount Stacey paid.

Solving equations using inverse operations

Discussion

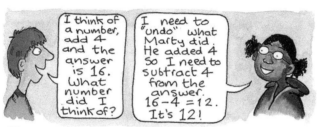

I think of a number, add 4 and the answer is 16. What number did I think of?

I need to "undo" what Marty did. He added 4 So I need to subtract 4 from the answer. $16 - 4 = 12$. It's 12!

Marty   **Rani**

Is Rani correct? **Discuss**.

We can **solve** equations using **inverse** operations.

Mike ran 25 times round a track. Altogether he ran 16 km. How long is the track?

Solve means find the answer.

Let t be the length of the track.

$$25t = 16$$

$$\therefore t = \frac{16}{25}$$

$$t = \frac{64}{100}$$

$$t = \mathbf{0.64 \text{ km}}$$

The inverse of multiplying by 25 is dividing by 25.
Change to a denominator of 100 so we can write as a decimal.

$$\frac{16}{25} \overset{\times 4}{\underset{\times 4}{=}} \frac{64}{100}$$

$\therefore$ means therefore.

$t = 0.64$ km is the **solution** to the equation.

Verify the answer by substituting it back into the equation.

$25t = 25 \times \mathbf{0.64}$ substitute 0.64 for t
$25t = 16$ ✓

Verify means check.

Worked Examples

Solve   **a** $\frac{n}{4} = 6$ **b** $24 = a - 17$

Answers

a $\frac{n}{4} = 6$

    $\therefore n = 6 \times 4$ The inverse of dividing by 4 is multiplying by 4.

    $n = \mathbf{24}$

b $24 = a - 17$

    $a - 17 = 24$ It is easier to have the unknown on the left.

     $a = 24 + 17$ The inverse of subtracting 17 is adding 17.

     $a = \mathbf{41}$

Exercise 13

1   Solve these equations.

a $y + 9 = 12$	**b** $x + 8 = 14$	**c** $7 + w = 21$	**d** $30 = 5 + p$
e $m - 4 = 6$	**f** $n - 2 = 10$	**g** $y - 12 = 4$	**h** $m - 14 = 8$
i $7x = 28$	**j** $6x = 54$	**k** $5a = 45$	**l** $24 = 3m$
m $\frac{x}{4} = 2$	**n** $\frac{x}{7} = 3$	**o** $\frac{n}{4} = 6$	**p** $\frac{x}{5} = 12$
q $m + 16 = 33$	**r** $\frac{w}{4} = 16$	**s** $4f = 64$	**t** $63 = p - 19$
u $48 = n - 17$	**v** $\frac{a}{7} = 14$	**w** $16m = 80$	

2   Write and solve an equation for these.

   a   Ross bought n CDs at 'Sounds'. He bought 2 more at 'CD Warehouse'.
      Altogether he bought 8 CDs.
      How many did he buy at 'Sounds'?

   b   Caroline bought n packets of muesli bars. They cost £3 each. The total cost was £12.
      How many did she buy?

   c   20 copies of the same book are stacked on top of one another. The stack is 35 cm high.
      What is the thickness of one book?

Review   Solve these equations.

a $x + 8 = 11$	**b** $y + 7 = 24$	**c** $9 + p = 23$	**d** $70 = 60 + p$
e $n - 3 = 7$	**f** $20 = a - 3$	**g** $6a = 54$	**h** $7a = 84$
i $\frac{n}{6} = 3$	**j** $\frac{n}{17} = 5$	**k** $72 = p - 27$	

Discussion

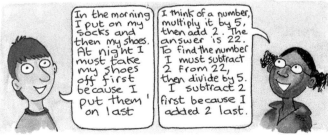

Marty   Rani

Discuss Marty's and Rani's statements.

To solve equations like $3n + 6 = 18$, we can use **inverse operations**.
We undo the operations by working backwards doing the inverse.

To get $3n + 6 = 18$

$n \rightarrow$ [multiply by 3] $\rightarrow 3n \rightarrow$ [add 6] $\rightarrow 3n + 6$

To solve the equation do the **inverse** operations in the reverse order.

$4 \leftarrow$ [divide by 3] $\leftarrow$ [subtract 6] $\leftarrow 18$

 Start with 18 and work backwards doing the inverse.

$n = \mathbf{4}$

Worked Example

Laura buys 4 Christmas presents for £n each and a card for £2.
The total cost is £34. Write and solve an equation to find the cost of each present.

Answer

$4n + 2 = 34$ n is multiplied by 4 then 2 is added.

$\quad 4n = 34 - 2$ The inverse of adding 2 is subtracting 2.

$\quad 4n = 32$

$\quad\;\; n = \frac{32}{4}$ The inverse of multiplying by 4 is dividing by 4.

$\quad\;\; n = \mathbf{8}$

We don't add 2 to the n first because this would give $4(n + 2)$.

Each Christmas present cost **£8**.

Verify the answer.

$4n + 2 = 4 \times \mathbf{8} + 2$ Substitute 8 for n.
$\qquad\;\; = 34$ ✓

Discussion

● Gemma keyed this on her calculator.

7 × 8 = + 6 = − 6 = ÷ 8 =

Without keying this, predict what the answer will be. **Discuss.**

Would Gemma get the same answer as she did above if she keyed

7 × 8 = + 6 = ÷ 8 = − 6 =   ? **Discuss.**

Look at the first keying sequence again.
What would happen if Gemma missed keying the first = ? **Discuss.**
What if she missed keying the second = ?
What if she missed keying all of the = except the last one? **Discuss.**

● Amy was asked to solve $3x + 2 = 14$.
She wrote $3x + 2 = 14$
$\qquad\qquad\qquad = 14 - 2$
$\qquad\qquad\qquad = \frac{12}{3}$
$\qquad\qquad\qquad = 4$
What is wrong with this? **Discuss.**

Algebra

1 What is the inverse of these?
- **a** multiplying by 2 then adding 4
- **b** dividing by 6 then adding 7
- **c** multiplying by 4 then subtracting 2
- **d** adding 2 then multiplying by 5

2 a I think of a number.
When I multiply by 4, then add 2, the answer is 26.
What is the number?

    b I think of a number.
I divide by 2, then add 3. The answer is 8.
What is the number?

    c I think of a number.
I add 5 then multiply by 2.
The answer is 22.
What is the number?

    d I think of a number.
I add 2·7 then multiply by 5.
The answer is 24·5.
What is the number?

3 Solve these equations.

- **a** $2x + 5 = 13$
- **b** $3y - 6 = 9$
- **c** $5w + 3 = 23$
- **d** $7y - 12 = 2$
- **e** $4p + 5 = 33$
- **f** $6m - 7 = 29$
- **g** $5m + 7 = 37$
- **h** $4n - 9 = 27$
- **i** $8b - 6 = 66$
- **j** $12 + 9a = 75$
- **k** $20 + 7x = 76$
- **l** $14 = 2 + 3n$
- **m** $11 = 4a - 1$
- **n** $21 = 3t - 6$
- **o** $17 = 3 + 2d$
- **p** $66 = 3w + 6$
- **q** $5x + 7 = 67$
- **r** $4p + 3 = 83$
- **s** $6n + 3 = 75$
- **t** $\frac{2x}{3} = 4$
- **u** $\frac{3x}{2} = 9$
- **v** $\frac{4a}{3} = 1$
- **w** $\frac{4a}{5} = 2$

*4 Write and solve an equation for these. Use the letter given in brackets as the unknown.
- **a** Jesse had saved £29. Her aunt gave her £5 of this. The rest she got from working for a neighbour for £3 an hour. How many hours work (h) did she do?
- **b** Freda cycled four times around a track and then 800 m along the road.
The odometer on her bike showed she had cycled 2400 m altogether.
How long is the track (t)?
- **c** Joseph measured the length he wanted his curtains. He then added 7 cm. He multiplied this length by 4 and got 480 cm.
What length (l) does he want his curtains?

Review 1 Solve these equations.
- **a** $2n + 4 = 10$
- **b** $3y + 4 = 22$
- **c** $4x - 7 = 25$
- **d** $9p - 7 = 56$
- **e** $57 = 7m + 8$
- **f** $31 = 5p + 6$
- **g** $4x + 3 = 63$

Review 2 Write and solve an equation to find the answer to these.
- **a** Melanie delivered 7 packs of pamphlets minus the 4 pamphlets she had left over. There were n pamphlets in each pack. Altogether she delivered 66 pamphlets. How many pamphlets were in each pack?
- **b** Ashad earned £5 for every car he cleaned. His boss gave him an extra £10 for working hard. Altogether Ashad earned £35. How many cars did he clean?

Discussion

7P was asked to solve this equation. $3(x + 4) = 27$.

Damian solved it like this.

$3(x + 4) = 27$
$3x + 12 = 27$ remove brackets first
$3x = 27 - 12$
$3x = 15$
$x = \frac{15}{3}$
$x = \mathbf{5}$

Rhian solved it like this.

$3(x + 4) = 27$
$(x + 4) = \frac{27}{3}$ inverse of ×3 is ÷3
$x + 4 = 9$
$x = 9 - 4$
$x = \mathbf{5}$

Discuss Damian's and Rhian's methods.

Exercise 15   **Only use a calculator if you need to.**

1 Solve these equations.
   a $3(x + 1) = 12$ **b** $6(y - 3) = 18$ **c** $4(p - 2) = 24$ **d** $8(w - 6) = 32$
   e $10(d + 4) = 40$ **f** $5(x + 2) = 20$ **g** $2(a - 7) = 18$ **h** $7(y + 3) = 42$
   i $20 = 5(f + 3)$ **j** $36 = 9(y - 4)$ **k** $2(2x + 1) = 6$ **l** $3(2y - 7) = 9$
   m $3(3b + 2) = 33$ **n** $3(4t - 1) = 27$ ∗**o** $4(3x + 1) = 37$ ∗**p** $5(4b - 3) = 21$
   ∗**q** $5(6x + 1) = 41$ ∗**r** $8(5y - 2) = 84$ ∗**s** $2(5 + x) - 12 = 4$ ∗**t** $4 + 5(2x - 1) = 9$

2 Write and solve an equation for these.
   a 'Timing' took £5 off some watches. Jon bought 3 identical watches.
      The total cost was £60. What was the original price (p) of the watches?
   b Toffee Delights cost £n. Chocolate Delights cost £2 more than this.
      Bianca bought 9 packets of Chocolate Delights. They cost £45 altogether.
      How much do Toffee Delights cost?
   ∗**c** Miles, Miriam and Marlene each had a bag of books.
      They did not know how many books were in each bag.
      Miles had 3 more than Miriam. Marlene had 4 times as many as Miles.
      Miriam had n books and Marlene had 36.
      How many books did Miriam have?

Review 1   Solve these equations.
a $4(x + 2) = 8$ **b** $7(y - 3) = 28$ **c** $5(d + 5) = 30$
d $24 = 6(p - 3)$ **e** $2(2x + 3) = 22$ **f** $4(2y - 5) = 28$
∗**g** $5(4m - 3) = 33$ ∗**h** $4(3b + 7) = 49$

Review 2   Write and solve an equation for this.
At a furniture shop a chair costs £n. A table costs £25 more.
Peter bought 5 tables for his new café. Altogether they cost £200.
How much does a chair cost?

Worked Example
In this diagram, the number in each box is found by adding
the two numbers above it.

What are the missing numbers in this diagram?

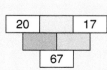

Answer

Let n be the number in the yellow box.

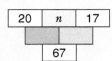

20
n
17
67

red box $= n + 20$
blue box $= n + 17$

$67 = \text{red box} + \text{blue box}$
$67 = n + 20 + n + 17$
$= 2n + 37$

$2n + 37 = 67$

$\therefore 2n = 67 - 37$ **The inverse of adding 37 is subtracting 37.**

$2n = 30$

$\therefore n = \frac{30}{2}$ **The inverse of multiplying by 2 is dividing by 2.**

$\mathbf{n = 15}$

Verify the answer.

$2n + 37 = 2 \times \mathbf{15} + 37$ **Substitute 15 for n**

$= 67$ ✓

The diagram can be filled in as shown.

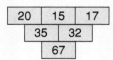

20
15
17
35
32
67

Exercise 16

1 Write and solve an equation to find the value of n. Use this to find the missing numbers. **a** has been started for you.

The number in each box is the sum of the two numbers above it.

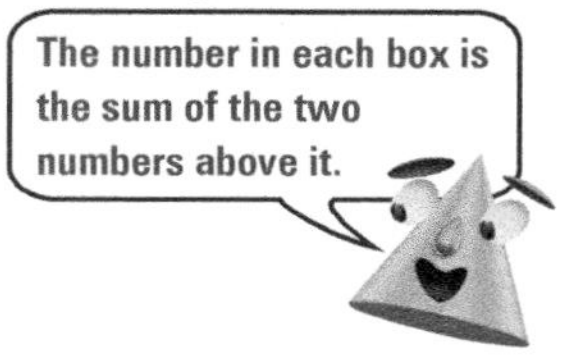

a

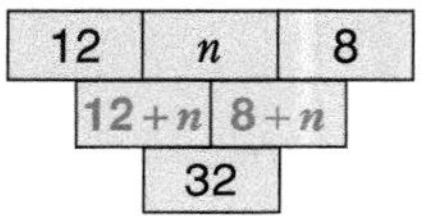

12
n
8
12+n
8+n
32

$12 + n + 8 + n = 32$

b

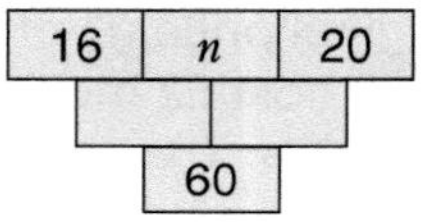

16
n
20
60

c

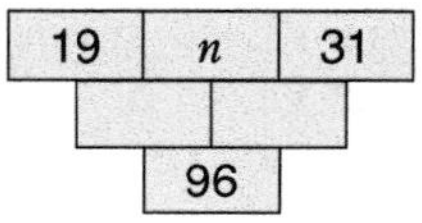

19
n
31
96

d

2
7
n
27

e

3
n
5
20

***f**

1½
n
4
12

***g**

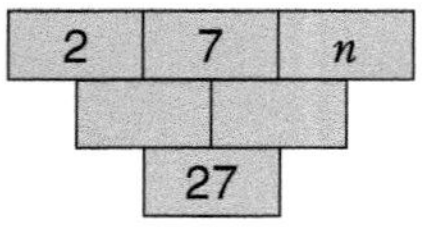

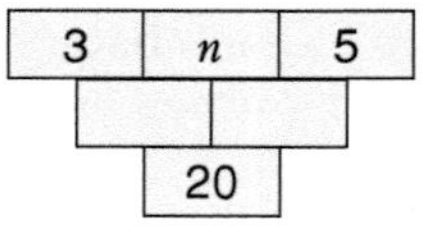

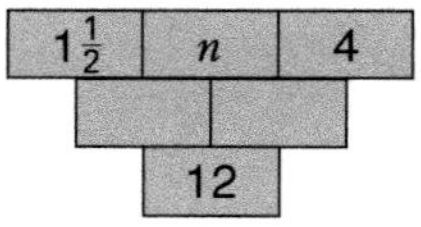

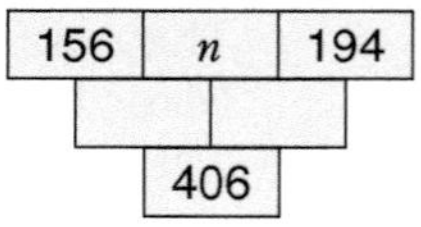

156
n
194
406

***h**

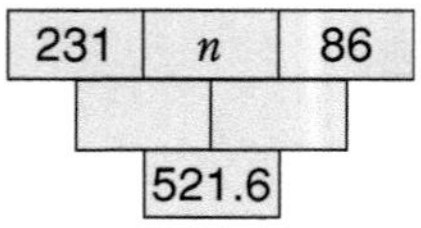

231
n
86
521.6

***i**

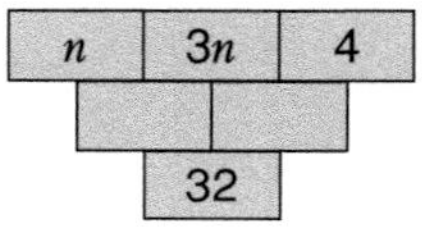

n
3n
4
32

***j**

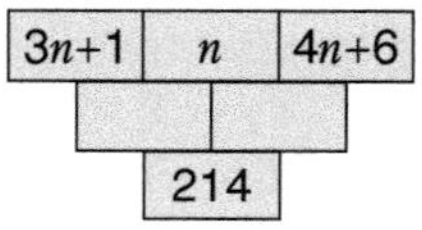

3n+1
n
4n+6
214

2 Write and solve an equation to find x.

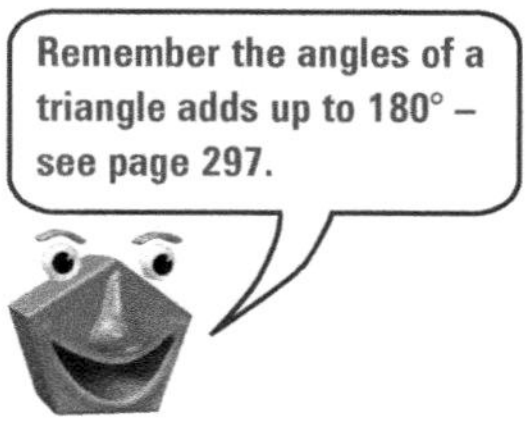

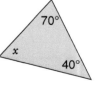

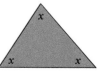

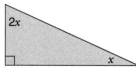

Remember the angles of a triangle adds up to 180° – see page 297.

a

70°
x
40°

b

x
x
x

c

2x
x

d

3x
2x
30°

e

2x
4x
3x

f

2x
5x
x

g

2x
(x–30°)
x

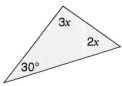

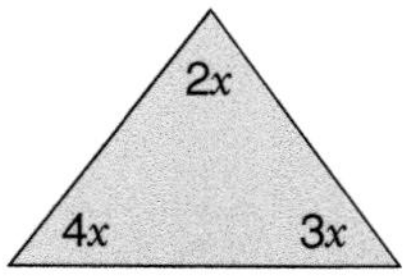

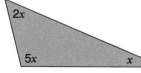

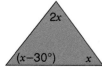

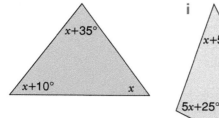

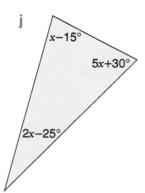

3 Write and solve an equation for these.
  a There are 32 pupils in a class.
     There are 6 more boys than girls.
     How many girls are in the class?
  b There are 30 scones altogether on two plates.
     The second plate has 6 fewer scones than the first plate.
     How many scones are there on each plate?
  c In a yacht race, the second leg is twice as far as the first. The total distance for both legs is 204km.
     How far is the first leg?
  d A building has a flag-pole on the top. The building is 12 times as tall as the flag-pole and together they are 65 m high.
     How high is the flag-pole?
  e A field is twice as long as it is wide. Its perimeter is 900 metres. What is its width?

4 $m + n = 7$ is an equation with two unknowns.
   If $m = 4$ and $n = 3$, the equation would be true.
   What other values of m and n make this equation true?

5 What value could the unknowns have?
   a $a + c = 32$ **b** $16 - p = q$ **c** $\frac{m}{n} = \frac{1}{2}$ **d** $gf = 24$

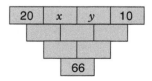

*6 The number in each box is the sum of the two numbers above it.
   a What positive values could x and y have?
   b What must the sum of x and y be?

*7 This shape is made from 5 square tiles.
   The length of a side of a tile is t cm.
   a Write an expression for the area of the shape.
   b The area of the shape is 720 cm².
      What is the perimeter of the shape?

   Another cross is made from square tiles of side s cm.
   c Write an expression for the perimeter of the shape.
   d If the perimeter is 120 cm, what is the area?

Review 1 The number in each box is the sum of the two numbers above it.
Find the value of n.

a 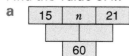   **b**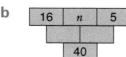

Review 2 Find x.

a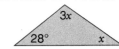

b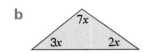

* c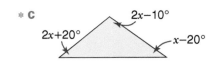

Review 3 Write and solve an equation for each of these. Let n be the unknown.

a In a library, one shelf holds 3 times as many books as another. Together they hold 128 books. Write and solve an equation to find the number of books on the smaller shelf.

* b Jayne had a 16 cm length of liquorice. She shared this with a friend. The piece she gave her friend was 2 cm less than twice the length she kept for herself. Write and solve an equation to find the length of the piece Jayne kept for herself.

Investigation

Arithmagons

This is an arithmagon.
The number in the square is the sum of the numbers in the circles on either side of it.

What could the numbers A, B, C, D be?
Is there more than one answer?

Investigate. Think about negative numbers and decimals as well.

Manoli wrote $A + B = 9$
$A + B + C + D = 17$
$B = 9 - A$

> Manoli's 1st and 3rd relationship are equivalent.

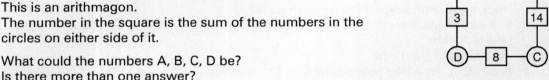

Write some more relationships between A, B, C and D.
Are any of the relationships you have written equivalent?
Check that the relationships are true for all the sets of possible values for A, B, C and D you have found.
When writing this investigation, we had to find numbers for the boxes that worked.
How might we have found them? **Investigate**.

Solving equations by transforming both sides

Discussion

● This set of scales is balanced.

The mass of 4 oranges and 3 bananas equals the mass of 2 oranges and 4 bananas.
If we take a banana from the right-hand pan what will happen?
What could we do to the left-hand pan to make the scales balance? **Discuss**.
What would you need to do to the pan on the other side to rebalance the scales if
a 2 oranges are taken from the left-hand pan
b 3 bananas are taken from the right-hand pan?

continued ...

● This set of scales is balanced.

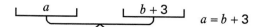

$a = b + 3$

If we subtract 3 from both sides, will the new equation be true?

$a - 3 = b + 3 - 3$

$a - 3 = b$

We could check by substituting.
Choose values for a and b that make $a = b + 3$ true.
$a = 9$ and $b = 6$ make $a = b + 3$ true.

Substitute these into the new equation, $a - 3 = b$.
$9 - 3 = 6$ ✓ It is true.

What if we now add 7 to both sides?
Is $a + 4 = 6 + 7$ true for $a = 9$ and $b = 6$?
Discuss.

$a + 4 = b + 7$

What if we now double each side?
Discuss.

$2(a + 4) = 2(b + 7)$

What if we now add 5 to both sides? **Discuss**.
Check by substituting.

$2(a + 4) + 5 = 2(b + 7) + 5$

If you do the same to both sides, do you always get an equivalent equation? **Discuss**.

●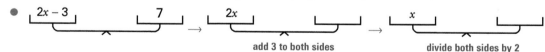

add 3 to both sides   divide both sides by 2

Discuss what goes in the empty pans.

Draw balance diagrams to solve these equations.

$3x = 12$ $\frac{x}{2} = 5$ $4x - 1 = 7$ $\frac{5x}{2} = 10$ $3n + 8 = 26$

How did you work out what order to do things in? **Discuss**.

We can solve equations by **doing the same to both sides** of the equation.

Worked Examples
Solve   a $4x - 7 = 15$ b $\frac{a}{5} + 2 = 8$

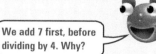
We add 7 first, before dividing by 4. Why?

Answers

a $\quad 4x - 7 = 15$
$\quad 4x - 7 + 7 = 15 + 7$ add 7 to both sides
$\quad\quad\quad 4x = 22$
$\quad\quad\quad \frac{4x}{4} = \frac{22}{4}$ divide both sides by 4
$\quad\quad\quad\quad x = 5\cdot5$

b $\quad\quad \frac{a}{5} + 2 = 8$
$\quad \frac{a}{5} + 2 - 2 = 8 - 2$ subtract 2 from both sides
$\quad\quad\quad \frac{a}{5} = 6$
$\quad\quad\quad \frac{a}{5} \times 5 = 6 \times 5$ multiply both sides by 5
$\quad\quad\quad\quad a = 30$

Algebra

Exercise 17

Solve these equations by doing the same to both sides.

1. a $n + 6 = 8$ b $x - 5 = 4$ c $3a = 33$ d $\frac{n}{3} = 5$
   e $\frac{a}{6} = 1$ f $\frac{x}{2} = 5$ g $3 + n = 14$ h $11 + n = 23$
   i $5b = 20$ j $12n = 144$ k $70p = 140$

2. a $2a + 1 = 7$ b $3a - 2 = 10$ c $3n + 4 = 19$ d $3 + 4a = 15$
   e $1 + 2a = 13$ f $2 + 5a = 17$ g $5a - 6 = 19$ h $4m - 6 = 18$
   i $5a - 3 = 17$ j $20 = 3m - 4$ k $26 = 5n + 11$

3. a $\frac{2x}{3} = 4$ b $\frac{3x}{2} = 9$ c $\frac{4a}{3} = 4$ d $\frac{5n}{2} = 10$
   e $\frac{4a}{5} = 2$ f $\frac{2n}{5} = 3$ g $\frac{4b}{3} = 5$

4. a $\frac{x}{2} + 1 = 5$ b $\frac{n}{3} - 1 = 3$ c $\frac{a}{5} + 3 = 5$ d $2 + \frac{n}{2} = 3$
   e $3 + \frac{n}{4} = 5$ f $5 + \frac{x}{3} = 21$ g $\frac{b}{4} - 5 = 27$

5. a $3n + n + 2 = 10$ b $5p + 4 + 2p = 18$ c $8x + 6 - 3x = 16$
   d $4s + 3 - 2s - 1 = 4$ e $8y - 6 - 3y = 29$

> Collect the like terms first.

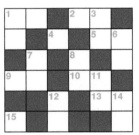

T

Review   Use a copy of the cross number. Fill it in by solving the equations.

Across
1. $n - 7 = 4$
2. $\frac{n}{7} = 3$
5. $n - 20 = 12$
9. $n - 10 = 3$
10. $\frac{n}{3} = 14$
13. $2n - 1 = 41$
15. $\frac{n}{3} + 4 = 5$

Down
1. $\frac{3n}{4} + 1 = 10$
3. $n + 3 = 16$
4. $3n = 12$
6. $\frac{2n}{3} = 16$
7. $2 + n = 15$
8. $n - 1 = 13$
11. $2n = 44$
12. $1 + 2n = 21$
14. $2n - 3 = 19$

Discussion

Trudy and Mike have the same number of sweets.
Trudy has 3 boxes of sweets and 2 loose sweets.
Mike has 2 boxes of sweets and 22 loose sweets.
There are n sweets in each box.
We can write an equation for this.

$$3n + 2 = 2n + 22$$

↑ Trudy's sweets   ↑ Mike's sweets

We can solve this equation by doing the same to both sides. We begin by subtracting the same number of unknowns from both sides so that one side then has no unknowns.

$$3n + 2 = 2n + 22 \qquad \text{subtract } 2n \text{ from both sides}$$
$$3n + 2 - \mathbf{2n} = 2n + 22 - \mathbf{2n}$$

> We choose to subtract $2n$ rather than $3n$ because $2n$ is a smaller amount.

How would we continue to solve this equation? **Discuss.**
Use this method to solve these.

$$5b + 1 = 4b + 3 \qquad 9x - 8 = 5x + 4$$

Exercise 18

1 Solve these equations.

a	$3x - 4 = 2x + 6$		b	$4y + 2 = 3y + 4$		c	$7w - 10 = 6w + 4$
d	$5a + 2 = 3a + 6$		e	$8n - 3 = 6n + 5$		f	$2b - 5 = 10 - b$
g	$2p + 5 = 10 + p$		h	$5m - 9 = 2m + 3$		i	$4 + 3x = 12 - x$
j	$5n - 4 = n + 8$		k	$6x - 8 = 4x + 2$		l	$2x + 5 = 15 - 2x$
*m	$6x + 3 = 4x - 1$		*n	$7y + 8 = 5y - 4$		*o	$12 - x = 3x - 4$
*p	$6x + 8 = 8x + 2$		*q	$x - 3 = 2x + 5$		*r	$2x + 5 = 4x + 7$

*2 Write and solve an equation to find the value of n.

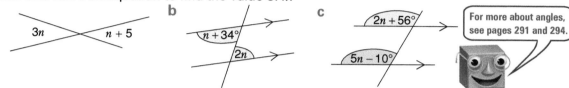

a $3n$ $n + 5$

b $n + 34°$ $2n$

c $2n + 56°$ $5n - 10°$

For more about angles, see pages 291 and 294.

Review Solve these equations.

a	$5a = 4a + 2$	b	$4p - 7 = 5 + 3p$	c	$7x + 3 = 5 + 5x$
d	$3n + 2 = n + 6$	*e	$3n + 8 = 6 - n$	*f	$2x + 25 = 7 + 4x$

 Puzzle

Three brothers were travelling together.
They had with them a briefcase which contained a number of ten pound notes.
On the third night of their journey they stayed with their cousin.
During the evening, the youngest brother opened the briefcase. He gave one ten pound note to his cousin, then took his third share of the remaining notes.
Later during the evening, the oldest brother did the same; again giving one ten pound note to his cousin and then taking a third share of the remaining notes.
Much later in the evening, the third brother did the same; again giving one ten pound note to his cousin, then taking a third share of the remaining notes.

What was the smallest number of ten pound notes in the briefcase?

Summary of key points

A $p - 7$, $8n$, $7t + 4$ are all **expressions**.
The letter stands for an **unknown**.

$2b + 7 = 16$ is an **equation**. b has a particular value.

$n + m = 12$ is also an **equation**. n and m can have any values as long as they add to 12.

$C = \frac{3}{5}m$ is a **formula**. C is the sale price of an item which originally cost £m.
If we know m, we can work out C.

Equations and formulae have an equals sign. Expressions do not.
In a formula the letters stand for something specific.

We **write expressions** without multiplication or division signs.

Examples $3 \times n = 3n$ $5 \times (x + y) = 5(x + y)$ $(n + m) \div p = \frac{n + m}{p}$

$a \times a \times a \times a = a^4$ $5 \times 6a = 30a$

 The operations used in algebra **follow the same rules as numbers**.

Examples

$x + y = y + x$ $pq = qp$ $m + (n + p) = (m + n) + p$ $a(bc) = (ab)c$

$3 + 4 = 4 + 3$ $8 \times 9 = 9 \times 8$ $5 + (6 + 4) = (5 + 6) + 4$ $3 \times (4 \times 5) = (3 \times 4) \times 5$

 We can **simplify** expressions by **collecting like terms**.

Examples $a + a + a = 3a$ $3n + 6n = 9n$ $8b - 3b = 5b$

$4m + 6 - 3m - 2 = 4m - 3m + 6 - 2$

$= m + 4$

$3x + 8y - 2x - y = 3x - 2x + 8y - y$

$= x + 7y$

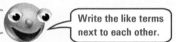

Write the like terms next to each other.

 We can **simplify** expressions by **cancelling**.

Examples $\frac{8b}{2} = 4b$ Divide numerator and denominator by the common factor 2.

$\frac{6a}{a} = 6$ Divide numerator and denominator by the common factor a.

$\frac{m^2}{m} = m$ Divide numerator and denominator by the common factor m.

 To **write an expression without brackets** we use the rules of arithmetic.

Examples $4(x + 7) = 4x + 28$ $5(2y - 3) = 10y - 15$

	x	$+7$
4	$4x$	$+28$

	$2y$	-3
5	$10y$	-15

 We **evaluate an expression** by **substituting** values for the unknown.

We follow the **order of operations** rules.

If $a = 2$, $b = 3$ and $c = 5$ then

$3a = 3 \times 2$ $2c - b = 2 \times 5 - 3$ $\frac{c}{a} + 3 = \frac{5}{2} + 3$ $30 - c^2 = 30 - 5^2$

$= 6$ $= 10 - 3$ $= 2 \cdot 5 + 3$ $= 30 - 25$

$= 7$ $= 5 \cdot 5$ $= 5$

 A **formula** is a rule for working something out.

Example $T = \frac{D}{S}$ is the formula for finding the time taken, T, to travel distance, D, at speed, S.

If $D = 20$ and $S = 4$ then $T = \frac{20}{4}$

$= 5$.

 We can **write expressions, formulae and equations** for practical situations.

J We can **solve equations** using inverse operations.

Example $2x - 3 = 15$

> Start with the right-hand side of the equation and work backwards doing the inverse.

$2x = 15 + 3$ The inverse of subtracting 3 is adding 3.

$2x = 18$

$x = \frac{18}{2}$ The inverse of multiplying by 2 is dividing by 2.

$x = \mathbf{9}$

We can also solve equations by doing the same to both sides of the equation.

Example $\frac{n}{6} + 4 = 13$

$\frac{n}{6} + 4 - \mathbf{4} = 13 - \mathbf{4}$ subtract 4 from both sides

$\frac{n}{6} = 9$

$\frac{n}{6} \times \mathbf{6} = 9 \times \mathbf{6}$ multiply both sides by 6

$n = \mathbf{54}$

Test yourself

1 Which of these is   **a** an equation   **b** an expression   **c** a formula?   ◆**A**
   A $3(f - 7)$
   B $F = \frac{9}{5}C + 32$ where F is temperature in °F and C is the temperature in °C
   C $2x + 3 - x = 8$

2 Simplify these.   ◆**B**
   a $3 \times y$ **b** $b \times 4$ **c** $c \times c$ **d** $4 \times (x + 3)$ **e** $(y - 4) \times 6$
   f $p \times p \times 2$ **g** $3 \times b \times b$ **h** $5y \times 5y$ **i** $(s + t) \div r$ **j** $5 \times 4m$ **k** $7 \times 8r$

3 Simplify these.   ◆**C** ◆**D**
   a $x + x + x + x$ **b** $8a + 3a$ **c** $10b - 5b$ **d** $10m + 3m + 2m$
   e $5n + 3n - n$ **f** $12p - 4p + 3p$ **g** $10q - 3q - 2q$ **h** $6p + 8p + 4$
   i $3y + 2y + 7 - 2$ **j** $3x + 5 + 2x - 1$ **k** $9m + 8 - 3m - 5$ **l** $3a + 4b - 2a - 2b$
   m $5x + 2y - 3x - 2y$ *∗***n** $8p - 2q - 3p + 3q$ *∗***o** $4m + 7n - 4m - 2n - 4n$

4 Marion's house is a rectangle.
   Write an expression for the perimeter of
   Marion's house.
   Simplify your expression.   ◆**C** ◆**D**

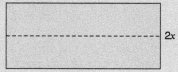
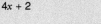 $2x$

$4x + 2$

5 Simplify these expressions by cancelling.   ◆**E**
   a $\frac{5w}{w}$ **b** $\frac{20n}{5}$ **c** $\frac{28x}{7}$ **d** $\frac{30m}{m}$ **e** $\frac{21a}{14}$ **f** $\frac{2b^2}{b}$

6 Write without brackets.   ◆**F**
   a $5(x + 3)$ **b** $7(n - 4)$ **c** $8(2a + 4)$ **d** $4(2y - 3x)$

7 Match each of these with an expression from the box.   ◆**B** ◆**C**
   a $2x + x + 2x$ **b** $2x + 4 + x$ **c** $\frac{4x}{x}$
   d $2(x + 2)$ **e** $2 \times x \times x$ **f** $\frac{16x^2}{4x}$ ◆**D** ◆**E**
   g $(x + y) \times 3$ **h** $(x + y) + 2$

A. $4x$	B. $3x + 3y$
C. $2x^2$	D. $5x$
E. $2 + (x + y)$	F. $2x + 4$
G. $3x + 4$	H. 4

   ◆**F**

8 If $x = 3$, $y = 2$ and $z = 6$ find the value of
  a xy **b** $3x - 2y$ **c** $xy + z$ **d** $4(z - y)$ **e** $\frac{z}{y} + x$ **f** $4(2x + y)$
  g $3x^2$ **h** $\frac{y^2}{2}$ **i** $\frac{2x + y}{2y}$ **j** $\frac{z^2}{x}$ **k** $3y^2 + 2x^2$

9 a If $x + y = 6$, write 3 other facts that must be true.
  b What values could x and y have?

10 $C = 80 + 65h$ is the formula for the cost to hire a wedding car where C is the cost,
  in pounds, and h is the number of hours hired.
  Find the cost to hire a car for
  a 4 hours   **b** 8 hours   **c** 5·5 hours.

11 The formula $E = \frac{1}{2}mv^2$ gives the kinetic energy (energy of movement) of an
  object of mass, m, and velocity v. E is measured in Joules.
  What is E if
  a $m = 20$ kg, $v = 3$ m/s   **b** $m = 500$ kg, $v = 10$ m/s?

12 Andy, Terry and Joshua each have a bag of counters.
  They do not know how many counters are in each bag.
  Terry has 2 more counters than Andy.
  Joshua has 4 times as many counters as Andy.
  a Andy calls the number of counters in his bag n.
    Write an expression using n to show the number of counters in
  *   **i** Terry's bag   **ii** Joshua's bag.
  b Terry calls the number of counters in his bag m.
    Write an expression using m to show the number of counters in
    i Andy's bag   **ii** Joshua's bag.
  c Joshua calls the number of counters in his bag p.
    Write an expression using p to show the number of counters in Terry's bag.

13 *Write an equation for each of these. Let the unknown be m.
  a When the length of a fence is divided by 6 the answer is 4 metres.
  b Three times James' height is 450 cm.
  c When the length of a pool was doubled and then 4 m was added, the final length was 20 m.
  d A book is wrapped in cardboard. This adds 1 cm to its thickness, Eight of these wrapped books make a stack 88 cm high.

14 Write an equation for this.
  Two numbers, p and q add to 136.

15 Solve these equations.
  a $b + 9 = 21$ **b** $x + 7 = 20$ **c** $5x = 50$ **d** $\frac{y}{9} = 4$
  e $2n + 6 = 16$ **f** $8p + 12 = 76$ **g** $\frac{5x}{4} = 8$ **h** $5(x + 4) = 30$
  i $7(y - 2) = 28$ **j** $9(3x + 1) = 36$

16 Write and solve an equation to find the answer to these.

  a I think of a number.
   When I multiply by 4 and add 9 the answer is 37.
   What is the number?

  b Jimmy bought 5 CDs at 'Sounds Right'.
   The total cost after the discount was £51.
   How much was each CD before the discount?

  c Find the value of x.

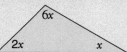

  *d There were a number of goats in a field. One of them had only 3 legs. The total number of legs was 127.
   How many goats were in the field?

17 a The number in the square is found by adding the two
   numbers in the circle either side of it.
   Write and solve an equation to find the value of n.
   Copy the diagram and fill in the missing numbers.

  *b What could x and y be?

18 Solve these equations by doing the same to both sides.

  a $5p + 6 = 26$ **b** $7 + \frac{y}{4} = 13$ **c** $4n + 9 + 2n = 21$ **d** $6m + 3 - 3m = 18$

  e $5x + 3 = 4x + 12$ **f** $7x - 8 = 3x + 4$ **g** $8x + 12 = 10x - 4$

9 Sequences and Functions

You need to know

✓ unknowns   page 169

✓ counting on and back   page 170

Key vocabulary

consecutive, continue, decrease, difference pattern, finite, function, function machine, general term, generate, increase, infinite, input, mapping, nth term, output, predict, relationship, rule, sequence, term

▶▶ Wheel of fortune

If we start at 0 and keep adding 4 we get this pattern.

  0, 4, 8, 12, 16, 20, 24, 28, 32, 36, 40, 44, ...

The pattern formed by the last digit of these numbers is

  0, 4, 8, 2, 6, 0, 4, 8, 2, 6, 0, 4, ...

If we join these last digits in order on a number wheel, we get the pattern shown.

What pattern do the last digits of these make on a number wheel?

1   Start at 0 and keep adding 2
2   Start at 0 and keep adding 6
3   Start at 0 and keep adding 3
4   Start at 0 and keep adding 7

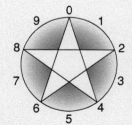

Keep writing down numbers until you get a repeating pattern.

Sequences

2, 4, 6, 8, 10, 12

This is a **number sequence**. It is a set of numbers in a given order.
Each number in the sequence is called a **term**.

2, 4, 6, 8, 10, 12

1st term
or term
number 1

4th term
or term
number 4

Terms next to each other are *consecutive* terms.
4 and 6 are consecutive terms. We usually put
commas between consecutive terms.

A sequence can have a **finite** or **infinite** number of terms.

The sequence of odd numbers, 1, 3, 5, 7, 9, ...
is **infinite**. It continues forever.

These dots tell us the sequence
continues forever.

The sequence of two-digit multiples of three, 12, 15, 18, 21, ..., 99
is **finite**. It stops at 99.

These dots tell us the sequence
continues up to 99.

Discussion

- 7 MN were asked to write examples of sequences found in everyday life.

I wrote down
the numbers of
the houses on
my side of
the street.
1, 3, 5, 7, 9 ...

I wrote down
the times the
sun rises
every day.
0649, 0647, 0645,
0644, 0642, 0640.

The rule for Paul's sequence is 'start at 1 and count on in steps of 2'.
Sirah's sequence has a more complex rule (one that is harder to find.)

Think of other sequences from everyday life. Think of some with simple rules and
some with a rule that is hard to find. **Discuss**.

I wrote down
the midday
temperatures
for May.
21°c, 24°c,
18°c, 19°c ...

I wrote down
the winning
numbers in
the school
raffle.
3, 59, 26, 83,
92, 36 ...

James' sequence has an irregular pattern and no rule.
Maisy's sequence has no rule and no pattern. It is called a set of random numbers.
Think of other sequences from everyday life with an irregular pattern. **Discuss**.
Think of other sequences from everyday life that are a set of random numbers.
Discuss.

- Is Paul's sequence finite or infinite? **Discuss**.
What about Sirah's, James' and Maisy's?

Counting on and counting back

Leslie started doing press-ups.
She wrote down the number she did each morning.

This sequence starts at 1 and increases by 1, 2, 3, 4, 5, ...

Worked Example
Write down the first six terms of these sequences.
a Start at 10 and count forwards by 1, 2, 3, 4, ...
b Start at 50 and count backwards by 2, 3, 4, 5, ...

Answer
a

The first six terms are **10, 11, 13, 16, 20, 25**.

b

The first six terms are **50, 48, 45, 41, 36, 30**.

Exercise 1

1 Write down the first six terms of these sequences.
   a Start at 5 and count on in steps of 0·5.
   b Start at 2 and count back in steps of 0·2.
   c Start at 5·5 and count on in steps of 1·5.
   d Start at ⁻2 and count back in steps of 2.
   e Start at 2 and count back in steps of 5.
   f Start at 0·3 and count on in steps of 0·6.
   g Start at ⁻3 and count back in steps of 5.
  * h Start at ⁻1·01 and count on in steps of 0·7.
  * i Start at 1·68 and count on in steps of 0·8.
  * j Start at 2 and count back in steps of 0·45.

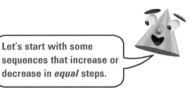
Let's start with some sequences that increase or decrease in *equal* steps.

2 In which of the sequences you made in question **1** are the terms ascending (getting bigger)?

3 After a run Patty took her pulse every 4 minutes. She took it the
   first time at 10·05 a.m. Write down the times of the next 5 times
   she took her pulse.

4 Write down the first six terms of these sequences.
   a Start at 0 and count forwards by 1, 2, 3, 4, ...
   b Start at 5 and count forwards by 2, 3, 4, 5, ...
   c Start at 100 and count backwards by 1, 3, 5, 7, ...
   d Start at 2 and count forwards by 2, 4, 6, 8, ...
  * e Start at 10 and count backwards by 2, 3, 4, 5, ...

5 Predict the next three terms of these sequences.
  a 6, 12, 18, 24, 30, ...   **b** 7, 12, 17, 22, 27, ... **c** 0·3, 0·6, 0·9, 1·2, 1·5, ...
  d 5, 5·2, 5·4, 5·6, 5·8, ... **e** 78, 70, 62, 54, 46, ... **f** 10, 5, 0, ⁻5, ⁻10, ...
  g 2, ⁻1, ⁻4, ⁻7, ⁻10, ..., **h** 4, 0, ⁻4, ⁻8, ... **i** ⁻64, ⁻55, ⁻46, ⁻37, ⁻28 ...
  j 0·8, 0·4, 0, ⁻0·4, ⁻0·8, ... **k** 1, 2, 4, 7, 11, 16, ... **l** 3, 5, 8, 12, 17, ...
  m 2, 3, 6, 11, 18, 27, ... ***n** 4·7, 5·9, 7·1, 8·3, ... ***o** ⁻0·24, ⁻0·47, ⁻0·7, ⁻0·93, ...
  ***p** ⁻2·16, ⁻0·46, 1·24, 2·94, ... ***q** 0·032, ⁻0·228, ⁻0·488, ⁻0·748, ...

6 Trent described the sequence in question **5a** as
'The sequence starts at 6 and increases in steps of 6.
Each term is a multiple of 6.' Describe the sequences in
question **5b, c, d** and **k**.

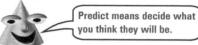

> Predict means decide what you think they will be.

7 Maryanne used her graphical calculator to display the sequence 0·2, 0·6, 1, 1·4, 1·8, 2·2.
She keyed

  [0·2] [EXE]
  [Ans] [+] [0·4]
  [EXE] [EXE] [EXE] [EXE] [EXE]

Maryanne then keyed [4] [EXE]
    [Ans] [+] [5]
    [EXE] [EXE] [EXE] [EXE] [EXE]

> You could key this into your calculator.

  a Describe the sequence Maryanne will get on her screen.
  b How many times would she have to press [EXE] after the [5] to get a number bigger
    than 50?

8 **a** Write down what you need to key on a graphical calculator to get a sequence which
    begins at 0·3 and goes up in steps of 0·3.
  b Without using the calculator, predict which of these numbers will be in the sequence.
    How can you tell?

    3·0 3·9 4·0 4·2 5·1

9 Jeff's calculator screen looked like this. He described the sequence as
'The sequence begins at 0·3 and increases in steps of 0·3.'
  a Describe the sequence that has terms 10 times larger than these.
  b Describe the sequence that has terms 100 times larger than these.

	0.3
Ans+0.3	
	0.6
	0.9
	1.2

10 A spider was crawling up the wall. Every minute it crawled 4 cm further than the minute
before. It crawled 5 cm in the first minute.
  a Write a sequence to show how far the spider crawled each minute.
  b How far did the spider crawl in the fifth minute?

11 Emily joined a running programme. Each week she increased the distance she ran by 1, 2,
3, 4, 5 ... kilometres.
In the first week she ran 1 kilometre.
  a Write down how far she ran each week during the first 6 weeks.
  b In which week did she run 29 kilometres?

*12 Ann cycled a total of 85 km over a period of 5 days.
Each day she cycled 5 km more than the day before.
  a Find a sequence for how far she cycled each day.
  b How many kilometres did Ann cycle on the first day?

Review 1 Write down the first six terms of these sequences.
a Start at 2 and count forwards by 1, 2, 3, 4, ...
b Start at 1 and count forwards by 3, 5, 7, 9, ...
c Start at 60 and count backwards by 2, 4, 6, 8, ...

Review 2 Write down the next three terms of these sequences.
a 3, 8, 13, 18, 23, ...   b 0·6, 1, 1·4, 1·8, 2·2, ...   c 2, ⁻6, ⁻14, ⁻22, ...

Review 3 Describe in words the sequence in Review **2a**.

Review 4 Predict if 3·9 will be a term of the sequence generated by this keying sequence.

[0·5] [EXE]
[Ans] [+] [0·3]
[EXE] [EXE] [EXE] [EXE] [EXE] ...

Review 5 Victoria decided to practise her goal shooting. Each practice, she shot 5 more goals than the practice before. On her first practice she shot 12 goals.
a Write down a sequence to show the number of goals Victoria shot during each practice.
b How many goals did she shoot on the sixth practice?

[T]

Practical

You will need a graphical calculator.
Use your graphical calculator to generate each of the sequences in question **5** of exercise **1**.

```
6
              6
Ans+6
              12
              18
```

Showing sequences with geometric patterns

The **numbers** in a sequence can be shown using diagrams.

Square Numbers

1   4   9   16   25

Triangular Numbers

1   3   6   10   15

Multiples of 3

3   6   9   12   15

Powers of 2

2   4   8   16

Discussion

Look at the diagrams in the screen on the previous page.

● The multiples of 3 have been arranged in rectangles of dots. Can multiples of any number be arranged in rectangles? **Discuss**.
  Try to arrange the multiples of 4, 5, 6, 7, ... in rectangles.

● How else could the triangular numbers be drawn using dots? **Discuss**.

● Why are the powers of two drawn as shown? **Discuss**.

● How could you draw dots to show this sequence? **Discuss**.
  2, 6, 12, 20, 30

Exercise 2

1 i Predict the next three terms of these sequences.
     a 8, 16, 24, ...   b 4, 9, 16, ...   c 121, 100, 81, ...
     d 10, 100, 1000, ...   e 128, 64, 32, ...   f 1, 3, 6, 10, ...
   ii For each of the sequences in **i** say whether it
     A ascends by equal steps   B ascends by unequal steps
     C descends by equal steps   D descends by unequal steps.
  * iii Describe the terms in the sequences in **i**. The answer to **a** is 'multiples of 8'.

Review Repeat parts **i**, **ii** and **iii** of question **1** for these sequences.
a 81, 72, 63, ...   b 36, 49, 64, ...   c 2, 4, 8, ...

Investigation

Pascal's Triangle

Row	Pascal's triangle	Sum of row
0	1	1
1	1 1	2
2	1 2 1	4
3	1 3 3 1	8
4	1 4 6 4 1	16

Copy these rows of Pascal's triangle.
Find the pattern and continue Pascal's Triangle up to at least row 15.

What pattern do the second numbers in each row give?
What pattern do the third numbers in each row give?
What is the sum of the second and third numbers in each row?
What pattern do the sums of rows give?
Which rows contain only odd numbers?
What else do you notice? **Investigate**.

> If you write down more rows you can investigate patterns of multiples.

Writing sequences from rules

A sequence can be written down if we know the first term and the rule for finding the next term. This is called a **term-to-term rule**.

Example   **1st term** 4   **rule for finding the next term** add 3
The sequence is   4, ₊₃ 7, ₊₃ 10, ₊₃ 13, ₊₃ 16, ₊₃ 19, ...

Each term is three more than the one before.

Example   **1st term** 1000   **term-to-term rule** halve
The sequence is   1000, ₊₂ 500, ₊₂ 250, ₊₂ 125, ₊₂ 62·5, ...

Each term is half the one before.

Exercise 3

1   The first term and the rule for finding the next term are given.
    Write down the first five terms of these sequences.
    You could use a graphical calculator if you wish.

   a   **first term** 2000   b   **first term** 100 000   c   **first term** 1
       rule halve   **rule** divide by 10   **rule** multiply by 4

   d   **first term** 50 000   e   **first term** 25   f   **first term** 0·1
       rule divide by 5   **rule** subtract 2·5   **rule** add 0·1

   g   **first term** 5   h   **first term** 1   i   **first term** 0·4
       rule subtract 0·25   **rule** add 101   **rule** subtract 0·05

   j   **first term** 2   k   **first term** 5
       rule multiply by 0·1   **rule** divide by ⁻0·1

2   Which of the sequences in question **1** increase or decrease by unequal steps?

3   We could describe the sequence generated in question **1a** as 'the first term is 2000 and
    each term is half of the one before'.
    Describe the sequences in Question **1b, c** and **e** in this way.

4   Some of the terms of these sequences are smudged.
    Find the missing terms.

   a   3, ▢, ▢, ▢, 19   **rule** add 4
   b   67, ▢, ▢, 46, ▢   **rule** subtract 7
   c   1600, ▢, ▢, ▢, 6·25   **rule** divide by 4
   d   1, ▢, 25, ▢ , ▢   **rule** multiply by 5
   e   2, ▢, ▢, ▢, 32, ▢   **rule** double
   f   1, ▢, ▢, ▢, 25, ▢   **rule** add 3, 5, 7, 9, ...

5 Write down the first six terms of these sequences.

	1st term	**term-to-term rule**
a	6	add 4
b	2	subtract 3
c	1	multiply by 2
d	3	multiply by ⁻1
e	0	add consecutive numbers starting with 1
f	100 000	divide by 10
g	1	divide by ⁻1
h	3	add consecutive odd numbers starting with 1
i	2	add consecutive even numbers starting with 2
j	100	subtract consecutive numbers starting with 1
k	1, 3	add the two previous terms
l	2, 3	add the two previous terms
*m	1	multiply the previous term by 2 then add 3
*n	1, ⁻5	add the two previous terms
*o	⁻3	multiply the previous term by ⁻1 then add 1·5

Add 1, 2, 3, 4, ...

6 Write down the first six terms of the sequences I am thinking of.
  a I am thinking of a sequence.   **b** I am thinking of a sequence.
  The 3rd term is 8.   The 4th term is 80.
  The rule is 'add 3'.   The rule is 'multiply by 2'.
  c I am thinking of a sequence.
  The 3rd term is 11.
  The rule is 'add consecutive odd numbers starting at 1'.

7 The rule for a sequence is 'to find the next term, add 4'.
  Two possible sequences with this rule are 1, 5, 9, 13, ... and 23, 27, 31, 35, ...
  a Write down another possible sequence with this rule.
  b Is it possible to find a sequence with the rule 'add 4' for which
    i all terms are multiples of 4   **ii** all terms are multiples of 8
    iii all terms are even numbers   **iv** all terms are negative numbers
    v none of the terms is whole number?
    If it is possible, give an example.

8 Mr Patel asked his class to write down a rule for a sequence.
  He then asked them to choose a first term and write down the first six terms.
  What rule and first term might these people have chosen?
  a Jessica's sequence has all even terms.
  b Barry's sequence has all odd terms.
  c The terms of Jake's sequence are all multiples of 4.
  d The terms of Amanda's sequence all end in 0.
  e The terms of Oliver's sequence all end in the digit 6.

*9 Mrs James put this sequence on the board.
    She asked her class to show how it continued.
    Penny wrote down 1, 2, 3, 4, 5, 6, ...
    Blair wrote down 1, 2, 3, 5, 8, 13, ...
  a Explain how Penny and Blair got their sequences.
  b How might the sequence 1, 2, 4, ... continue?
    Give more than one way.
  c How might the sequence 3, 8, 13, ... continue?
    Give more than one way.

Algebra

Review 1 Write down the first five terms of these sequences.
You may use a graphical calculator.
a **first term** 40 000 **rule** divide by 2   b **first term** 2   **rule** multiply by 4
c **first term** 1000 **rule** subtract 25   d **first term** ⁻7·5 **rule** add 1·2

Review 2 Find the smudged terms.
a 4, ⬭, ⬭, ⬭, 48 **rule** add 11   b 80, ⬭, ⬭, 10, ⬭, ⬭ **rule** halve
c 243, 81, ⬭, ⬭, ⬭ **rule** divide by 3   d 0·4, ⬭, ⬭, 0·1, ⬭ **rule** subtract 0·1

Review 3 Write down the first five terms of these sequences.

	1st term	term-to-term rule
a	400 000	divide by 5
b	⁻1	multiply by 2
c	1	add consecutive odd numbers starting with 1
d	1, 1	add the two previous terms
e	⁻4	multiply the previous term by ⁻2 then add 2·5.

Review 4
i The term-to-term rule for a sequence is *add 6*.
   What might the first term be if
   a all terms are multiples of 6   b all terms are odd
   c there are at least three negative terms?
ii The term-to-term rule for a sequence is *subtract 10*.
   What might the first term be if the terms of the sequence
   a are all multiples of 5   b all end in 5?

Investigations

Adding terms

What are the missing terms in these if the term-to-term rule is 'add the two previous terms'.

3, ____, 8, ____, 21
8, ____, ____, ____, 22
4 ____, ____, ____, 26
____, ____, ____, ____, 24

Is it possible to use the same rule as above to find the missing terms for these?
If so, write down the missing terms.

4, ____, ____, 9
10, ____, ____, ____, 24

Try fractions.

* New sequences from old

The sequence 1, 4, 7, 10, 13, ... has first term 1 and rule *add 3*.
We can generate a new sequence from this one using the term-to-term rule
'add each term to all the previous terms in the original sequence'.
The new sequence is   1, 5, 12, 22, 35,...

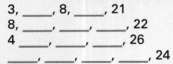

1+4   1+4+7   1+4+7+10   1+4+7+10+13

From 1, 5, 12, 22, 35, ... we can generate the sequence 2, 10, 24, 44, 70, ...
How was this done?

How is 8, 14, 20, 26 generated from 2, 10, 24, 44, 70, ...?

What other sequences could you generate from 1, 4, 7, 10, 13, ...?
What if you began with 2, 4, 6, 8 ... **or** 2, 4, 8, 16, **or** 1, 4, 9, 16, ... **or** 1, 1, 2, 3, 5, 8, ... **or** ...?

Writing sequences given the *n*th term

We can show the sequence 3, 4, 5, 6, ... in a table.

Term number, n	1	2	3	4	5	...
Term			3	4	5	6	7	...

The term-to-term rule is '**first term** 3   **rule** add 1'.
We can also see that each term is the term number, n, plus **2**.

> 1st term ($n = 1$) = $1 + 2 = 3$
> 2nd term ($n = 2$) = $2 + 2 = 4$
> 3rd term ($n = 3$) = $3 + 2 = 5$
> $\vdots$
> nth term $(n) = n + 2$

The **rule for the *n*th term** is $n + 2$.

We often call the first term T(1), the second term T(2), the third term T(3), ..., the *n*th term T(n).
For the sequence above $\mathbf{T}(n) = n + 2$.

Worked Example
The rule for the *n*th term of a sequence is T(n) = $4n - 1$.
a Write the first four terms of this sequence.
b Find T(24).

Answer
a   1st term ($n = 1$) = $4 \times 1 - 1$
                       = 3
   2nd term ($n = 2$) = $4 \times 2 - 1$
                       = 7
   3rd term ($n = 3$) = $4 \times 3 - 1$
                       = 11
   4th term ($n = 4$) = $4 \times 4 - 1$
                       = 15
   The first four terms are **3, 7, 11, 15**.
b To find the 24th term, substitute **24** into $4n - 1$.
       $4n - 1 = 4 \times 24 - 1$
               = 95
       T(24) = **95**

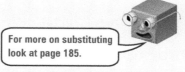

For more on substituting look at page 185.

Exercise 4

1 The *n*th term of a sequence is given. Write the first five terms.
   a $n + 1$ b $n - 1$ c $2n + 1$ d $3n$ e $2n - 0.5$
   f $20 - n$ g $3n - 4$ h $100 - 5n$ i $4n - 2$ j $80 - 3n$
   k $4 - 3n$ l $66 - 6n$ m $n + \frac{1}{2}$ n $n \times 0.2$

2 Write down T(20) for each of the sequences in question **1**.

3 Describe each of the sequences in question **1** in words.
   The answer to **1a** could be 'the integers greater than 1' or '**first term** 2   **rule** add 1'.

4

  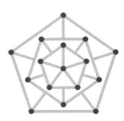

shape 1   **shape 2**   **shape 3**

The number of dots in the nth shape of this pattern is given by $5n + 1$.
Write down the number of dots in
a shape 5   b shape 10   c shape 20.

T

5 This table shows the sequence for $T(n) = 2n + 1$.
Use a copy of these tables.
Fill in the missing terms.

Term number	1	2	3	4	5	6	...	n
Term	3	5	7	9	11	13	...	$2n+1$

a

Term number	1	2	3	4	5	6	...	n
Term	0	1					...	$n-1$

b

Term number	1	2	3	4	5	6	...	n
Term	7	9					...	$2n+5$

c

Term number	1	2	3	4	5	6	...	n
Term	4						...	$3n+1$

d

Term number	1	2	3	4	5	6	...	n
Term	28						...	$30-2n$

T

6 a Use a copy of this table.
    Fill it in for the multiples of 3.
  b Use another copy.
    Fill it in for the multiples of 5.
 ∗**c** Use another copy.
    Fill it in for two more than the multiples of 4.

Term number	1	2	3	4	5	6	...	n
Term							...	

Review 1 The nth term of a sequence is given. Write the first five terms.
a $n + 6$ b $6n$ c $2n - 0{\cdot}5$ d $80 - 8n$ e $2n - 5$
f $n + 1\frac{1}{2}$ g $n \times 0{\cdot}3$

Review 2 Write down $T(25)$ for each of the sequences in Review **1**.

Review 3 Describe each of the sequences in Review **1** in words.

T

Review 4 Use a copy of this table.
Fill it in for the multiples of 4.

Term number	1	2	3	4	5	6	...	n
Term							...	

T

 Practical

You will need a spreadsheet and a graphical calculator.

1   4, 8, 12, 16, 20, ...

The terms of this sequence are the multiples of 4.
The rule for finding the next term is 'add 4'.
The nth term of the sequence is $4n$.

Using a spreadsheet we can find the first 20 terms of the sequence in two
ways.

Using the term-to-term rule

	A	B	C	D	E	
1	Term number	1	= B1+1	= C1+1	= D1+1	= E
2	Term	4	= B2+4	= C2+4	= D2+4	= E

Using the expression for the nth term

	A	B	C	D	E	
1	Term number	1	= B1+1	= C1+1	= D1+1	= E
2	Term	= B1*4	= C1*4	= D1*4	= E1*4	= F

Use a spreadsheet to find the first 20 multiples of these in two different ways.

| 5 | 7 | 11 | 15 | 24 | 29 |

Use a spreadsheet to find

the 24th multiple of 15
the 18th multiple of 24
the 100th multiple of 29.

2 Use a spreadsheet to find the first 20 terms of these sequences.

100, 95, 90, 85, 80, ...
7, 11, 15, 19, 23, ...

Use the term-to-term rule.

T 3 Janice was asked to generate this sequence on her graphical calculator.

100 000, 50 000, 25 000, 12 500, ...

She keyed (100 000) (EXE) (Ans) (÷) (2) (EXE) (EXE) (EXE) (EXE)

Use a graphical calculator to generate these sequences.

128, 64, 32, 16, ...
1, 4, 16, 64, ...
1, ⁻1, 1, ⁻1, 1, ⁻1, ...

Write down the rules that you used.

* 4 Use a spreadsheet to find the first 20 terms of a sequence with nth term $2n + 8$.
Now use the term-to-term rule to find the same sequence.

Investigation

Shuffles

Put 16 chairs in 4 rows of 4.
Seat 15 students in the chairs as shown.
Move the student in seat 1 to seat 16 in as few moves as possible.

1	2	3	4
5	6	7	8
9	10	11	12
	14	15	16

Follow these rules.
- The person in front of, behind or next to the empty chair may move to it.
- No one else may move.
- No one may move diagonally.

Each time a person moves this is counted as one move.

What is the fewest number of moves possible? **Investigate**.

What if the chairs were put in 5 rows of 5 and 24 students were seated.
Leave seat 21 empty and move the student in seat 1 to seat 25.
What if ...

Sequences in practical situations

Discussion

Jamie made this pattern with matchsticks.

1 square **2 squares** **3 squares** **4 squares**

Number of squares	1	2	3	4	...
Number of matchsticks	4	7	10	13	...

Jamie was asked how many matchsticks would be needed for 30 squares.

Is Jamie right?

How else could Jamie have worked it out?

Jamie worked out that for n squares the number of matchsticks needed is $n \times 3 + 1$ or $3n + 1$. Is he right?

Each added square needs 3 matchsticks but the first square needs an extra matchstick. That's $30 \times 3 + 1 = 91$.

Worked Example
Jasmine made these shapes with floor tiles.

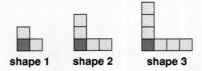

shape 1   **shape 2**   **shape 3**

Shape number	1	2	3	4	...
Number of tiles	3	5	7	?	...

a Draw shape 4. How many tiles are in this shape?

b Describe the pattern and how it continues.

c Explain how you can find the number of tiles in the nth shape.

d Write an expression for the number of tiles in the nth shape.

Look carefully at how the shapes are changing.

Answer

a   There are **9** tiles.

b **The first L-shape has 3 tiles.**
   Two more tiles are added each time the next shape is drawn.

c The first shape has two green tiles, the second four green tiles, the third six green tiles and so on. The number of green tiles in each shape is two times the shape number. Each shape also has one red tile. **The number of tiles altogether in each shape is 2 × shape number +1.**

d **$2n + 1$.**

Exercise 5

1   Ellen drew a fish. She made a pattern with the bubbles.

   group 1   group 2   group 3   group 4

a   Copy and complete this table.
b   Ellen drew 10 groups of bubbles altogether.
Explain how you could find the number of
bubbles in the 10th group without drawing the whole pattern.

Group of bubbles	1	2	3	4	5
Number of bubbles					

c   Explain how you could find the number of bubbles in the nth group.
d   Which of these expressions gives the number of bubbles in the nth group?
A $n+1$ B $n+2$ C $2n$ D $2n+1$

2   Vince's mowing has a V made of circles as its logo.

size 1   size 2   size 3

a   Copy and complete this table.
b   Describe the sequence made by the number of circles.

Size of V	1	2	3	4
Number of circles				

c   Will one of the Vs have 29 circles? Give a reason for your answer.
d   Explain how you would find the number of circles needed for the nth V.
e   Write an expression for the number of circles needed for the nth V.

3

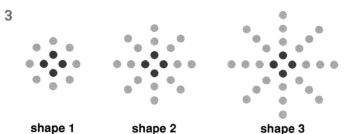

shape 1   shape 2   shape 3

Pritesh drew these shapes on her bedroom wall.
a   Draw shape 4.
Copy and complete this table.
b   Describe the sequence made by
the total number of circles.

Shape number	1	2	3	4	5
Total number of circles	12	20			

c   Explain how you could find the total number of circles needed for the nth shape.
d   Write an expression for the number of circles needed for the nth shape.
e   How many circles are needed for shape 20?
f   Does one of the shapes have 124 circles? Explain.

4   The Light Company makes a wall panel with tiny blue and yellow lights.

model 1   model 2   model 3

a   Copy and complete this table.

Model number	1	2	3	4
Number of blue lights				
Number of yellow lights				
Total number of lights				

225

b   Describe the sequence made by the number of
    i blue lights   ii yellow lights   iii lights in total.
c   Explain how you would find the number of blue lights in the model n.
d   Write an expression for the number of blue lights in model n.
e   Write an expression for the number of yellow lights in the nth model.
f   Add together the expressions you got in d and e.
*g Write an expression for the total number of lights in model n.
*h What do you notice about your answers to f and g?
    Explain why this is the case.

5   Naim made window hangers with coloured glass squares.

shape 1   shape 2   shape 3

a   How many red and how many orange squares will there be in shape 8?
    Explain how you worked this out.
b   Will one of the shapes have 29 orange squares? Give a reason for your answer.
c   Write an expression for the number of red glass squares in shape n. Explain how you
    found it.
d   Write an expression for the number of orange glass squares in shape n. Explain how
    you found it.
e   Add the expressions you got in c and d.
*f Write an expression for the total number of coloured glass squares in shape n. Explain
    how you found it.
*g What do you notice about your answers to e and f. Explain why this is the case.

6   Marjorie made these shapes for her bathroom wall with green and blue tiles.

shape 1   shape 2   shape 3

a   How many blue and how many green tiles will there be in shape 7?
    Explain how you worked this out.
b   Will one of the shapes have 81 blue tiles? Give a reason.
c   How many green tiles will there be in shape n?
*d Write an expression for the number of blue tiles in shape n.
*e Write an expression for the total number of tiles in shape n.

*7

1 hexagon   2 hexagons   3 hexagons

Julian made these patterns with matchsticks.
a   Find the missing numbers (?) in this table.

Number of hexagons	1	2	3	4	...	10	...	?	...	30
Perimeter	6	10	14	?	...	?	...	82	...	?

b   Julian had 87 matchsticks.
    Can he make a hexagon pattern like those above and have no matchsticks left over?

Review 1

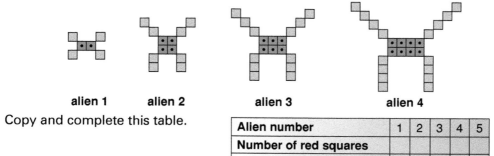

| Size 1 | Size 2 | Size 3 | Size 4 |

Size of cross	1	2	3	4	5
Number of squares					

a Copy and complete this table.
b Describe the sequence made by the number of squares.
c Will one of the crosses have 40 squares? Explain your answer.
d Explain how you could find the number of squares in cross size n.
e Write an expression for the number of squares in cross size n.

*Review 2   Melissa drew these 'alien' patterns for the cover of her space project.

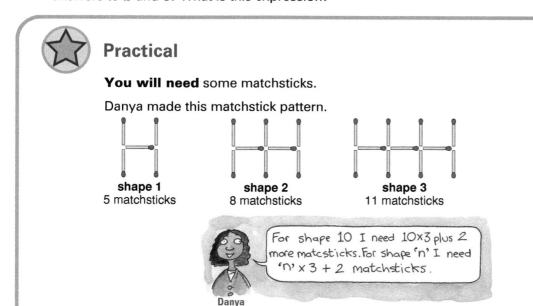

| alien 1 | alien 2 | alien 3 | alien 4 |

a Copy and complete this table.

Alien number	1	2	3	4	5
Number of red squares					
Number of green squares					
Total number of squares					

b Explain how you would find the number of red squares in alien n.
  Write an expression for this.
c Explain how you would find the number of green squares in alien n.
  Write an expression for this.
d How could you find an expression for the total number of squares in alien n from your answers to b and c? What is this expression?

★ Practical

You will need some matchsticks.

Danya made this matchstick pattern.

| shape 1 | shape 2 | shape 3 |
| 5 matchsticks | 8 matchsticks | 11 matchsticks |

> For shape 10 I need 10×3 plus 2 more matchsticks. For shape 'n' I need 'n' × 3 + 2 matchsticks.

Danya

Explore rows of other shapes made from matchsticks.
Find the rule for finding the number of matchsticks needed for the nth shape.

Finding the rule for the *n*th term

Discussion

● These tables show the difference patterns for the sequences $T(n) = 2n$ and $T(n) = 2n + 1$.

$T(n) = 2n$

Term number	1	2	3	4	5	...	n
T(n)	2	4	6	8	10	...	$2n$
Difference		2	2	2	2	...	

$T(n) = 2n + 1$

Term number	1	2	3	4	5	...	n
T(n)	3	5	7	9	11	...	$2n+1$
Difference		2	2	2	2	...	

Draw similar tables for these sequences.

$T(n) = 2n - 2$ $T(n) = 3n$ $T(n) = 3n - 2$ $T(n) = 3n + 4$

$T(n) = 4n$ $T(n) = 4n - 4$ $T(n) = 4n + 3$

What do you notice about the difference and the number multiplying *n*? **Discuss.**
What would the difference be if you drew tables for these sequences? **Discuss.**

$T(n) = 5n - 2$ $T(n) = 7n + 6$ $T(n) = 20n - 4$ $T(n) = 8n - 3$

$T(n) = 0·5n + 1$ $T(n) = an + b$

● Mr James asked his class if anyone could explain how to find the 20th term of
5, 9, 13, 17, 21, ...

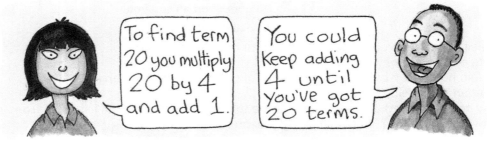

Rebecca   **Paul**

Are Rebecca and Paul both correct? **Discuss.**
Whose way would be quicker? **Discuss.**
How could Rebecca and Paul find the *n*th term? **Discuss.**

Rebecca worked out the *n*th term of some other sequences.
This is what she wrote. Explain how she got these answers. **Discuss.**

4   7   10   13   *n*th term is **3** times *n* plus 1 or $3n + 1$
  +3   +3   +3

8   13   18   23   *n*th term is **5** times *n* plus 3 or $5n + 3$.
  +5   +5   +5

To **find the rule for *n*th term**, draw a table to show the **difference pattern**.

Example   7, 11, 15, 19, 23, ...

Term number	1	2	3	4	5	...
T(*n*)	7	11	15	19	23	...
Difference		4	4	4	4	

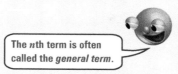

The *n*th term is often called the *general term*.

Each term is **4** times the term number, *n*, plus **?**.
The first term is 7.
This is $4 \times 1 + 3$.
Check that the other terms are $4 \times n + 3$.

Term number

$T(n)$

There is a *relationship* between the term number and the term.

The rule for the *n*th term is $4n + 3$.

Exercise 6

1 Write an expression for the *n*th term.
  a 2, 4, 6, 8, 10, ...   **b** 3, 6, 9, 12, 15, ...   **c** 6, 12, 18, 24, 30, ...
  d 6, 11, 16, 21, 26, ...   **e** 4, 9, 14, 19, 24, ...   **f** 3, 7, 11, 15, 19, ...
  g 0·3, 0·6, 0·9, 1·2, 1·5, ... **h** 0·5, 0·9, 1·3, 1·7, ...   **i** 110, 120, 130, 140, ...
  j 60, 75, 90, 105, 120, ...   **k** 1·5, 3·5, 5·5, 7·5, 9·5, ... *__l__ 100, 90, 80, 70, 60, ...
 *__m__ 30, 25, 20, 15, ...   *__n__ 40, 32, 24, 16, ...

Review   Write an expression for the *n*th term.
a 1, 3, 5, 7, 9, ...   **b** 4, 10, 16, 22, 28, ...   *__c__ 60, 50, 40, 30, 20, ...

Functions

We can find the **output** of a **function machine** if we are given the input.

Example

$7 \rightarrow$ [multiply by 4] $\rightarrow$ [add 1] $\rightarrow 29$ If the input is 7, the output is $7 \times 4 + 1 = 29$.

$x \rightarrow$ [multiply by 4] $\rightarrow$ [add 1] $\rightarrow y$ If the input is *x*, the output is $x \times 4 + 1 = y$ or $y = 4x + 1$.

Sometimes we show the inputs and outputs on a table.

Example

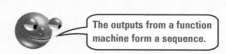

$x \rightarrow$ [add 3] $\rightarrow$ [multiply by 2] $\rightarrow y$

x	1	2	3	4	5	← inputs
y	8	10	12	14	16	← outputs

The outputs from a function machine form a sequence.

Algebra

Sometimes the inputs and outputs are written like this.

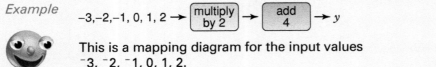

3, 7, 2, 4 → **add 3** → **multiply by 2** → 12, 20, 10, 14

First number to be input

First number to be output

Sometimes we show the inputs and outputs on a **mapping diagram**.

Example   −3,−2,−1, 0, 1, 2 → **multiply by 2** → **add 4** → y

This is a mapping diagram for the input values
⁻3, ⁻2, ⁻1, 0, 1, 2.

input →

output →

$$^-3 \times 2 + 4 = {}^-6 + 4 = {}^-2$$
$$^-2 \times 2 + 4 = {}^-4 + 4 = 0$$
$$^-1 \times 2 + 4 = {}^-2 + 4 = 2$$
$$0 \times 2 + 4 = 0 + 4 = 4$$
$$1 \times 2 + 4 = 2 + 4 = 6$$
$$2 \times 2 + 4 = 4 + 4 = 8$$

Exercise 7

1   Find the output for each of these if the input is 8.

a   x → **multiply by 2** → **add 1** → y

b   x → **multiply by 3** → **subtract 4** → y

c   x → **divide by 2** → **add 3** → y

d   x → **divide by 4** → **add 5** → y

e   x → **add 1** → **multiply by 2** → y

f   x → **subtract 5** → **multiply by 3** → y

2   Find the output for each of the function machines in question **1** if the input is
    i 12   **ii** ⁻4.

3   Describe, in words, what each of the function machines in question **1** does to x to get y.
    The answer to **a** is 'x **is multiplied by 2 and then 1 is added to get** y'.

4   Copy and fill in a table like this for each of these.

x	1	2	3	4	5
y					

a   x → **multiply by 2** → **subtract 1** → y

b   x → **add 3** → **multiply by 4** → y

c   x → **multiply by 3** → **add 6** → y

d   x → **subtract 1** → **multiply by 3** → y

5 Find the output for these.

a 3, 7, 8, 4.5 → | multiply by 2 | → | subtract 4 | → ___, ___, ___, ___

b 18, 6, ⁻6, 2.1 → | divide by 3 | → | add 1 | → ___, ___, ___, ___

c 0, ½, 6.5, ⁻3 → | add 4 | → | multiply by 2 | → ___, ___, ___, ___

d 7, 5½, 9.6, ⁻5 → | subtract 3 | → | divide by 2 | → ___, ___, ___, ___

T

6 Use a copy of these mapping diagrams.
Fill them in for the function machine and input given. **a** is started for you.

a 0, 1, 2, 3, 4 → | add 3 | → y

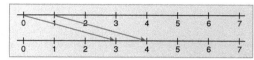

b 0, 1, 2, 3, 4 → | subtract 2 | → y

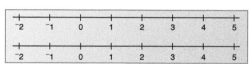

c 0, 1, 2, 3, 4 → | multiply by 2 | → | add 1 | → y

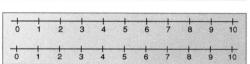

d 0, 1, 2, 3, 4 → | multiply by 2 | → | subtract 4 | → y

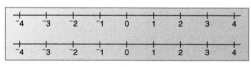

e ⁻4, ⁻3, ⁻2, ⁻1, 0 → | add 1 | → | multiply by 2 | → y

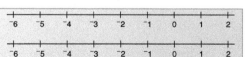

*7 Draw a mapping diagram for these function machines.

a 0, 1, 2, 3, 4 → | multiply by 3 | → | subtract 2 | → y   **b** ⁻2, ⁻1, 0, 1, 2 → | subtract 1 | → | multiply by 2 | → y

Review 1 Find the output for each of these if 4 is the input.

a x → | divide by 2 | → | add 3 | → y   **b** x → | multiply by 4 | → | subtract 2 | → y

c x → | subtract 1 | → | divide by 2 | → y   **d** x → | add 3 | → | multiply by 4 | → y

Review 2 Copy and fill in a table like this for the
function machines in Review **1b** and **d**.

x	1	2	3	4	5
y					

Review 3 Describe, in words, what each of the function machines in Review **1** does to x to
get y.

Review 4 Find the output.

$8, 6\frac{1}{2}, 0, ^-1 \rightarrow$ [subtract 4] $\rightarrow$ [multiply by 2] $\rightarrow$ ___, ___, ___, ___

T **Review 5** Use a copy of these mapping diagrams.
Fill them in for the function machine and input given.

a
$0, 1, 2, 3, 4 \rightarrow$ [multiply by 2] $\rightarrow$ [subtract 1] $\rightarrow y$

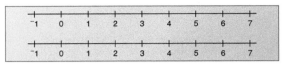

b
$^-3, ^-2, ^-1, 0, 1 \rightarrow$ [add 2] $\rightarrow$ [multiply by 2] $\rightarrow y$

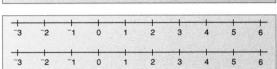

$x \rightarrow$ [multiply by 3] $\rightarrow$ [add 2] $\rightarrow y$

The rule for this function machine is 'multiply by 3 then add 2'.
We can write this as $x \times 3 + 2 = y$ or $y = 3x + 2$,
or using a mapping arrow ($\rightarrow$). $x \rightarrow x \times 3 + 2$ or $x \rightarrow 3x + 2$.
$y = 3x + 2$ and $x \rightarrow 3x + 2$ are both **functions**.

> We read this as x maps onto 3 times x plus 2.

Worked Example
a Write the rule for this function machine as $y =$ ___

$x \rightarrow$ [subtract 3] $\rightarrow$ [multiply by 4] $\rightarrow y$

b Write the rule for this function machine as $x \rightarrow$ ___

$x \rightarrow$ [divide by 2] $\rightarrow$ [add 4] $\rightarrow y$

Answer
a 3 is subtracted from x and the result is multiplied by 4.
$y = (x - 3) \times 4$ or $y = \mathbf{4(x - 3)}$
b x is divided by 2 and then 4 is added.
$x \rightarrow \frac{x}{2} + 4$

> We put a bracket around $x - 3$ because all of it must be multiplied by 4.

Exercise 8

1 Write the rules for these function machines as $y =$ ___.

a $x \rightarrow$ [multiply by 2] $\rightarrow$ [add 4] $\rightarrow y$

b $x \rightarrow$ [multiply by 8] $\rightarrow$ [subtract 7] $\rightarrow y$

c $x \rightarrow$ [divide by 4] $\rightarrow$ [subtract 3] $\rightarrow y$

d $x \rightarrow$ [add 3] $\rightarrow$ [multiply by 4] $\rightarrow y$

2 Write the rules for the function machines in question **1** as $x \rightarrow$ ___.

3 a

$x \longrightarrow$ [add 4] $\longrightarrow$ [multiply by 2] $\longrightarrow y$

Copy and fill in this table.

x	1	2	3	4	5
y					

Write a function for this function machine.

b The order of the operations is changed.

$x \longrightarrow$ [multiply by 2] $\longrightarrow$ [add 4] $\longrightarrow y$

Copy and fill in this table

x	1	2	3	4	5
y					

Write a function for this function machine.

c What happens when the order of the operations is changed?

*__**d** Marty said he had found a function machine which gave the same output when he changed the order of the operations. Is this possible? If so, give an example.

4

$x \longrightarrow$ [multiply by 4] $\longrightarrow$ [multiply by 2] $\longrightarrow y$

Mrs Patel asked her class to write the rule for this function machine as $x \longrightarrow$ ____.
Gemma wrote $x \longrightarrow x \times 4 \times 2$.
Anna wrote $x \longrightarrow 8x$.
Are both Gemma's and Anna's answers correct? Explain.

5 Write the function for these as $x \longrightarrow$ ____. Simplify your answer as much as possible.

a $x \longrightarrow$ [multiply by 4] $\longrightarrow$ [multiply by 7] $\longrightarrow y$

b $x \longrightarrow$ [add 4] $\longrightarrow$ [add 6] $\longrightarrow y$

c $x \longrightarrow$ [add 9] $\longrightarrow$ [subtract 4] $\longrightarrow$ [add 2] $\longrightarrow y$

d $x \longrightarrow$ [multiply by 8] $\longrightarrow$ [divide by 2] $\longrightarrow y$

6 A function machine changes the number n to the number $2n - 3$. $n \longrightarrow 2n - 3$.
What number will it change these to?
a 4   **b** 5   **c** 7   **d** 23   **e** 0   **f** ⁻1   **g** $\frac{1}{2}$ **h** $4\frac{1}{4}$

Review 1   Write the rules for these function machines as $y =$ ____.

a $x \longrightarrow$ [multiply by 4] $\longrightarrow$ [add 10] $\longrightarrow y$

b $x \longrightarrow$ [add 4] $\longrightarrow$ [multiply by 3] $\longrightarrow y$

Review 2   Write the rules for these as a mapping, $x \longrightarrow$ ____

a $x \longrightarrow$ [divide by 3] $\longrightarrow$ [add 7] $\longrightarrow y$

b $x \longrightarrow$ [subtract 8] $\longrightarrow$ [multiply by 4] $\longrightarrow y$

Review 3   A function machine changes the number p to $3p + 2$. $p \longrightarrow 3p + 2$.
If this number is input, what is the output?
a 2   **b** 6   **c** 10   **d** 21   **e** 0   **f** ⁻3   **g** $3\frac{1}{3}$

Review 4   Which function matches these function machines?

a $x \longrightarrow$ [multiply by 3] $\longrightarrow$ [multiply by 5] $\longrightarrow y$

b $x \longrightarrow$ [add 20] $\longrightarrow$ [subtract 15] $\longrightarrow y$

A $y = x + 15$
B $y = 6x$
C $y = 15x$
D $y = x + 5$

Review 5 $x \rightarrow$ [multiply by 3] $\rightarrow$ [add 2] $\rightarrow y$ **machine 1**

a   Write a function for this function machine.
b   Hugh changed the order of the operations and put 'add 2' first. He called this **machine 2**.
     He input 5 into both machines. Will the output be the same? Explain your answer.
* c   Hugh drew a function machine which did give the same
       output when he changed the order of the operations. $x \rightarrow$ [?] $\rightarrow$ [?] $\rightarrow y$
       What might Hugh have drawn?

Finding the function given the input and output

   ### Guess my rule – a game for a group

You will need pencil and paper and some copies of this function machine.

$x \rightarrow$ [   ] $\rightarrow$ [   ] $\rightarrow y$

To play
● Choose a leader.
● This person puts an operation in each box without showing anyone.
● Each person in the group takes turns to call out a number for x.
● The leader works out the value of y for each and tells this to the group.
● The first person to work out the rule is the new leader.

Example   Jill was the leader. She wrote $x \rightarrow$ [multiply by 3] $\rightarrow$ [add 2] $\rightarrow y$

The numbers called were 2, 0, 4, 10.
Jill told the group the answers were 8, 2, 14, 32.

Worked Example
Find the rule for these function machines.
a
     3, 5, 8, 2, $\rightarrow$ [?] $\rightarrow$ 15, 25, 40, 10
b
     4, 0, 7, 3 $\rightarrow$ [multiply by 2] $\rightarrow$ [?] $\rightarrow$ 11, 3, 17, 9
c
     5, 3, 2, 1 ,4, $\rightarrow$ [?] $\rightarrow$ [?] $\rightarrow$ 16, 10, 7, 4, 13

Answer
a   Each of the numbers has been multiplied by 5. $x \rightarrow$ **5x**
b   Put the input numbers in order first.

     0, 3, 4, 7 $\rightarrow$ [multiply by 2] $\rightarrow$ [?] $\rightarrow$ 3, 9, 11, 17

> Remember to change
> the order of the
> output as well.

Find the output from the first box.

     0, 3, 4, 7 $\rightarrow$ [multiply by 2] $\rightarrow$ 0, 6, 8, 14 $\rightarrow$ [?] $\rightarrow$ 3, 9, 11, 17

We can now see that the rule for the second box is 'add 3'.

     0, 3, 4, 7 $\rightarrow$ [multiply by 2] $\rightarrow$ [add 3] $\rightarrow$ 3, 9, 11, 17

The rule for the function machine is $x \rightarrow x \times 2 + 3$ or $x \rightarrow$ **2x + 3**.

c Put the numbers in order first.

1, 2, 3, 4, 5 → [?] → [?] → 4, 7, 10, 13, 16

Because the input numbers are consecutive we can find the rule using a difference table.

Input (x)	1	2	3	4	5
Output (y)	4	7	10	13	16
Difference		3	3	3	3

> Remember: from finding the nth term of a sequence that the difference is the number that multiplies n – see page 229.

This means the rule is $x \longrightarrow 3x + c$ where c is a number.
From the first number input, we see that $c = 1$.

1 → [multiply by 3] $\overset{3}{\rightarrow}$ [add 1] → 4

Check the other input values.

2 → [×3] $\overset{6}{\rightarrow}$ [+1] → 7 ✓   3 → [×3] $\overset{9}{\rightarrow}$ [+1] → 10 ✓

4 → [×3] $\overset{12}{\rightarrow}$ [+1] → 13 ✓   5 → [×3] $\overset{15}{\rightarrow}$ [+1] → 16 ✓

The rule is $x \longrightarrow 3x + 1$

Exercise 9

1 Write down the rule for each of these.

a 33, 52, 21, 103 → [?] → 27, 46, 15, 97

b 24, 16, 100, 10 → [?] → 6, 4, 25, 2.5

c 3, 7, 9, 0.5 → [?] → 15, 35, 45, 2.5

d 100, 35, 65, 45 → [?] → 80, 15, 45, 25

2 Find the missing operations. Write the rule as a mapping, $x \longrightarrow$ ___ .

a 3, 2, 1, 4 → [multiply by 2] → [?] → 5, 3, 1, 7,

b 0, 1, 5, 2 → [multiply by 3] → [?] → 1, 4, 16, 7

c 4, 1, 6, 2 → [multiply by 3] → [?] → 10, 1, 16, 4

d 10, 3, 8, 0 → [?] → [add 4] → 24, 10, 20, 4

3 Find the function for these.

a 3, 1, 4, 5, 2 → [?] → [?] → 7, 3, 9, 11, 5

b 5, 1, 3, 2, 4 → [?] → [?] → 9, 1, 5, 3, 7

c 1, 5, 3, 2, 4 → [?] → [?] → 5, 17, 11, 8, 14

d 2, 3, 5, 1, 4 → [?] → [?] → 10, 12, 16, 8, 14

e 2, 4, 3, 1, 5 → [?] → [?] → 3, 9, 6, 0, 12

*f 1, 4, 5, 3, 2 → [?] → [?] → 2.5, 4, 4.5, 3.5, 3

*g 4, 3, 5, 1, 2 → [?] → [?] → 1, 0.5, 1.5, −0.5, 0

Algebra

4 Find the function for these. You may need to use trial and improvement.

  a 1, 20, 3, 4 → [?] → [?] → 4, 99, 14, 19

  b 3, 10, 4, 6 → [?] → [?] → 18, 53, 23, 33

Trial and improvement means make a good guess then test it. If you are wrong, *improve* your guess and test again.

  c 1, 6, 8, 2 → [?] → [?] → 7, 27, 35, 11

  d 16, 20, 24, 7 → [?] → [?] → 9, 11, 13, 4.5

  e 12, 9, 15, 21 → [?] → [?] → 6, 5, 7, 9

5 a 4, 1, 8, 3 → [?] → [?] → 7, 4, 11, 6

Joel was asked to fill in the missing operations.
He wrote

  4, 1, 8, 3 → [+2] → [+1] → 7, 4, 11, 6

Jade said that the two operations could be replaced with one.

  4, 1, 8, 3 → [+3] → 7, 4, 11, 6

Is Jade correct?

  b What could go in the boxes?

  5, 3, 1, 4 → [?] → [?] → 30, 18, 6, 24

  c What single operation could replace the two operations in **b**?

  5, 3, 1, 4 → [?] → 30, 18, 6, 24

6 7TP was asked to find the rule for this function machine.

  5, 0, 3, 2 → [?] → [?] → 12, 2, 8, 6

Beth wrote $x \rightarrow 2x + 2$ 5, 0, 3, 2 → [×2] → [+2] → 12, 2, 8, 6

Samuel wrote $x \rightarrow 2(x + 1)$ 5, 0, 3, 2 → [+1] → [×2] → 12, 2, 8, 6

Who is correct? Explain your answer.

***7** Find two different ways of filling in the function machines.

  a 4, 2, 0, 3 → [?] → [?] → 15, 9, 3, 12

  b 5, 10, 1, 7 → [?] → [?] → 24, 44, 8, 32

Review 1   Find the function for these.

  a 24, 12, 18, 6 → [?] → 18, 6, 12, 0

  b 1, 3, 5, 4, 2 → [?] → [?] → 5, 13, 21, 17, 9

  c 1, 3, 5, 4, 2 → [?] → [?] → 1, 7, 13, 10, 4

  d 24, 8, 2, 9 → [?] → [?] → 19, 11, 8, 11.5

Review 2

a What could go in the boxes?

5, 1, 7, 4 → ☐ ? → ☐ ? → 40, 8, 56, 32

b What single operation could replace the two operations?

5, 1, 7, 4 → ☐ ? → 40, 8, 56, 32

* Review 3 Find two different ways of filling in the function machine.

8, 3, 2, 1 → ☐ ? → ☐ ? → 45, 20, 15, 10

Investigation

Magic squares

A is magic square. Each number in **A** has been mapped onto a new number in **B** using the mapping $x \rightarrow x + 4$.

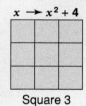

A

11	4	9
6	8	10
7	12	5

$x \rightarrow x + 4$

B

15	8	13
10	12	14
11	16	9

Copy these three diagrams.
Begin with the numbers in square **A** above and map each number onto a new number using the mapping given.

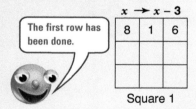

The first row has been done.

$x \rightarrow x - 3$

8	1	6

Square 1

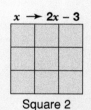

$x \rightarrow 2x - 3$

Square 2

$x \rightarrow x^2 + 4$

Square 3

Are squares 1, 2 and 3 magic squares?
What if the function is $x \rightarrow \frac{x}{2} + 1$ or $x \rightarrow 3x - 2$ or $x \rightarrow 2x^2 - 2$.
Is the new square always a magic square? **Investigate**.

Finding the input given the output

Remember
To 'undo' an operation we use the **inverse operation**.

The inverse of adding is subtracting and the inverse of subtracting is adding.
The inverse of multiplying is dividing and the inverse of dividing is multiplying.

Algebra

Worked Example

$$x \rightarrow \boxed{\begin{array}{c}\text{add} \\ 6\end{array}} \rightarrow y$$

What was the input if 8 is the output?

Answer
We find the input using inverse operations.
The inverse of 'add 6' is 'subtract 6'.
We draw an inverse function machine

$$x \leftarrow \boxed{\begin{array}{c}\text{subtract} \\ 6\end{array}} \leftarrow y$$

We use this to find the input

The directions of the arrows change.

$$2 \leftarrow \boxed{\begin{array}{c}\text{subtract} \\ 6\end{array}} \leftarrow 8$$ Begin with the output and work in the direction of the arrows.

2 was the input.

Worked Example

$$x \rightarrow \boxed{\begin{array}{c}\text{multiply} \\ \text{by 3}\end{array}} \rightarrow \boxed{\begin{array}{c}\text{add} \\ 6\end{array}} \rightarrow y$$

What was the input if 21 is the output?

Answer
Draw the inverse function machines.

$$x \leftarrow \boxed{\begin{array}{c}\text{divide} \\ \text{by 3}\end{array}} \leftarrow \boxed{\begin{array}{c}\text{subtract} \\ 6\end{array}} \leftarrow y$$

The inverse of multiply by 3 is divide by 3.
The inverse of add 6 is subtract 6.

Use the inverse function machine to find the input.

$$5 \leftarrow \boxed{\begin{array}{c}\text{divide} \\ \text{by 3}\end{array}} \overset{15}{\leftarrow} \boxed{\begin{array}{c}\text{subtract} \\ 6\end{array}} \leftarrow 21$$

$21 - 6 = 15$
$15 \div 3 = 5$

5 was the input.   start with the output and work in the direction of the arrows

Worked Example
Find the inverse function of $x \rightarrow 2x + 3$.

Answer

$$x \rightarrow \boxed{\begin{array}{c}\text{multiply} \\ \text{by 2}\end{array}} \rightarrow 2x \rightarrow \boxed{\begin{array}{c}\text{add} \\ 3\end{array}} \rightarrow 2x + 3$$

The inverse is found by working backwards doing the inverse operations.

$$\frac{x-3}{2} \leftarrow \boxed{\begin{array}{c}\text{divide} \\ \text{by 2}\end{array}} \leftarrow x-3 \leftarrow \boxed{\begin{array}{c}\text{subtract} \\ 3\end{array}} \leftarrow x$$

Start with x.

Note that $x - 3$ is all divided by 2
The inverse function is $x \rightarrow \dfrac{x-3}{2}$

Exercise 10

1 Find the input for these.

a
—, —, —, — → [add 5] → 12, 20, 36, 72

b
—, —, —, — → [subtract 4] → 7, 24, 42, 63

c
—, —, —, — → [multiply by 4] → 28, 40, 100, 240

d
—, —, —, — → [divide by 3] → 9, 4, 6, 0

Draw an inverse function machine to help.

2 Find the input.
Draw the inverse function machines to help.

a
— → [multiply by 2] → [add 3] → 13

b
— → [subtract 2] → [multiply by 6] → 60

c
— → [divide by 2] → [add 5] → 13

d
— → [multiply by 3] → [add 8] → 23

e
— → [divide by 6] → [subtract 4] → 5

3 Find the input for these.

a
—, —, — → [multiply by 3] → [add 4] → 34, 7, 16

b
—, —, — → [divide by 4] → [add 3] → 13, 6, 9

c
—, —, — → [add 8] → [divide by 2] → 6, 10, 7.5

d
—, —, — → [multiply by 4] → [subtract 7] → 5, 9, 33

e
—, —, — → [divide by 2] → [add 5] → 9, 15, 55

f
—, —, — → [subtract 3] → [multiply by 5] → 10, 20, 45

4 A function machine changes the number n to the number $3n + 2$.
What numbers must be input to get these output numbers?
a 11   b 5   c 20   d 62   e ⁻19   f $6\frac{1}{2}$ *g ⁻5

5 What are the missing operations in these?

a
1, 2, 3, 4, 5 → [add 2] → [?] → 1, 2, 3, 4, 5

b
1, 2, 3, 4, 5 → [subtract 6] → [?] → 1, 2, 3, 4, 5

c
1, 2, 3, 4, 5 → [multiply by 2] → [?] → 1, 2, 3, 4, 5

d
1, 2, 3, 4, 5 → [divide by 6] → [?] → 1, 2, 3, 4, 5

6 i
3, 4, 8, 12 → [add 4] → [multiply by 2] → [ ] → [ ] → 3, 4, 8, 12

The numbers have stayed the same.
Which of these gives the missing operations?

A 3, 4, 8, 12 → [add 4] → [multiply by 2] → [divide by 2] → [subtract 4] → 3, 4, 8, 12

B 3, 4, 8, 12 → [add 4] → [multiply by 2] → [subtract 4] → [divide by 2] → 3, 4, 8, 12

ii What are the missing operations in these?
Check to make sure you have the operations in the right order.

a   8, 4, 7, 2 → [multiply by 3] → [add 2] → [ ] → [ ] → 8, 4, 7, 2

b   9, 6, 12, 18 → [divide by 3] → [subtract 4] → [ ] → [ ] → 9, 6, 12, 18

c   4, 8, 1, 7 → [add 3] → [divide by 2] → [ ] → [ ] → 4, 8, 1, 7

d   7, 8, 4, 2 → [subtract 6] → [multiply by 3] → [ ] → [ ] → 7, 8, 4, 2

7 In a game, orange numbers were found by multiplying a green number by 2 then adding 1. Carrie wrote this function.

green number × *2 + 1* = *orange number*

To find a green number from an orange numbers the function must be inverted.

green number = (*orange number* − *1*) ÷ *2*

Invert these functions to show how to get a green number from an orange numbers.

a (green number − 2) × 3 = orange numbers
b green number × 4 − 2 = orange numbers
c green number ÷ 2 + 3 = orange numbers

8 Find the inverse function of these.

a $x \longrightarrow 2x + 1$ **b** $x \longrightarrow 3x + 2$ **c** $x \longrightarrow 3x - 1$ **d** $x \longrightarrow \frac{x}{2} + 2$
e $x \longrightarrow 2(x + 3)$ **f** $x \longrightarrow 2(x - 1)$ **g** $x \longrightarrow 4(x - 3)$ **h** $x \longrightarrow \frac{x + 3}{2}$
i $x \longrightarrow \frac{x - 1}{3}$ *j** $x \longrightarrow x^2 + 2$ *k** $x \longrightarrow x^2 - 3$

Review 1   Write down the input for each of these.

a ___ , ___ , ___ → | subtract 7 | → 16, 24, 33   **b** ___ , ___ , ___ → | divide by 4 | → 3, 8, 1

c ___ , ___ , ___ → | add 8 | → 23, 42, 106   **d** ___ , ___ , ___ → | multiply by 5 | → 25, 45, 100

Review 2   Find the input.
Draw inverse function machines to help.

a ___ → | multiply by 2 | → | add 7 | → 19   **b** ___ → | subtract 3 | → | divide by 4 | → 9

c ___ , ___ , ___ → | subtract 8 | → | multiply by 3 | → 15, 21, 30

d ___ , ___ , ___ → | divide by 2 | → | add 4 | → 12, 54, 40

Review 3   What are the missing operations?

a 5, 1, 6, 10 → | add 4 | → | ? | → 5, 1, 6, 10

b 3, 10, 7, 0 → | multiply by 3 | → | add 2 | → | ? | → | ? | → 3, 10, 7, 0

c 8, 9, 4, 3 → | divide by 4 | → | add 1 | → | ? | → | ? | → 8, 9, 4, 3

d 5, 14, 7, 24 → | subtract 4 | → | multiply by 3 | → | ? | → | ? | → 5, 14, 7, 24

Review 4   Find the inverse function for these.

a $x \longrightarrow 2x - 3$ **b** $x \longrightarrow \frac{x}{4} + 1$ **c** $x \longrightarrow 2(x - 4)$ **d** $x \longrightarrow \frac{x - 2}{3}$

Summary of key points

A A **sequence** is a set of numbers in a given order.

Each number is called a **term**.

Examples 8, 15, 22, 29, 36, ... These three dots mean the sequence continues forever. It is *infinite*.

↑ 1st term   ↑ 5th term

4, 8, 12, 16, ... 100 This sequence is *finite*. It starts at 4 and ends at 100.

B We can **write a sequence** by **counting on** or **counting back**.

Examples Starting at 7 and counting forwards by steps of 1, 2, 3, 4, ... we get

7, 8, 10, 13, 17, ...

This sequence ascends in unequal steps.

C We can **write a sequence** if we know the **first term** and the **rule for finding the next term** (term-to-term rule).

Example **1st term** ⁻6   **term-to-term rule** add 2

⁻6, ⁻4, ⁻2, 0, 2, 4, ...

Each term is 2 more than the one before.

D We can **write a sequence** if we know the **rule for the nth term**.

Example If the rule for the nth term is $T(n) = 3n - 1$, the sequence is

2, 5, 8, 11, 14, 17, 20, ...

$T(n)$ means the nth term.

E **Sequences in practical situations**

shape 1   shape 2   shape 3   shape 4

Shape number	1	2	3	4
Number of squares	4	7	10	13

In the nth shape there are $3n + 1$ squares.

There are 3 arms on each shape. The nth shape has $3 \times n$ squares on the arms plus one in the middle.

F We can find a **rule for the nth term** by drawing a difference table.

Example For the sequence 5, 8, 11, 14, 17, ...

Term number	1	2	3	4	5
$T(n)$	5	8	11	14	17
Difference		3	3	3	3

The difference is the number that multiplies x.

Each term is **3** times the term number, n, plus 2.

$T(n) = 3n + 2$

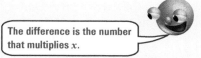

241

G This is a **function machine**. $x \rightarrow$ [add 3] $\rightarrow$ [multiply by 2] $\rightarrow y$

If 5 is the input, the output is $(5 + 3) \times 2 = 16$.

$5 \rightarrow$ [add 3] $\xrightarrow{8}$ [multiply by 2] $\rightarrow 16$

We can write the **function** for this machine as $y = (x + 3) \times 2$ or $y = 2(x + 3)$.

It can also be written as a **mapping**, $x \rightarrow 2(x + 3)$.

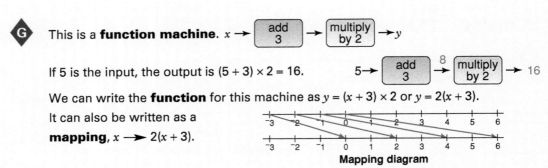

Mapping diagram

H If we are given the input and output we can **find the rule** for the machine.

Example $3, 5, 10, 20 \rightarrow$ [ ] $\rightarrow 9, 15, 30, 60$

The rule for this machine is 'multiply by 3'.

I We use **inverse operations** to find the input if we are given the output.

Example $\rightarrow$ [multiply by 3] $\rightarrow$ [subtract 4] $\rightarrow 20$

We draw an inverse function machine.

$8 \leftarrow$ [divide by 3] $\xleftarrow{24}$ [add 4] $\leftarrow 20$ **Start with the output.**

The input was **8**.

J To find the **inverse function** work backwards doing inverse operations.

Example $x \rightarrow 3x - 4$

$x \rightarrow$ [multiply by 3] $\rightarrow 3x \rightarrow$ [subtract 4] $\rightarrow 3x - 4$

To find the inverse, draw the inverse function machine.

$\dfrac{x+4}{3} \leftarrow$ [divide by 3] $\leftarrow x+4 \leftarrow$ [add 4] $\leftarrow x$ **Start with x.**

The inverse function of $x \rightarrow 3x - 4$ is $x \rightarrow \dfrac{x+4}{3}$.

Test yourself

1 Write down the first six terms of these sequences.   **A B**
   a Start at 1 and count back in steps of 3.
   b Start at 5 and count forwards by 1, 2, 3, 4, ...

2 The Patel family went on a touring holiday.   **B**
Each day they travelled 10 fewer kilometres than the day before.
On the first day they travelled 140 km.
   a Write down a sequence to show the daily distance travelled.
   b How far did the Patel family travel on the sixth day?

3 a Write down the next three terms of these sequences.
  i 25, 36, 49, 64, ...   **ii** 256, 128, 64, ...   **iii** 1, 3, 6, 10, ...
  b For each of the sequences in **a** say whether it
   A ascends by equal steps   **B** ascends by unequal steps
   C descends by equal steps   **D** descends by unequal steps.

4 Find the missing terms.
  a 5000, ☐, ☐, 40, ☐   **term-to-term rule** divide by 5
  b 3, ☐, ☐, 12, ☐   **term-to-term rule** add consecutive numbers starting with 2
  c 1, 2, ☐, ☐, 8, 13, ☐   **term-to-term rule** add the two previous terms

5 Write down a possible first term and the term-to-term rule for a sequence which has
  a all odd terms   **b** all terms multiples of 20   **c** all terms ending in the digit 9.

6 The nth term of a sequence is given. Find the first five terms.
  a $n + 5$ **b** $2n - 3$ **c** $20 - 2n$ **d** $n \times 0.4$

7 Write down $T(15)$ of each of the sequences in question **6**.

8 Use a copy of this table. Fill it in for the multiples of 6.

Term number	1	2	3	4	5	6	...	n
T(n)	6						...	

9

Bruce made these 'bridge' shapes with blocks.

Bridge 1   **Bridge 2**   **Bridge 3**

  a Draw bridge 4. How many blocks are in this shape?
  b Copy and complete this table

Bridge number	1	2	3	4	5
Number of blocks					

  c Describe the sequence made by the number of blocks.
  d Explain how you would find the number of blocks in the nth bridge.
  e Write an expression for the nth term.
  f Will one of the bridges have 38 blocks? Give a reason for your answer.

10 Write an expression for the nth term of these.
  a 1, 3, 5, 7, 9, ...   **b** 100, 118, 136, 154, 172, ...

11 Find the output for these.
  a 5, 8, ⁻2, 2.5 → [multiply by 3] → [add 4] → ___, ___, ___, ___
  b 9, 17, 2, 5.6 → [subtract 4] → [multiply by 2] → ___, ___, ___, ___

12 Use a copy of this. Fill in the table for this function machine.

x → [multiply by 2] → [add 7] → y

x	1	2	3	4	5
y					

13 Use a copy of this mapping diagram.
Fill it in for this function machine.

$^-3, ^-2, ^-1, 0, 1, 2 \rightarrow$ [add 2] $\rightarrow$ [multiply by 2] $\rightarrow$

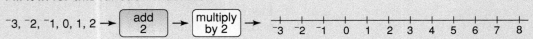

14 Copy and complete the rules for these functions machines.

a $x \rightarrow$ [multiply by 3] $\rightarrow$ [add 4] $\rightarrow$

$x \rightarrow$ _____

b $x \rightarrow$ [subtract 3] $\rightarrow$ [multiply by 4] $\rightarrow y$

$y =$ _____

c $x \rightarrow$ [divide by 4] $\rightarrow$ [add 3] $\rightarrow y$

$y =$ _____

d $x \rightarrow$ [add 2] $\rightarrow$ [multiply by 5] $\rightarrow$

$x \rightarrow$ _____

15

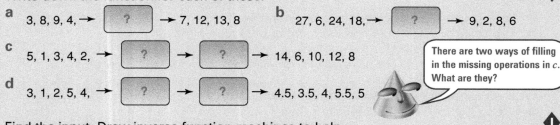

Gemma   **Ricky**

a The rule for Gemma's function machine is $x \rightarrow 2x - 4$. $x \rightarrow$ [multiply by 2] $\rightarrow$ [?] $\rightarrow y$
Copy and complete Gemma's function machine.

b Copy and complete Ricky's function machine. $x \rightarrow$ [?] $\rightarrow$ [?] $\rightarrow y$

c If 10 was put into both Gemma's and Ricky's function machines would they both give
the same output? Explain.

16 Write down the function for each of these.

a $3, 8, 9, 4, \rightarrow$ [?] $\rightarrow 7, 12, 13, 8$

b $27, 6, 24, 18, \rightarrow$ [?] $\rightarrow 9, 2, 8, 6$

c $5, 1, 3, 4, 2, \rightarrow$ [?] $\rightarrow$ [?] $\rightarrow 14, 6, 10, 12, 8$

d $3, 1, 2, 5, 4, \rightarrow$ [?] $\rightarrow$ [?] $\rightarrow 4.5, 3.5, 4, 5.5, 5$

> There are two ways of filling in the missing operations in c. What are they?

17 Find the input. Draw inverse function machines to help.

a ___, ___ $\rightarrow$ [multiply by 3] $\rightarrow$ [add 4] $\rightarrow 22, 34$

b ___, ___ $\rightarrow$ [divide by 3] $\rightarrow$ [subtract 5] $\rightarrow 5, 3$

18 A function machine changes n to the number $3n - 4$.
a What output would the function machine give if $^-2$ was the input?
b What number must be input to get 0·5 as the output?

19 What are the missing operations?

$7, 2, 1, 6 \rightarrow$ [add 7] $\rightarrow$ [divide by 2] $\rightarrow$ [?] $\rightarrow$ [?] $\rightarrow 7, 2, 1, 6$

20 Find the inverse function of these.
a $x \rightarrow 3x - 2$ **b** $x \rightarrow \frac{x}{4} - 7$ **c** $x \rightarrow 3(x + 4)$ **d** $x \rightarrow 5(x - 3)$
e $x \rightarrow \frac{x - 5}{3}$ **f** $x \rightarrow \frac{x + 2}{5}$ ***g** $x \rightarrow x^2 + 4$

10 Graphs

You need to know

✓ coordinates page 273

Key vocabulary

axes, axis, coordinate pair, coordinate point, coordinates, equation (of a graph), graph, intersect, origin, relationship, straight-line graph, *x*-axis, *x*-coordinate, *y*-axis, *y*-coordinate

Working out the plot

1   These two 'pictures' were made by joining the dots at (2, 2) (2, 10) (10, 2) (10, 10) (4, 4) (4, 8) (8, 4) (8, 8) in different ways.
What other 'pictures' can be made by joining these points?
You will need to draw some copies of the grids and plot the given points.

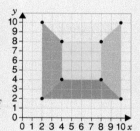 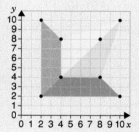

2

Gareth drew this picture of some of the buildings in his town. He plotted one point on this grid for each building.
Match each building with a point on the grid.

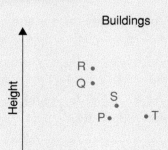

Buildings

Coordinate pairs

(1,3) is a **coordinate pair**

x-coordinate   *y*-coordinate

We say 'the coordinate pairs *satisfy* the rule $y = x + 2$.

(1,3)   (2,4)   (3,5)
+2   +2   +2

In each of these coordinate pairs, the **y-coordinate** is two more than the **x-coordinate**.
The rule is $y = x + 2$.

Worked Example
These coordinate pairs satisfy the rule $y = 12 - 2x$.
Fill in the missing numbers.
(2, ___), (1, ___), (0, ___), ($^-$1, ___)

Remember the order of operation rules.

Answer
We substitute the given *x*-coordinates into $y = 12 - x$.
When $x = 2, y = 12 - 2 \times 2$ $x = 1, y = 12 - 2 \times 1$ $x = 0, y = 12 - 2 \times 0$ $x = {}^-1, y = 12 - 2 \times {}^-1$
                    $= 12 - 4$ $= 12 - 2$ $= 12 - 0$ $= 12 - {}^-2$
                    $= 8$ $= 10$ $= 12$ $= 14$

We now fill in the missing numbers.
(2, **8**), (1, **10**), (0, **12**), ($^-$1, **14**)

We can list the coordinate pairs in a table.

It is easier to substitute for the positive values first. You might be able to find the negative ones by looking for a pattern.

Example

x	2	1	0	$^-$1
y	8	10	12	14

Exercise 1

1 Copy and complete these coordinate pairs so they satisfy the rule $y = 3x$.
   a (1, ___) **b** (3, ___) **c** (5, ___) **d** (___, 21) **e** (___, 12)

2 Copy and complete these coordinate pairs so they satisfy the given rule.
   a $y = x + 1$ (2, ___), (1, ___), (0, ___), ($^-$1, ___), ($^-$2, ___)
   b $y = 10 - x$ (2, ___), (1, ___), (0, ___), ($^-$1, ___) ($^-$2, ___)
   c $y = x$ (3, ___), (2, ___), (1, ___), (0, ___), ($^-$1, ___), ($^-$2 ___), ($^-$3 ___)
   d $y = 2x$ ($^-$2, ___), ($^-$1, ___), (0, ___), (1, ___), (2, ___)
   e $y = 2x + 5$ (3, ___)(2, ___), (1, ___), (0, ___), ($^-$1, ___), ($^-$2, ___)
   f $y = 3x - 3$ (6, ___), (4, ___), (0, ___), ($^-$1, ___), ($^-$3, ___), ($^-$6, ___)
   g $y = 10 - 2x$ (3, ___), (2, ___), (1, ___), (0, ___), ($^-$1, ___), ($^-$2, ___), ($^-$3, ___)
   h $y = 20 - 4x$ (3, ___), (2, ___), (1, ___), (0, ___), ($^-$1, ___), ($^-$2, ___), ($^-$3, ___)

3 Copy and complete these tables of coordinate pairs for the given rules,

a $y = 2x - 1$

x	-2	-1	0	1	2
y					

b $y = 20 - 3x$

x	-1	4	5	10
y				

c $y = \frac{1}{2}x$

x	-4	-2	0	2	4
y					

4 The coordinate pairs (3, 6), (4, 8), (10, 20) all satisfy the rule $y = 2x$. Does (8, 16) satisfy the rule? Explain your answer.

5 Which of these coordinate pairs satisfy the rule $y = 2x + 4$?
    a (1, 6)   **b** (10, 6)   **c** (20, 44)   **d** (-4, 0)   **e** (-1, 2)   **f** $(\frac{1}{2}, 5)$

6 Which of these coordinate pairs satisfy the rule $y = 2x - 3$?
    a (2, 1)   **b** (4, 1)   **c** (10, 23)   **d** (6, 9)   **e** $(\frac{1}{2}, -2)$

Review 1 Copy and complete these coordinate pairs so they satisfy the given rule.
a $y = x + 4$ (0, ___), (1, ___), (2, ___), (3, ___), (4, ___)
b $y = 2x - 4$ (6, ___)(4, ___), (2, ___), (0, ___), (-2, ___), (-4, ___), (-6, ___)
c $y = 6 - 5x$ (-2, ___), (-1, ___), (0, ___), (1, ___), (2, ___)

Review 2 Copy and complete these tables of coordinate pairs so they satisfy the given rules.

a $y = 5x + 1$

x	1	2	3	4	5
y					

b $y = 5 - 3x$

x	-2	-1	0	1	2
y					

c $y = \frac{1}{2}x - 2$

x	-4	-2	0	2	4
y					

Review 3 Which of these coordinate pairs satisfy the rule $y = 3x - 5$?
a (3, -5)   **b** (0, -5)   **c** (-1, -8)   **d** (3, 4)   **e** $(\frac{1}{2}, -3\frac{1}{2})$

Drawing straight-line graphs

Coordinate pairs can be **plotted** on a grid.

Worked Example
The values on this table satisfy the rule $y = 2x + 1$.
a Write down the coordinate pairs.
b Plot the **coordinate points** on a grid.
    Draw a straight line through the points.
    Label the line $y = 2x + 1$.
c Choose another point on the line.
    Write down the coordinate pair of your point.
    Check that this coordinate pair satisfies the rule $y = 2x + 1$.
d Does $y = 2x + 1$ go through the point (-4, -7)?

x	-2	0	2
y	-3	1	5

Answer

a (⁻2, ⁻3), (0, 1), (2, 5)

b The coordinate points are shown plotted.
   A straight line is drawn through them.

c Choose a point on the line, say (1, 3).
   If $x = 1$, $y = 2 \times 1 + 1$
   $= 2 + 1$
   $= 3$
   (1, 3) does satisfy the rule $y = 2x + 1$.

d Check that (⁻4, ⁻7) satisfies $y = 2x + 1$
   If $x = ⁻4$ $y = 2 \times ⁻4 + 1$
   $= ⁻8 + 1$
   $= ⁻7$
   $y = 2x + 1$ **does go through the point** (⁻4, ⁻7).

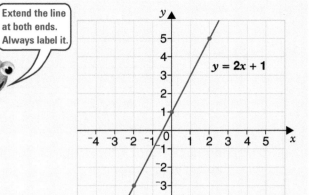

Extend the line at both ends. Always label it.

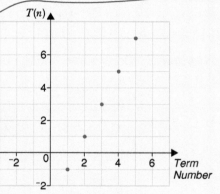

m and *c* stand for numbers.

In the above worked example, $y = 2x + 1$ is the **equation of the line**.
The coordinate pairs of *all* points on the line satisfy the equation of the line.
The coordinate pairs of points that do not lie on the line do not satisfy the equation of the line.
A graph with equation of the form $y = mx + c$ is always a straight line.

If we graph a **linear sequence**, we get a set of points in an 'imagined' straight line.

A linear sequence ascends or descends in equal steps.

Example

Term Number	1	2	3	4	5
$T(n)$	⁻1	1	3	5	7

It is impossible to have a term number of $1\frac{1}{2}$ or $2\frac{1}{4}$ or 3·7, ... so we do not draw a line through the points.

Exercise 2

1 a $y = x + 1$
      Copy and complete the table of values for this rule.

x	⁻3	⁻1	0	2
y	⁻2			

   b Copy and complete this sentence.
      Four points on the line $y = x + 1$ are (⁻3, ⁻2), (⁻1, __), (0, __), (2, __).

   c Use a copy of this grid.
      On the grid, plot these four points.
      Draw and label the line $y = x + 1$

Draw your line carefully.

   d Does the point (⁻2, ⁻1) lie on this line?
   e Does $y = x + 1$ go through the point (⁻3, ⁻4)?
   f Will the point ($\frac{1}{2}$, $1\frac{1}{2}$) lie on this line?
   g Will the point (18, 17) lie on this line? Explain.

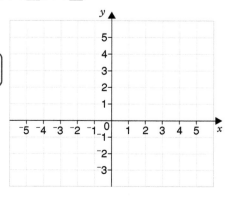

2 **a** (0, __), (2, __) (4, __)
   Copy and complete these coordinate pairs for the
   rule $y = 1 - x$.
   b Use a copy of this grid.
   Draw and label the line with equation $y = 1 - x$.
   c Which of these points lie on the line?
   (3, 4), (0, 1) (4, ⁻3)
   d Will $y = 1 - x$ go through the point (7, ⁻6)?
   e Choose two other points which lie on the line.
   Write down the coordinate pairs for each. Do these
   coordinate pairs satisfy the rule $y = 1 - x$?

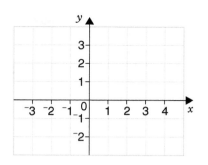

3 **a** Copy and complete this table for the equation
   $y = 2x - 2$.

x	⁻3	⁻2	⁻1	1	2
y					

   b Write down the coordinate pairs from the table.
   c Plot the points on a grid with x values from ⁻4 to 4 and y values from ⁻10 to 4.
   d Does $y = 2x - 2$ go through the point $(⁻\frac{1}{2}, 2)$? Explain.
   e Will the point (24, 46) lie on this line? Explain.
   f $y = 2x - 2$ goes through each of these points.
   What are the missing coordinates?
   (⁻4, ___), (___, ⁻8), (⁻0·5, ___), (___, ⁻4·5)

4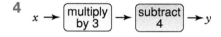

x	0	1	2	3
y				

There is more about
function machines
on page 229.

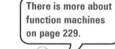

   a Copy and complete this table for the rule given by the function
   machine.
   b Write down the coordinate pairs from the table.
   c On a grid, plot the four points.
   Draw a straight line through them.
   d Will the point $(\frac{1}{2}, ⁻2\frac{1}{2})$ lie on the line?

5

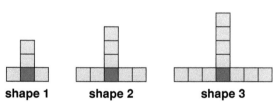

shape 1   shape 2   shape 3

   a Copy and complete this table.

Shape number	1	2	3	4
Number of shapes	5	8		

   b Use a copy of this grid.
   On your grid, plot the coordinate
   pairs from the table.
   The first pair is plotted for you.
   Do the points lie in a straight line?

Algebra

6 The nth term of a sequence is given by $T(n) = 2n - 3$.

  a Copy and complete this table for the sequence.

  b Plot the sequence on a grid.

  c Is 9 a term of the sequence? Explain.

  d Copy and complete this table for the equation $y = 2x - 3$.

  e Draw and label the graph of $y = 2x - 3$.

  f Explain why we do not draw a line through the points plotted in **b** but we do through the points plotted in **e**.

Term number	1	2	3	4	5
$T(n)$					

x	1	3	5
y			

T

Review 1

a Copy and complete the table for $y = 2x - 2$.

b Write down the coordinate pairs for the points.

c Use a copy of this grid.
On the grid, plot the three points.
Draw and label the line with equation $y = 2x - 2$.

d Write down the coordinate pairs for two other points that lie on the line. Do the coordinate pairs satisfy the rule $y = 2x - 2$?

e Explain how you can tell from the graph that the coordinate pair (3, 4) does satisfy the rule $y = 2x - 2$.

f Does $y = 2x - 2$ go through the point (18, ⁻16)?

g Will the point $(4\frac{1}{2}, 7)$ lie on the line? Explain.

x	⁻1	0	2
y			

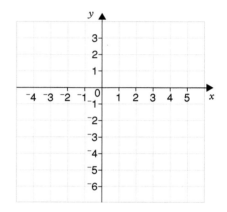

Review 2

This table shows the sequence for the number of matchsticks in a shape.

a Plot the sequence on a grid.
Do the points lie in a straight line?

b Is 22 a term of the sequence? Explain.

c Would it be sensible to join the points with a straight line? Explain.

Shape number	1	2	3	4	5
Number of matchsticks	2	4	6	8	10

Practical

You will need a graphical calculator.
Draw three straight lines on your calculator screen through each of these points.

  a (0, 3)   **b** (0, ⁻2)   **c** (0, ⁻4)   **d** (2, 2)   **e** (⁻4, ⁻4)

  * Draw a line on your calculator screen that goes through these points.

    a (⁻4, 0) and (0, 4)   **b** (⁻2, 0) and (0, 4)   **c** (0, ⁻10) and (10, 0)

Investigation

Families of graphs

You will need graph paper or a graph plotter software package or a graphical calculator.

T

1 a Use a copy of these tables. Fill them in.

$y = x$

x	‾4	0	4
y			

$y = 2x$

x	‾3	0	3
y			

$y = 4x$

x	‾2	0	2
y			

Draw x and y axes with x-values from ‾5 to 5 and y-values from ‾12 to 12.
On this set of axes draw and label the graphs of $y = x$, $y = 2x$ and $y = 4x$.
Describe what happens to the graph as the number multiplying x increases.
Which point do all the graphs have in common?

b Draw tables like the ones above for $y = 3x$ and $y = 5x$.
On the same set of axes, draw and label the graphs of
$y = 3x$ and $y = 5x$.
Before you draw them, predict what they will look like.

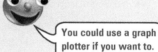

You could use a graph plotter if you want to.

T

2 a Use a copy of these tables. Fill them in.

$y = x + 1$

x	‾3	0	3
y			

$y = x + 4$

x	‾6	0	3
y			

$y = x + 5$

x	‾2	0	2
y			

Draw x-and y-axes with x-values from ‾8 to 8 and y-values from ‾4 to 10.
On this set of axes draw and label the graphs of $y = x + 1$, $y = x + 4$, $y = x + 5$.
What do you notice about the slopes of these graphs?

b Look at the graphs you drew in **a**.
What do you notice about where each graph intersects the
y-axis and its equation?

Intersects means crosses.

c Draw tables like the ones above for $y = x + 3$ and $y = x + 7$.
On the same set of axes, draw and label the graphs of $y = x + 3$ and $y = x + 7$.
Before you draw them, predict what they will look like.

T

3 Repeat **2a** and **2b** but replace the tables and equations with

$y = 8 - x$

x	0	2	5
y			

$y = 5 - x$

x	0	1	4
y			

$y = 4 - x$

x	0	1	3
y			

4 For each of the following, draw up a table of values for x and y.
Choose x-values between ‾4 and 4.
Plot the points on a grid. Draw and label the line for each.

$y = 2x$ $y = 2x + 1$, $y = 2x + 5$, $y = 2x - 3$, $y = 2x - 1$

What do you notice about the equation of the line and
a the slopes of the lines **b** where the lines cut the y-axis?

*5 The graph of a straight line is given by $y = mx + c$.
What does m represent? What does c represent?
Predict what the graphs of these would look like.

You could check using a graph plotter.

$y = 3x - 2$ $y = 4x + 2$ $y = x - 4$ $y = 10 - x$ $y = 5 - 2x$

$y = mx + c$ is the equation of a straight line.

m represents the **gradient** or slope. The gradient is a measure of the steepness of a line. The greater the value of m the steeper the line. Lines with the same value of m are parallel.

Examples $y = 2x + 3$ and $y = 2x - 4$ are parallel.
$y = 4x + 2$ has a steeper slope than $y = 2x + 4$.

If m is positive, the gradient is positive.
If m is negative, the gradient is negative.

Examples $y = 3x + 2$ has a **positive** gradient.
$y = 5 - 2x$ has a **negative** gradient.

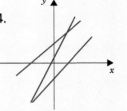

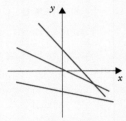

These lines have a
positive gradient
(m is positive)

These lines have a
negative gradient
(m is negative)

In $y = mx + c$, c tells us where the straight line crosses the y-axis.
This is often called the y-**intercept**.
The y-intercept has coordinates $(0, c)$ where c is given by the equation.

Examples $y = 4x - 3$ crosses the y-axis at $(0, {}^-3)$
$y = 10 - 2x$ crosses the y-axis at $(0, \mathbf{10})$

*c is often called
the constant.*

Exercise 3

1   Which graph will have the steepest slope? Explain why.
   A $y = 2x + 4$ **B** $y = 3x - 2$ **C** $y = \frac{1}{2}x + 6$ **D** $y = 4x - 1$

2   Will these have a positive or negative slope?
   a $y = x + 3$ **b** $y = 2x$ **c** $y = x - 4$ **d** $y = 6 - x$
   e $y = 12 - 7x$ **f** $y = 2x + 7$ **g** $y = 5 - 4x$

3   Which of these graphs will cut the y-axis at $(0, 3)$?
   A $y = x$ **B** $y = x - 3$ **C** $y = 3 - x$ **D** $y = 3x + 2$

4   Which of these graphs will cut the y-axis at $(0, {}^-5)$?
   A $y = -5x$ **B** $y = 5 - x$ **C** $y = x - 5$ **D** $y = x + 5$

5   Will $y = 2x - 4$ have a steeper slope than $y = x$?

6   Which number in these equations represents the gradient?
   a $y = 2x$ **b** $y = 3x$ **c** $y = 3x + 2$ **d** $y = \frac{1}{2}x - 4$

   e $y = \frac{1}{2}x$ **f** $y = \frac{1}{3}x + 2$ **g** $y = 4 - 2x$ **h** $y = 3 - 5x$

   i $y = 2x - 1\cdot5$ **j** $y = {}^-3x$ **k** $y = \dfrac{x - 4}{2}$

7   Where do the lines in question **6** cross the y-axis?

8   Which two of these lines are parallel?
      $y = 3 - 2x$ $y = 2x + 3$ $y = 3x + 2$ $y = 2x - 3$ $y = 2 - 3x$

9   Sketch the graph of $y = 2x - 1$ by looking at the equation. Do not plot points.

Review 1 Choose from the box.

$$y = 2x + 1 \qquad y = 3 - x \qquad y = x - 3 \qquad y = \tfrac{1}{2}x - 2 \qquad y = 3x + 1 \qquad y = -x$$

Which graph(s) will   **a**   have the second steepest slope
                               b   have a negative slope
                               c   cut the y-axis at $(0, {}^-3)$
                          *d**   be parallel to $y = 3 - x$?

Review 2 Write down the gradient of each of these lines.
a $y = 20 - 2x$ **b** $y = 3x - 1$ **c** $y = {}^-4x$ **d** $y = 2\tfrac{1}{2} + \tfrac{1}{2}x$

Review 3 Write down the coordinates of the y-intercept for each of the lines in Review **2**.

Lines parallel to the x- and y-axes

$y = a$ is the equation of a straight line **parallel to the x-axis**.
It cuts the y-axis at a. The x-coordinate can be any value at all.

Examples $y = 3$ is parallel to the x-axis and cuts the y-axis at 3.
               $y = {}^-5$ is parallel to the x-axis and cuts the y-axis at ${}^-5$.

$x = b$ is the equation of a straight line **parallel to the y-axis**.
It cuts the x-axis at b. The y coordinate can be any value at all.

Examples $x = 2$ is parallel to the y-axis and cuts the x-axis at 2.
               $y = {}^-2$ is parallel to the y-axis and cuts the y-axis at ${}^-2$.

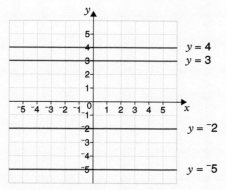

lines parallel to the x-axis

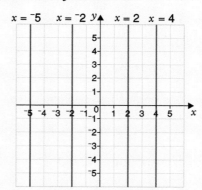

lines parallel to the y-axis

Exercise 4

1   Draw a grid with x- and y-values from ${}^-6$ to 6.
On your grid, draw and label these graphs.
    a $x = 4$ **b** $y = {}^-2$ **c** $x = {}^-4$
    d $y = 2$ **e** $x = 1 \cdot 5$ **f** $y = {}^-3 \cdot 5$

2   Write down the equation of a line which is
    a   parallel to the y-axis   **b**   parallel to the x-axis.

3 Draw a grid with x- and y-values from $^-6$ to 6.
Copy and complete this table of values for $y = x - 2$.
Draw the graph of $y = x - 2$ on your grid.
Draw the graph of $y = 4$ on the same grid.
Write down the coordinates of the point where the two graphs meet.

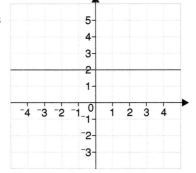

x	$^-2$	0	1	2	4
y					

*4 Draw the graphs of $y = 2x - 1$ and $y = ^-3$ on the same grid.
Write down the coordinates of the point where the two graphs meet.

*

5 The equation of any straight line can be written in the form
$y = mx + c$.
Write the equation of the line on this grid in the form $y = mx + c$.

Review 1   Which of these graphs will be lines
a parallel to the x-axis
b parallel to the y-axis?

$y = 4$ $x = 3$ $y = -1\frac{1}{2}$ $x = 2\frac{1}{2}$ $y = -4$ $x = -3$

Review 2   Draw a grid with x- and y-values from $^-6$ to 6.
On your grid draw the graphs of $y = x - 4$ and $y = 3$.
Write down the coordinates of the point where the two graphs intersect.

Reading real-life graphs

Worked Example
Mr Taylor's class sat a test.
It was out of 40.
He drew this graph to convert the marks to
percentages.
a What percentage does each small division
on the vertical axis represent?
b Kyle got 32 out of 40. What percentage is
this?
c Todd got 52%. What mark out of 40 did
Todd get?

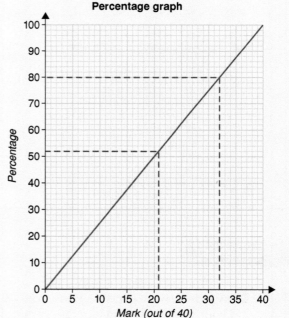

Answer
a 5 small divisions represent 10%.
One small division must represent $\frac{10\%}{5} = $ **2%**.
b Find 32 on the horizontal axis.
Draw a dotted line from 32 vertically up to
the graph.
From there draw a horizontal dotted line out
to the other axis.
The percentage can then be read. It is **80%**.
c Find 52% on the vertical axis.
Draw a horizontal dotted line from 52 to the graph.
From there, draw a vertical dotted line down to the other axis.
The mark out of 40 can then be read. It is about **21**.

Exercise 5

1 Janet drew this graph to show the charge for swimming lessons.
   a Do the values between the points have any meaning? Explain.
   b How much does it cost for the first lesson?
   c How much does it cost for
      i 2 lessons   ii 4 lessons?

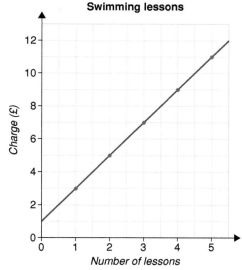

Swimming lessons

2

Energy conversions

   a How many Calories does each small division on the horizontal axis represent?
   b How many kilojoules does each small division on the vertical axis represent?
   c Convert these to Calories.
      i 50 kJ   ii 150 kJ   iii 210 kJ   iv 320 kJ
   d Convert these to kilojoules.
      i 60 Calories   ii 74 Calories   iii 46 Calories   iv 92 Calories

   Read the answer to the nearest division.

   e Allaf ate a pear (250 kJ) and a banana (350 kJ).
     How many Calories were in these altogether?
   f Debbie ate a piece of cheese (50 Calories) and a sandwich (100 Calories).
     How many kilojoules were in these altogether?
   g Jake is on a special diet. He must eat 420 kJ for breakfast and 350 kJ for lunch.
     How many Calories is this?
  *h A milkshake has 800 Calories. How many kilojoules is this?

3 This graph gives the length of two types of candles as they burn.

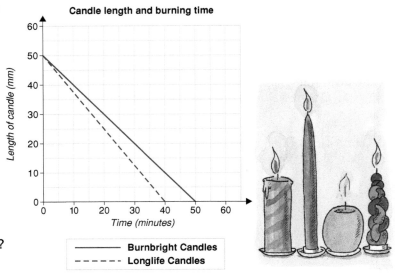

Candle length and burning time

a Sarah burned a Burnbrite candle for 20 minutes. What length would it be after that?

b Mrs Chen burned a Longlife candle for 40 minutes. What length would it be after that?

c Jeff burned a candle for 30 minutes. It was then 20 mm long. What sort of candle was it?

d Peter burned a Longlife candle until it was 25 mm long. For how long did he burn it?

——— Burnbright Candles
– – – – Longlife Candles

4 This graph gives the minimum distance that should be between cars at different speeds.

a Ed is driving in bad weather at 40 miles per hour. What is the minimum distance he should be from the car in front?

b Fay is driving in good weather at 55 miles per hour. What is the minimum distance she should be from the car in front?

c Mr Shaw is driving 70 metres behind another car. The weather is bad. What is the maximum speed at which he should be driving?

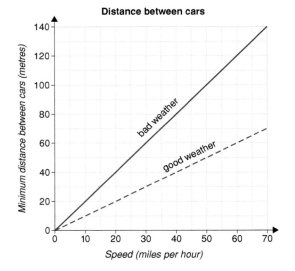

Distance between cars

5 Andrea heated a liquid, then let it cool. While it was cooling, she took the temperature every minute.
Andrea drew this graph.

a What was the temperature after 1 minute?

b After how many minutes was the temperature 24°C?

c To what temperature did Andrea heat the liquid?

d How much did the temperature drop in the first 5 minutes of cooling?

e What do you estimate the temperature to have been after $2\frac{1}{2}$ minutes?

*f After how many minutes do you estimate the temperature was 50°C?

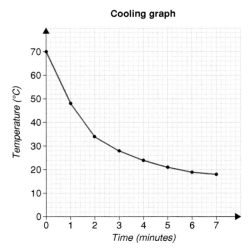

Cooling graph

*6 Sandy and Suzie ran a 1500 metre race.
This distance-time graph shows the race.

a What goes in the gaps?
Up until 900 m _____ was in the lead.
At 900 m Sandy and Suzie were equal.
For the next _____ minutes. _____ was in the lead.
At _____ m they were equal again.

b Who won the race?
c What was her time?
d How much longer did the other girl take?

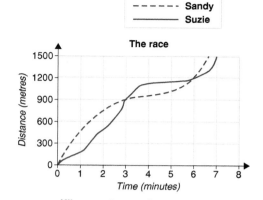

Review 1

a Work out what each small division on each of the axes represents?
b Convert these to pounds.
  i 35 kg   ii 10 kg
c Convert these to kilograms.
  i 60 lb   ii 100 lb
d Anita weighed 55 kg.
  How many lb is this?
e Anita stands on the scales holding her cat. The cat weighs 10 lb.
  What do the scales read, in kg?

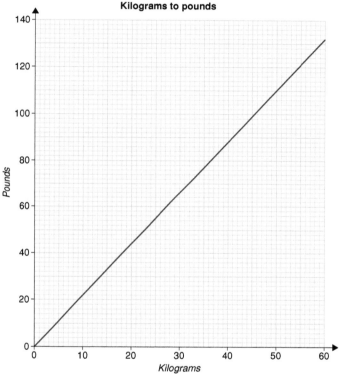

Kilograms to pounds

Review 2 This graph shows how much petrol two cars use.

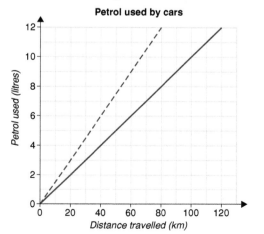

Petrol used by cars

a Tom's car went 60 km.
  How many litres of petrol did it use?
b Kim's car went 100 km.
  How many litres of petrol did it use?
c Tom's car used 6 litres of petrol one day.
  How far did it go?
d How much more petrol does Tom's car use than Kim's car when each car travels 60 km?

Review 3 At the end of each week Laura measured the height of a plant. This is the graph Laura drew.

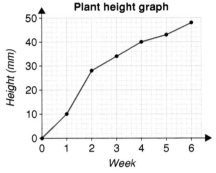

a How tall was this plant at the end of the 3rd week?
b How much did this plant grow in 6 weeks?
c How long did it take for the plant to be 40 mm tall?
d How many millimetres did this plant grow in the 2nd week?
e Estimate how tall the plant was after $4\frac{1}{2}$ weeks.
*f Estimate how long it took for the plant to be 35 mm tall.

Plotting real-life graphs

Pierre comes from France. In France all the road signs are in kilometres.
Pierre wants to draw a graph so he can quickly convert miles to kilometres.
He knows that to convert miles to kilometres you multiply by about 1·6.
He wrote this formula. **kilometres = miles × 1·6**
He chose some values for miles and used the formula to fill in this table.
He plots the points on a grid and draws a line though them.

miles	0	25	50
km	0	40	80

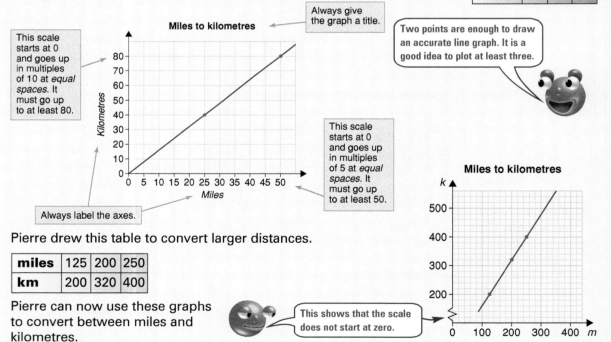

Always give the graph a title.

This scale starts at 0 and goes up in multiples of 10 at *equal spaces*. It must go up to at least 80.

Two points are enough to draw an accurate line graph. It is a good idea to plot at least three.

This scale starts at 0 and goes up in multiples of 5 at *equal spaces*. It must go up to at least 50.

Always label the axes.

Pierre drew this table to convert larger distances.

miles	125	200	250
km	200	320	400

Pierre can now use these graphs to convert between miles and kilometres.

This shows that the scale does not start at zero.

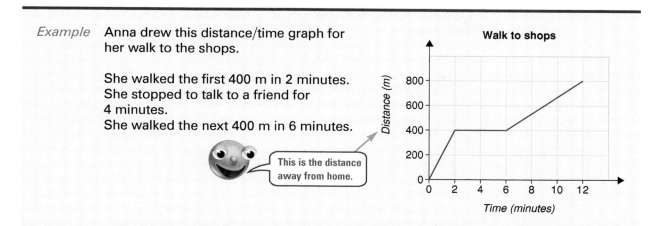

Example Anna drew this distance/time graph for her walk to the shops.

She walked the first 400 m in 2 minutes.
She stopped to talk to a friend for 4 minutes.
She walked the next 400 m in 6 minutes.

This is the distance away from home.

Discussion

Discuss how to draw the distance/time graph for Peter's journey.

Going home is like going backwards.

Peter left home and drove the 5 km to his friend's house in 5 minutes. He stayed at his friends house for 8 minutes.
He then drove the 5 km home in 8 minutes.

Exercise 6

T

1 Maria wants to draw a graph to convert pounds to kilograms. She knows that 5 kg is about 11 lb and that 20 kg is about 44 lb.

a Copy and fill in this table.

kg	0	5	20
lb	0		

b Use a copy of this grid.
   Plot the points from the table on it. Draw a straight line through the points. Label the axes and give the graph a title.

c Maria bought a 40 lb bag of chaf for her horse. Estimate how many kg this is.

d Maria paid £8·25 for a bag of pony nuts at 55p per kilogram. Estimate how many lb of nuts were in the bag.

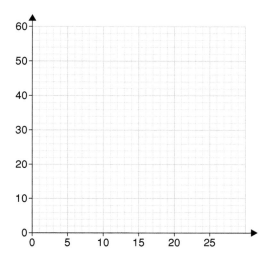

Algebra

2 Jones Electrical charge a call out fee of £25 and then an hourly rate of £15.

   a Copy and complete this table.

Hours	0	4	12
Charge (£)	25		

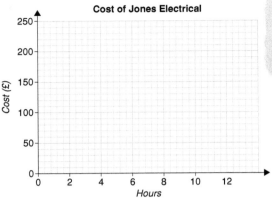

   b Use a copy of this grid. Plot the points from the table on the grid.

   c Draw a straight line through the points.

   d Use the graph to estimate the charge for 9 hours work.

   e Mrs Cassidy was charged £175 for a job. Estimate from your graph how many hours it took.

Cost of Jones Electrical

(graph with vertical axis Cost (£) marked 0, 50, 100, 150, 200, 250 and horizontal axis Hours marked 0, 2, 4, 6, 8, 10, 12)

3 Trevor is going on a holiday to New Zealand. He uses this exchange rate to convert pounds to NZ dollars.   £1 = NZ$3.

Trevor wants to draw a graph to help him convert from £0 to £50 to New Zealand dollars.

   a Copy and complete this table.

£	10		50
NZ$			

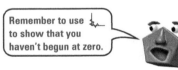

Choose a suitable value for here.

   b Draw a set of axes. Have pounds on the horizontal axis and NZ$ on the vertical axis.

   c Draw and label your graph.

   d Draw another graph to convert from £200 to £300 to NZ$. Do not begin at (0, 0).

Remember to use ⟿ to show that you haven't begun at zero.

4 $d = 4 + 0.5s$ gives the stopping distance d, in metres, for a car travelling at a speed s, in km/h, on a particular road.

   a Copy and complete this table.

s	10	50	100
d		29	

   b Copy and complete these coordinate pairs (10, ___), (50, 29), (100, ___).

   c Draw the graph that relates the stopping distance to the speed.

   d Use your graph to estimate the distance a car, travelling at 70 km/h, takes to stop.

   e It took 40 metres for Jenny to stop, from the time she put her foot on the brake. Use your graph to estimate the speed at which Jenny was travelling.

5 A mobile phone company charges £15 per month plus 15p per minute for peak time calls.

   a Write a formula for the cost of calls, C, at 15 pence per minute, m.

   $C =$ _____

   b Draw a graph to show the cost of between 0 and 200 minutes of peak time calls. Have time on the horizontal axis and cost on the vertical axis.

   c Last month Gerrard used 180 peak time minutes. Estimate the charge for this.

T **6** Use a copy of this grid.
Andrea took her niece, Lucy, for a walk.
They walked 800 m to the playground. It
took 5 minutes. They stayed at the
playground for 15 minutes. They then
walked on to the shops which are 1600 m
from home. This took 5 minutes.
They stayed at the shops for 5 minutes.
They then walked home in 10 minutes.
Finish the graph for Andrea's and
Lucy's walk.

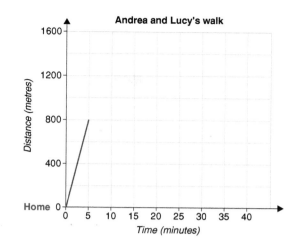

T **7** Use a copy of this grid.
Draw a distance/time graph for Daniel's journey.
Daniel left home and drove 40 km in half an hour.
He then stopped for $\frac{1}{2}$ hour to have morning tea.
He drove 100 km in the next hour and a half.
He realised he had missed his friend's house so
he drove back 40 km. This took him $\frac{1}{2}$ hour.

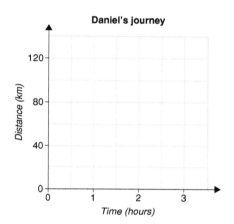

Review 1   A shop hires stereos. The cost is £50 plus £5 for every hour.
a Copy and complete this table.
b Use a copy of this grid.
   Choose a suitable scale for the
   vertical axis.
   Plot the points from the table on the grid.
   Draw a straight line through the points.
   Label the vertical axis and give the graph a title.
c Raewyn hired a stereo for 9 hours. Use your graph to estimate the
   cost of this.

Hours	0	4	8
Charge (£)			

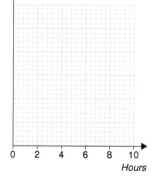

Review 2 $a = 2 \cdot 5 \, h$ gives an approximate relationship between area in hectares, h, and
acres, a.
a Copy and complete this table.

Area in hectares	2	6	10
Area in acres			25

b Draw a graph to convert between acres and hectares.
   Have area in hectares on the horizontal axis.
c One of Jim's field's is 12 acres. Estimate how many hectares this is.

Review 3 Use a copy of this grid.
Mahmud went on a training run.
He ran 12 km in the first hour.
He rested for $\frac{1}{2}$ an hour.
He then jogged the 12 km home in $1\frac{1}{2}$ hours.
Draw a distance/time graph for Mahmud's run.

Training run

⭐ **Practical**

You will need a spreadsheet.

Crocodile hunting in Australia? *Skiing in Switzerland?*
Rodeo riding in Canada? *Trekking in Nepal?*
Cruising the Greek Islands?

Decide where you would like to go for an overseas holiday.
Find the exchange rate between British pounds (£) and the currency you would
use on your holiday.
Use a spreadsheet to convert pounds to the currency of the country.

Example

	A	B	C	D
1	Pounds	£10	= B1+10	= C1+10
2	Australian Dollars	= B1*2.5	= C1*2.5	= D1*2.5

This spreadsheet changes
£ to Australian dollars.
The exchange rate is
£ = A $2·5.

Use the graph function of the spreadsheet to draw a graph
to change pounds into the currency you chose.

Interpreting and sketching real-life graphs

Discussion

depth

time

This graph shows the depth of water in a washing machine during one complete
cycle. Which parts of this graph show the machine filling, washing, rinsing, spinning
and draining? **Discuss.**

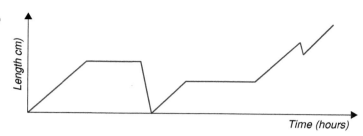

This graph shows the length of a scarf that Anna is knitting. Write a story for this graph. **Discuss** your story.

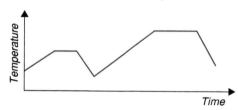

What might this be the graph of? Write a story for this graph. **Discuss** your story.

- What might the previous graph represent if temperature is replaced with distance or height or £? Write a story. **Discuss**.

A pot of rice, filled with water is cooked. This graph shows the amount of water left as the rice cooks.

As the rice cooks it absorbs water. The longer it cooks the less water is left. The amount of water left and cooking time are called **variables**.

The **relationship** between the variables is 'as cooking time increases the amount of water left decreases'.

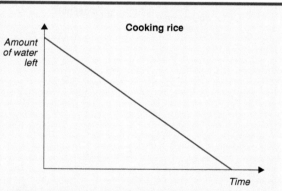

Worked Example

At the 'Just Juice' factory juice flows steadily from a large tank into different-shaped bottles.

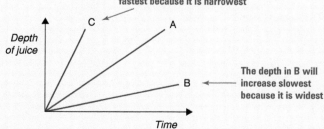

Sketch a graph of the depth of juice against time for each bottle. Sketch all three on the same grid.

Answer

The depth in C will increase fastest because it is narrowest

The depth in B will increase slowest because it is widest

263

Algebra

1 Explain these graphs.
For each, say how the variable on the *y*-axis is related to the variable on the *x*-axis.
Use words like increases or decreases.

a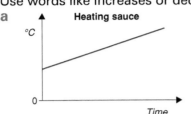

b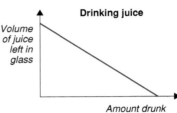

2 Which of these matches the graph shown?
   A A flashlight flashes once.
   B Cinema lights are dimmed slowly.
   C A car indicator flashes many times.
   D Car headlights flash three times.

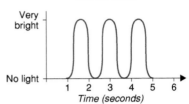

3 Explain these graphs.

a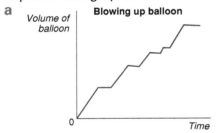

b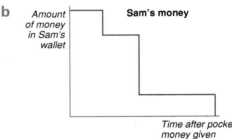

c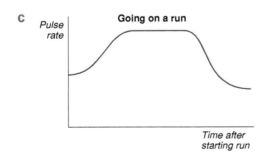

d

4 This graph shows the amount of petrol in the fuel tank of Tamara's car during a journey on a motorway in Europe.
   a How much petrol was in the tank at the beginning of the journey?
   b What total distance did Tamara travel?
   c How many times did she stop for petrol?
   d Assuming that Tamara filled the tank up each time she stopped, what was the capacity of the tank?
   e Did the fuel gauge read $\frac{1}{2}$, $\frac{1}{4}$ or $\frac{3}{4}$ at the beginning of the journey?
   f How much petrol was used altogether on this journey?
   g Could Tamara have completed the journey if she had only filled the tank up once?
   *h What was the petrol consumption, in kilometres per litre, for Tamara's journey?

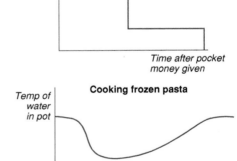

*5 Pieta's roof has a hole in it. Rain drips steadily from the hole into containers **A** and **B**. The line graph showing the depth of water against time for **A** is shown. Trace the graph. Sketch the line graph for **B** on the same set of axes.

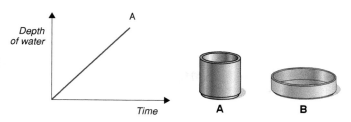

Review 1 Explain these graphs.
For **a**, say how the variables on the axes are related. Use words like increases, decreases.

a
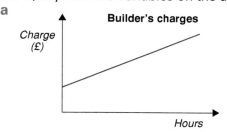

b
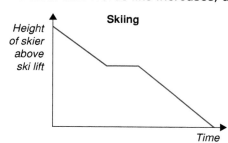

Review 2 This graph shows how much petrol was in the tank of a car at the end of the day.
On which days did petrol get put into the car?

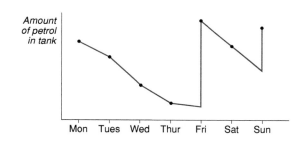

Review 3

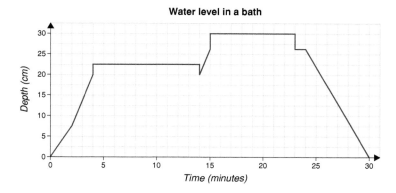

Ramon baths his two dogs, Jasper and Jess. He baths Jess first, then Jasper. The graph shows the water level in the bath.
a Which dog is bathed for the longer time?
b Which dog is larger, Jess or Jasper?
c How long after Jasper is taken out of the bath, does Ramon begin to empty it?

Discussion

- Some pupils at a science camp heated four identical containers with different volumes of water in each. They were all heated by the same burner. The volumes varied from 200 mℓ to 500 mℓ.

  They measured the temperature of the water after 2 minutes. This graph shows their results.

  Discuss these questions.

  1 Should the points have been joined?

  2 Should the points be joined by a straight line?

  3 How many different volumes do you think should have been heated to get an accurate picture of the trend?

  4 About what temperature do you think these volumes of water would be after 2 minutes?

    a 350 mℓ   **b** 600 mℓ

  5 If you wanted some water to be 73 °C after 2 minutes, about how much water should you heat?

  6 Describe the relationship between the temperature after 2 minutes and the volume of water.

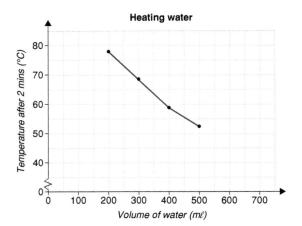

- Amy had six identical sponges.

  She poured a different volume of water onto each sponge.

  Each sponge completely absorbed the water poured onto it.

  Amy weighed each of the sponges.

  She drew a graph of mass of sponge against volume of water.

  Which of these is most likely to look like Amy's graph? **Discuss**.

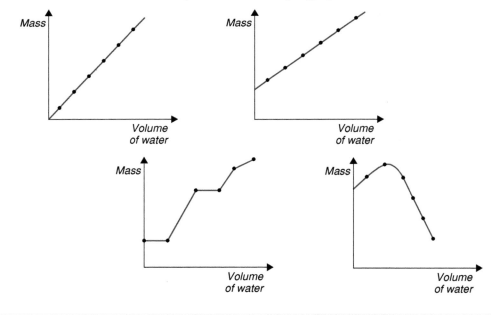

Summary of key points

 A (2, 3) is a **coordinate pair**.

Each of the coordinate pairs ($^-$1, 0), (0, 1), (1, 2), (2, 3) satisfy the rule $y = x + 1$.

Example ($^-$1, 0) satisfies the rule $y = x + 1$ because $0 = {^-1} + 1$.

↑   ↑
y x

 B To draw **a straight line graph**

    find some coordinate pairs that satisfy the rule

    plot them on a grid

    draw a straight line through the points.

Example $y = 2x - 1$

x	$^-$1	0	2
y	$^-$3	$^-$1	3

The coordinate pairs ($^-$1, $^-$3), (0, $^-$1) and (2, 3) are shown plotted.

The **equation of the line** in the example is $y = 2x - 1$.

The coordinate pairs of all points on the line satisfy the equation of the line.

The graph of a **linear sequence** gives a set of points in an 'imagined' straight line.

 C $y = mx + c$ is the equation of a straight line.

m represents the **gradient** or steepness of the line.

The greater the value of m, the steeper the slope.

If m is positive, the gradient is positive.

If m is negative, the gradient is negative.

positive gradient

negative gradient

c represents the **y-intercept**.

The coordinates of the y-intercept are $(0, c)$ where

c is given by the equation of the line.

Example $y = 2x - 1$ has a gradient of **2** and crosses the y-axis at $(0, {^-1})$.

D Lines **parallel to the x-axis** all have equation $y = a$.

a is any number.

Examples $y = 7$, $y = 1\frac{1}{2}$.

Lines **parallel to the y-axis** all have equation $x = b$.

b is any number.

Examples $x = 4$, $x = {^-3}\frac{1}{2}$.

Continued ...

We often draw straight-line graphs to represent **real-life situations**. We can read information from these graphs.

Example This graph shows monthly Internet charges.

The points (0, £10), (10, £17·50) and (20, £25) have been plotted and joined with a straight line.

We can estimate other values using the graph.

Example If I spend 5 hours on the Internet, the charge is about £13·50.

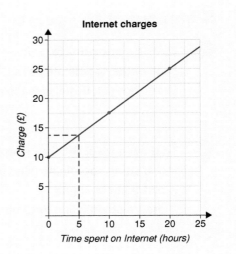

Internet charges

 We can **interpret graphs of real-life situations**.

Test yourself

1 Copy and complete these tables of coordinate pairs for the given rules.

a $y = 3x + 3$

x	$^-2$	0	2
y			

b $y = 2x - 4$

x	$^-1$	2	4
y			

c $y = 3 - 2x$

x	$^-1$	1	3
y			

[T]

2 **a** Write down the coordinate pairs from the table in question **1c** for $y = 3 - 2x$.

b Use a copy of the grid. Plot the three points. Draw and label the line with equation $y = 3 - 2x$.

c Does the point (4, $^-7$) lie on the line?

d Write down the coordinate pair of another point which lies on the line. Does the coordinate pair satisfy the equation $y = 3 - 2x$?

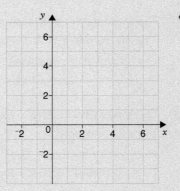

3 **a** $y = \underline{\quad} x + 4$ What number could go in the gap to give the graph of a line which is steeper than $y = 2x$?

b Which of these graphs has a positive gradient, $y = x + 3$ or $y = 5 - x$?

c Which of these graphs cuts the y-axis at (0, $^-3$)?

   A $y = x + 3$ **B** $y = 3 - x$ **C** $y = x - 3$ **D** $y = 3x$

4 Which of these are the equations of lines parallel to the x-axis?

   a $y = 4$ **b** $x = ^-4$ **c** $y = ^-3$ **d** $x = x$

T

5 The cost to hire a computer is given by the formula $C = 20w + 100$, where C is the cost in pounds and w is the number of weeks the computer is hired.

a Copy and complete this table.

Weeks (w)	2	6	10
Charge (£)			

b Use a copy of this grid. Plot the points from the table on the grid. Draw a straight line through the points.

c Explain what this symbol on the vertical axis means.

d How much does it cost to hire a computer for 9 weeks?

e Alison hired a computer and it cost her £240. For how many weeks did she hire it?

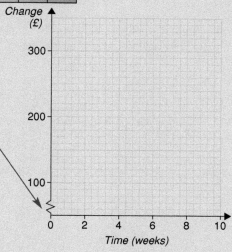

6 Richard sketched this graph to show how full his rain gauge was.

a One month it rained a lot more than it did the other months.
Which month do you think this was?

b When it is very sunny, the water in the rain gauge evaporates and so the level does down.
Which months do you think were very sunny?

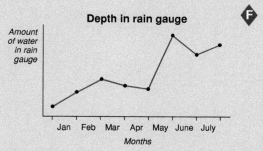

Angles

Angle is a measure of turn. We measure angles in degrees. There are 360° in a full turn.
A **reflex** angle is an angle between 180° and 360°.

One complete turn or 360°

Right angle or 90°

Straight angle or 180°

Acute angle less than 90°

Obtuse angle between 90° and 180°

Reflex angle between 180° and 360°

We use a **protractor** or angle **measurer** to measure and draw angles.

These are the arms of the angle

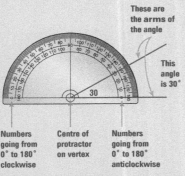

This angle is 30°

Numbers going from 0° to 180° clockwise

Centre of protractor on vertex

Numbers going from 0° to 180° anticlockwise

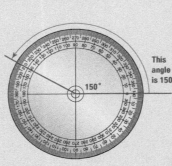

This angle is 150°

Clockwise is the direction clock hands move.

It is best to measure and draw reflex angles using a 360° protractor.

To measure a reflex angle with a 180° protractor measure the acute or obtuse angle and then subtract it from 360°.

Example   Obtuse angle is 138°
Reflex angle = 360° − 138°
= **222°**

Always **estimate** the size of the angle first.
Do this by comparing the acute or obtuse angle to a right angle.

measure this as 138°

Angle A is about $\frac{1}{3}$ of a right angle. It is about 30°.

A

The corner of a page is a right angle.

The three interior angles of a triangle add to 180°.

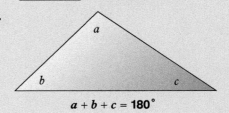

$$a + b + c = 180°$$

Angles at a point add to 360°.
$$d + e + f + g = \mathbf{360°}$$

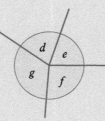

Practice questions 11, 34, 35, 36, 47, 50

Lines

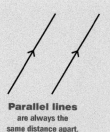

Parallel lines
are always the
same distance apart.

Perpendicular lines
intersect (cross)
at right angles.

Practice Questions 1, 51

2-D Shapes

Triangles

Scalene
no equal sides
no equal angles

Equilateral
3 equal sides
3 equal angles
3 lines of symmetry

Isosceles
2 equal sides
2 base angles equal
1 line of symmetry

equal
sides

base
angles

Right-angled

Quadrilaterals

A **quadrilateral** has four sides.

These are some special quadrilaterals.

Quadrilateral

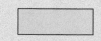

Square
4 equal sides
4 right angles

Rectangle
2 pairs of opposite sides equal
4 right angles

Parallelogram
Opposite sides
equal and parallel

Rhombus
a parallelogram
with 4 equal sides

Trapezium
1 pair of opposite
sides parallel

Kite
2 pairs of adjacent
sides equal
1 pair of equal angles

Arrowhead (delta)
2 pairs of adjacent
sides equal
1 pair of equal angles
1 reflex angle

Polygons

A 3-sided polygon is a triangle.   A 4-sided polygon is a quadrilateral.
A 5-sided polygon is a pentagon.   A 6-sided polygon is a hexagon.
A 7-sided polygon is a heptagon.   An 8-sided polygon is an octagon.
A **regular polygon** has all its sides equal and all its angles equal.

Examples

**Regular
pentagon**

**Regular
hexagon**

**Octagon
(irregular)**

A **diagonal** is a straight line drawn from one corner of a polygon to another.
The corners must not be next to each other.

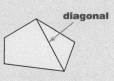

diagonal

Congruent shapes are the same size and shape.

Practice Questions 14, 16, 28, 48

3-D Shapes

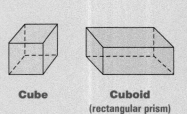
Cube   Cuboid (rectangular prism)   Triangular prism   Pyramid   Pyramid (pentagonal base)   Cylinder

Sphere   Cone   Hemisphere   Tetrahedron (triangular-based pyramid)   Octahedron   Dodecahedron

Faces, edges, vertices

A **face** is a flat surface.
This shape has 6 faces.
One of the side faces is shown in red.

An **edge** is a line where two faces meet.
This shape has 12 edges. One edge is shown in blue.

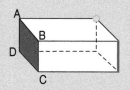

A **vertex** is a corner where edges meet. (**Vertices** is the plural of vertex.)
This shape has 8 vertices. One of the vertices has a yellow dot.

Practice Questions **5, 44**

Nets

A 2-D shape that can be folded to make a 3-D shape is called a **net**.

This net folds to make a cube.
The fold lines are dashed.

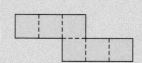

Example   This net folds to make a cuboid.

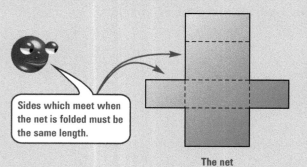

Sides which meet when the net is folded must be the same length.

The net   Folding the net   Completed cuboid

Practice Questions **39, 42**

Coordinates

We use **coordinates** to give the **position** of a point or place on a grid. The coordinates of the lighthouse are ($^-$1, $^-$2).
$^-$1 is the x-coordinate.
$^-$2 is the y-coordinate.
We always give the x-coordinate first.
The coordinates of the **origin** are (0, 0).

The *origin* is where the x-axis and y-axis intersect.

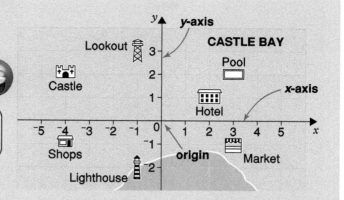

Practice Questions 8, 18

Symmetry

A shape has **reflection symmetry** if it can be folded along a line so that one half fits exactly onto the other half. The line is called a **line of symmetry**.

This shape has four lines of symmetry.

A **regular** polygon has the same number of lines of symmetry as it has sides.

Example   This regular pentagon has five sides and five lines of symmetry.

A shape has **rotation symmetry** if it fits onto itself **more than once** when it is rotated 360°.
The **order of rotation symmetry** is the number of times a shape fits onto itself when it is rotated 360°.
This shape fits onto itself four times in a 360° turn.
It has order of rotation symmetry 4.

Practice Questions 7, 27, 46, 49

Transformations

The blue shape has been **translated** 3 units to the right and 1 unit down to get the green shape.

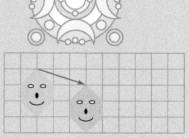

Translation

The red shape has been **reflected** in the mirror line, m, to get the orange shape.

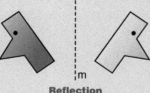

Reflection

This shape has been rotated by quarter turns (90° turns) about corner A.

$\frac{1}{4}$ turn = 90° $\frac{1}{2}$ turn = 180° $\frac{3}{4}$ turn = 270°

Practice Questions 10, 19, 22, 23, 40, 41

Rotations of 90°

Measures

Time

1 minute = 60 seconds
1 hour = 60 minutes
1 day = 24 hours
1 year = 12 months or 52 weeks and 1 day
1 year = 365 days (or 366 days in a leap year)

1 millennium = 1000 years
1 century = 100 years
1 decade = 10 years

April, June, September, November have 30 days.
January, March, May, July, August, October, December have 31 days.
February has 28 days except in a leap year when it has 29 days.

All years that are divisible by 4 are leap years, except centuries which are leap years only if they are divisible by 400.

a.m. time is from midnight until noon; **p.m.** time is from noon until midnight.

24-hour clocks begin at midnight. The hours are numbered as shown.

| 00:00 | 01:00 | 02:00 | ... | 12:00 | 13:00 | ... | 22:00 | 23:00 |
| midnight | 1 a.m. | 2 a.m. | | 12 p.m | 1 p.m | | 10 p.m | 11 p.m. |

24-hour clocks always have 4 digits.
1:20 p.m. is written as 13:20 or 1320 or 13·20.
4:35 a.m. is written as 04:35 or 0435 or 04·35.

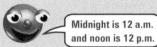

Midnight is 12 a.m. and noon is 12 p.m.

Metric measures

length
1 kilometre (km) = 1000 metres
1 metre (m) = 1000 millimetres (mm)
1 metre (m) = 100 centimetres (cm)
1 centimetre (cm) = 10 millimetres (mm)

capacity
1 litre (ℓ) = 1000 millilitres (mℓ)
1 litre (ℓ) = 100 centilitres (cℓ)
1 centilitre (cℓ) = 10 millilitres (mℓ)

mass
1 kilogram (kg) = 1000 grams (g)
1 tonne = 1000 kilograms (kg)

Examples   5 km = 5 × 1000 m   3 m = 3 × 100 cm   3 tonne = 3 × 1000 kg
             = 5000 m   **= 300 cm**   **= 3000 kg**

Imperial measures

length
8 km ≈ 5 miles
1 metre ≈ 3 feet

mass
1 kg ≈ 2·2 pound (lb)
1 oz ≈ 30 g

capacity
1 pint is about 600 mℓ (just over $\frac{1}{2}$ ℓ)
1 gallon ≈ 4·5 ℓ

Example   20 miles is about 32 km because **4** × 5 = 20
                                          4 × 8 = 32

Reading scales

To **read a scale** we must work out what each space on the scale stands for.

Example   On these scales, each space stands for 10 kg.
            The pointer is at about 92 kg or 90 kg to the nearest 10 kg.

Practice Questions 1, 2, 3, 4, 6, 9, 12, 13, 15, 17, 20, 21, 24, 25, 26, 29, 30, 31, 32, 33, 43, 52, 53, 54, 55

Perimeter and area

The distance right around the outside of a shape is called the **perimeter**.
Perimeter is measured in mm, cm, m or km.

The amount of surface a shape covers is called the **area**. Area is measured in square millimetres (mm^2), square centimetres (cm^2), square metres (m^2) or square kilometres (km^2).

We can count squares to find area. The area of each of these squares is 1 square centimetre. Since there are 12 squares in this rectangle, its area is 12 square centimetres or 12 cm^2.

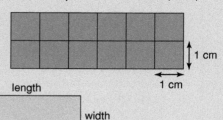
1 cm
1 cm

The **area of a rectangle** is found by multiplying the length by the width.

length
width

Example   Perimeter $= 2 \times 8 + 2 \times 4$
$= 16 + 8$
$= \mathbf{24\ cm}$
  Area $= 8 \times 4$
$= \mathbf{32\ cm^2}$

8 cm
4 cm

Practice Questions 37, 38, 45

Practice Questions   **Only use a calculator if you need to.**

1 **a** Estimate the lengths of these lines to the nearest centimetre.
Then measure them accurately.

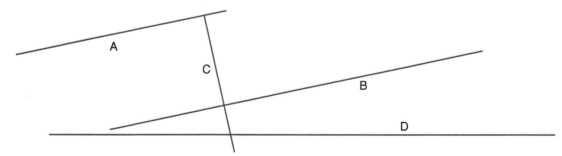

A
C
B
D

   b Which of the lines are parallel?
   c Name two pairs of lines that are perpendicular.

2 **a** Paul's birthday is on March 1st. Ray's is on May 3rd. How many days are there between Paul's and Ray's birthdays, not including their birthdays.
   b How many days are there between July 30th and September 20th, including both these days?
   c How many days were there between February 15th and March 15th in 1988, not including these days?
   d Oscar's birthday is the 15th February. His sister's is the 15th March. How many days are there **between** Oscar's birthday and his sister's in the year 2009?

3 What goes in the gap?
  a 1 hour 25 minutes = ___ minutes
  b 115 minutes = ___ hour ___ minutes
  c 124 seconds = ___ minutes ___ seconds
  d 30 days = ___ weeks ___ days
  e 5 weeks 6 days = ___ days
  f 2 centuries 10 years = ___ years
  g 1 millennium, 3 centuries, 2 decades, 4 years = ___ years

4 What is the total weight, in kg, of this fruit?

Apples   2.6 kg
Bananas   500 g
Lemons   1150 g

5 **a** How many edges does this shape have?
  b How many vertices does this shape have?
  c Three of the faces are rectangular.
     How many other faces are there?
     What shape are these?
  d Name this shape.

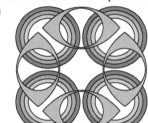

6 Write down two things you would measure in
  a km   **b** m   **c** tonnes   **d** g   **e** ℓ   **f** mℓ
  g cℓ   **h** miles   **i** pints   **j** pounds.

T

7 Use a copy of this.
Which of these patterns have reflection symmetry?
Draw the lines of symmetry on these patterns
  a   **b**   **c**

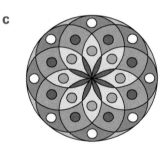

T

8 The places shown on this graph are the
places that 5-year-old Jake often visits.
  a Write down the coordinates of these.
     i school   **ii** post office
     iii bus station   **iv** granny
  b One Sunday morning, while his parents
     were still asleep, Jake decided to go out.
     From his home he walked to (⁻1, 2), then
     (1, 0), then (2, 1), then (2, ⁻2), then
     (⁻1, ⁻2), then (0, ⁻1) and then back
     home. Write down the names of all the
     places in the order that he visited them.
  c Use a copy of the graph. Beginning at
     his home, join each point on Jakes's
     walk with a straight line.
  d The diagram for **c** points to the place
     Jake likes best. Which place is this?

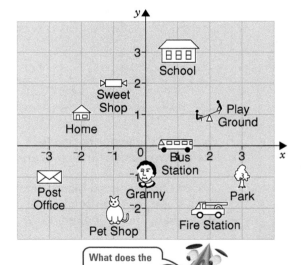

What does the arrow point to?

9 Alice's grandfather measured most things in imperial units.
The imperial unit he used is given. Which metric unit could he have used to measure them?
  a distance to his brother's in miles   **b** length of his bedroom in feet
  c weight of vegetables in pounds   **d** how much oil he needed in pints
  e mass of a bag of nails in ounces

10 Describe how each of these shapes have been translated.

a

b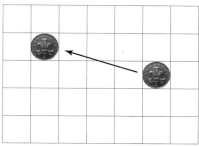

11 Write down the colour of the   **a** acute   **b** obtuse   **c** right
     d reflex angles in each of these.

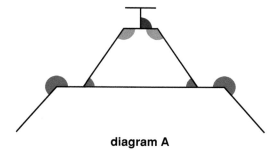

diagram A

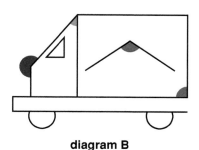

diagram B

12 A medicine bottle holds 0·3 litres.
Robert is to have 15 millilitres twice a day.
How many days will the medicine last?

13 Will the year 3000 be a leap year?

14
A trapezium	**B** parallelogram	**C** rhombus	**D** arrowhead (delta)
E kite	**F** equilateral triangle	**G** scalene triangle	
H isosceles triangle	**I** rectangle	**J** right-angled triangle	

Choose the best name from the box for each of these.

a 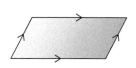   b   c 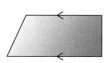   d

e    f   g   h

15 Max bought 2 m of wrapping paper to wrap gifts. From this he cut two lengths, one of 34 cm and the other of 85 cm.
How much paper did he have left?

T

16 Use a copy of this.
   a Draw all the diagonals on the hexagon.
   b How many diagonals does a hexagon have?
   c How many diagonals does an octagon have?

Draw an octagon to check.

277

17 Which is more, 10 lb of apples or 5 kg of apples?

T

18 a Write down the letters at each of the following
coordinates. You will get four words.
  1 (⁻4, 0) (1, 1) (⁻4, ⁻6) (⁻2, ⁻2)
  2 (4, 5) (0, ⁻6) (2, ⁻6)
  3 (⁻3, ⁻1) (2, ⁻6) (4, 5) (4, 5) (2, ⁻6) (3, 3) (1, 5)
  4 (7, 5) (⁻3, ⁻1) (2, 2) (0, ⁻6) (7, 5) (7, ⁻1) (2, ⁻6)
      (4, 5) (⁻4, ⁻6) (5, 1) (7, 5) (⁻3, ⁻1) (⁻3, ⁻1) (6, 5)

  b Do what the message in **a** says.
  What do you get?

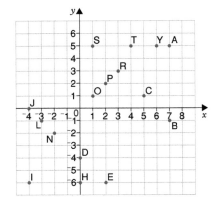

T

19 Use a copy of these.
Draw the reflections in the dotted mirror lines.

a

b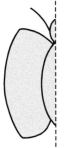

c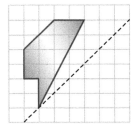

T

20

6·8	7·2	6800	720	820	72	6·8		7·2	680	7200	6800	112	6·8

7200	6800	680		720	12		112	7·2	680	720	6800

			D	
7·2	680	7200	82	6·8

Use a copy of this box.
What goes in the gap?

D 8·2 cm = 82 mm   **E** 0·68 km = ____ m   **A** 7·2 ℓ = ____ mℓ   **I** 0·72 kg = ____ g
N 0·12 ℓ = ____ cℓ   **T** 11·2 cℓ = ____ mℓ   **R** 6·8 kg = ____ g   **S** 0·68 cm = ____ mm
P 0·072 km = ____ m   **M** 8·2 m = ____ cm   **H** 0·072 m = ____ cm

21 This table shows the time-clock settings for Rari's central heating.
  a How long is the heating on in the morning?
  b How long is the heating on in the afternoon?
  c How long is the heating on for altogether during one day?

On	Off
0645	0840
1115	1300
1500	2130

22 Name the fish that is
  a a reflection of fish C
  b a translation of fish C
  c a rotation of fish C.

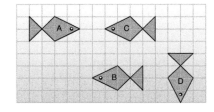

23 The green shape has been rotated anticlockwise about the origin to the yellow shape.
Write down the angle of rotation.
Choose from the box.

| 45° | 90° | 180° | 270° |

a

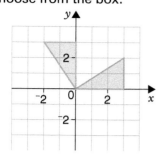

b

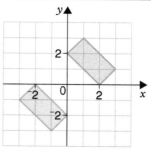

c

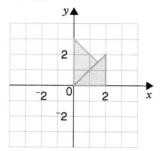

24 Last night the sun set at 19:30. It rose this morning at 06:45.
How many hours of darkness were there?

25 Keith entered a fun triathalon. He ran 16·5 km, cycled 38·4 km and
crawled 1560 m on his hands and knees.
How many kilometres did he race? How many metres?

26 Which of these would you expect to be
a the length of a table
    A 3 cm   **B** 2 m   **C** 25 cm   **D** 30 m
b The mass of an orange
    A $\frac{1}{2}$ kg   **B** 250 g   **C** 760 g   **D** 25 g
c The amount of milk in a glass
    A 20 mℓ   **B** 600 mℓ   **C** 200 mℓ   **D** 2 ℓ
d The distance between two cities
    A 150 mm   **B** 150 km   **C** 150 m   **D** 150 cm
e The length of a pencil
    A 15 cm   **B** 35 cm   **C** 3 cm   **D** 2 m
f The mass of a 3-year-old child
    A 450 g   **B** 5 kg   **C** 200 kg   **D** 18 kg
g The amount of water in a full bucket?
    A 2 ℓ   **B** 750 mℓ   **C** 10 ℓ   **D** 50 ℓ

27 Use a copy of this.
Shade eight more squares so that both dashed lines are lines of symmetry.
a   **b**   **c**

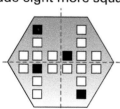 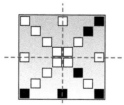 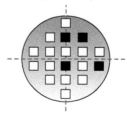

28 Nishi had 12 squares. She made three different rectangles with them.

    a How many different rectangles could she make with 24 squares?
    b How many squares does she need to make exactly five different rectangles?

29 What reading is given by each of the pointers?

a

b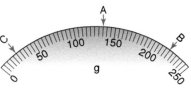

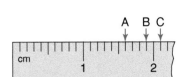

c

d

e

30 To keep in shape Charlotte swims 1 km twice a week. If the pool is 40 m long, how many lengths does Charlotte swim each week?

31 There is 200 ml of bath oil in a bottle. There is five and a quarter times as much in a larger bottle. How much bath oil is in the larger bottle?

32 How many grams of carrots must be added to 3·84 kg to make 6 kg?

33 Suzannah's dog weighed 44 kg. The vet said it had to lose at least 5 kg.
Suzannah's scales only weighed in pounds. About how many pounds does the dog have to lose?

34 Estimate the size of each of the marked angles.
Check your estimate by using a protractor to measure them.

a　　　　　　　　**b**　　　　　　　　**c**

35 Draw some large triangles. Make each triangle a different shape.
Measure the three angles in each triangle.
No matter what the shape of your triangle, the three angles should total 180°.

36 Draw a large four-sided shape.
Measure the four outside angles.
No matter what shape you draw these four angles should add to 1080°. Check that yours do.

37 Lucy used this pattern to make Christmas decorations.

Count squares to estimate the area of each decoration.

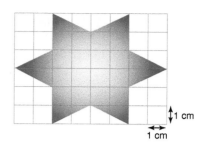

1 cm

1 cm

38 What is the area of these rectangles?

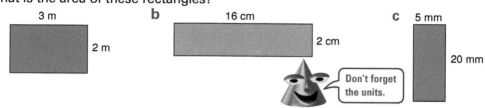

a 3 m

2 m

b 16 cm

2 cm

c 5 mm

20 mm

Don't forget the units.

[T]

39 Use a copy this.
Justin is making a box to display a butterfly.
He draws the net of the box like this.
The base of the box is shaded.

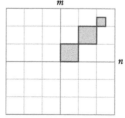

Justin wants to put a lid on the box.
He must add one more square to his net.
On each diagram below, show a different place to add the new square.
Remember the base of the box is shaded.

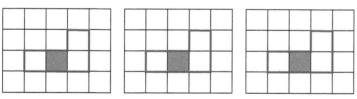

[T]

40 Use a copy of this.
Draw the reflection of these lines in mirror line *m*.
Then reflect all of the lines in mirror line *n*.

m

n

41 Which of the diagrams below shows this shape after
a a rotation of 90° clockwise about (0, 0)
b a rotation of 180° about (0, 0)
c a translation of 2 units left and 3 units down?

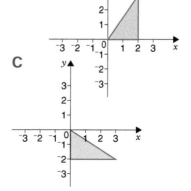

A

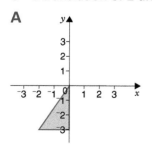

B

C

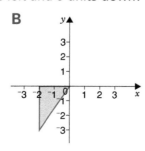

T

42 a Imagine folding each of these nets.
Which of them will fold to make a cube?

A   B   C   D

E   F   G

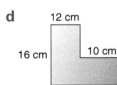

You could cut out copies
of the nets and fold them
to check.

H   I   J

b Draw another net for a cube. Do not use any of the nets given in **a**.

43 Change this recipe to metric units.

half a pint	water
2 oz	butter
4 oz	sugar
2	eggs
$\frac{1}{2}$ lb	flour

44 What is the smallest number of cubes you would need to add to these shapes to make them into cuboids?

a 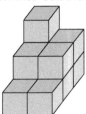   **b**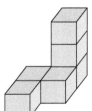

45 Find the perimeter of these.

a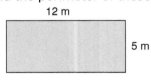
12 m
5 m

b
4 mm   6 mm
7 mm   6 mm

c
regular hexagon   3 cm

d
12 cm
16 cm   10 cm
9 cm

46 What is the order of rotation symmetry of these?

a   **b**   **c**   **d**

47 Find the value of *x*.

a

b

c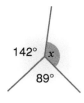

d

48 Write true or false for these.
   a   A rhombus is a parallelogram with four equal sides.
   b   A rectangle has four right angles and its opposite sides are equal.
   c   A rectangle is a square with opposite sides equal.
   d   A trapezium has one pair of opposite sides parallel.
   e   A kite has two pairs of adjacent sides equal.

49 Megan folded a piece of paper in half and then in half again.
   She made one straight cut. She opened the sheet of paper and
   it looked like this.

   Copy this folded sheet of paper. Draw on it
   where Megan made her cut.

50 a   Sketch a shape with 3 acute angles and 1 obtuse angle.
   b   Sketch a shape with 1 reflex angle, 1 right angle, 2 acute angles and 1 obtuse angle.

51 Which of these is true?
   A   A line which is perpendicular to a horizontal line must be vertical.
   B   A vertical line can never be perpendicular to another vertical line.
   C   Two horizontal lines can be perpendicular.

52 Suggest what units you would use to measure these.
   a   the thickness of a piece of paper
   b   the mass of a pin
   c   the amount of water in one drip from a tap

* **53** The digit 1 is often displayed on a digital clock. The digit 9 is not displayed as often. From
   1800 to midnight, for how many more minutes is at least one '1' displayed than at least
   one '9'?

* **54** 100 kg of potatoes are given to 100 people. Each adult gets 3 kg, each
   teenager gets 2 kg and each small child gets 0·5 kg. How many adults and
   teenagers and small children were there? (There are many answers to this
   problem. Find as many as you can.)

* **55** Rose needed to measure out exactly 2 ℓ of water. She had two containers,
   one that held 8 ℓ and one that held 5 ℓ. She began by filling the 5 ℓ container.
   How did she continue?

11 Lines and Angles

You need to know

✓ angles — acute, obtuse, right, straight, reflex   page 270
           — how to measure angles including reflex angles
           — how to estimate angles
           — inside angles of a triangle add to 180°
           — angles at a point add to 360°

✓ lines — parallel and perpendicular   page 271
           — measuring lines

✓ triangles — isosceles   page 271
           — equilateral
           — right-angled
           — scalene

Key vocabulary

adjacent, alternate angles, angle, angles at a point, angles on a straight line, base angles, corresponding angles, degree (°), equidistant, exterior angle, interior angle, intersect, intersection, isosceles triangle, line, line segment, parallel, perpendicular, plane, point, supplementary, vertex, vertically opposite angles, vertices

Straight or curved

1 Draw a right angle using a set square.
Draw each 'arm' of the angle 10 cm long.
Divide each 'arm' into cm lengths.
Number the cm lengths as shown.
Join 1 to 1, 2 to 2, 3 to 3, ...

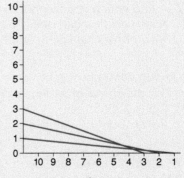

2 Draw another right angle with arms 10 cm long.
Divide each arm into $\frac{1}{2}$ cm lengths.
Number the $\frac{1}{2}$ cm lengths 1 to 20 in the same order as the diagram in 1.
Join 1 to 1, 2 to 2, 3 to 3, ...

3 Try and draw these.

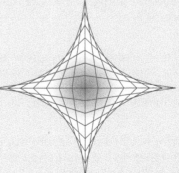

Naming lines and angles

This is a **line**. It goes on forever in both directions.
We say it has infinite length.

$\longleftrightarrow$

Is the horizon a line?

MN is a **line segment**.
It has end points M and N.
We say it has a finite length.

M ——————— N

We think of lines and line segments as having no measurable width.
However, when you *draw* lines they do have some width.

When we refer to lines and line segments we mean 'straight' lines, with no curves or kinks.

Two straight lines must either be
parallel or cross once (**intersect**).

A is the point where the lines intersect.

parallel lines   intersecting lines

Discussion

Above we said that two straight lines must either be parallel or cross once.

What if one of the lines was drawn on the wall and the other on the ceiling? Would this still be true? **Discuss.**

If one line is drawn on the ceiling and the other on the wall the lines are on different **planes**.
Lines on the same plane are on the same flat surface.

When two line segments meet at a point they make an **angle**.
The point where they meet is the **vertex**.
The angle is a measure of turn from one arm to the other.

vertex   F

G   arms or rays of angle

H

We name angles in two ways.

1 Using the letter at the vertex.
The angle shown is ∠G.

2 Using three letters, the middle one being the vertex.
The angle shown is ∠FGH or ∠HGF.
We could write this as FĜH or HĜF.

We must use three letters if there is more than one angle at the vertex.

E
F
G
H

Exercise 1

1 Name the red line segment.

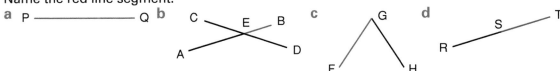

2 Use a single letter to name the marked angles.

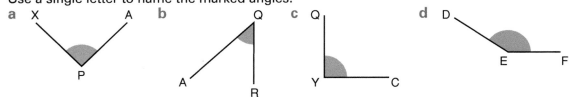

3 Use three letters to name the marked angles in question **2**.

4

SAL	M__	H__	O__	D__	A__	H__	T__	F__	T__
a	b	c	d	e	f	g	h	i	j

Copy this chart.
Name the marked angles on the diagrams below and fill in your chart.
The angle marked in **a** is ∠SAL so SAL is filled in as shown.

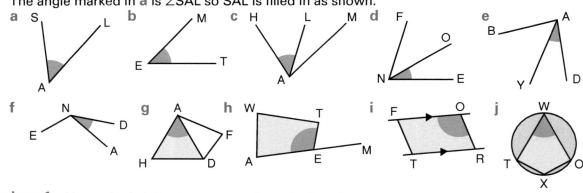

Review 1 Use a single letter to name each marked angle.

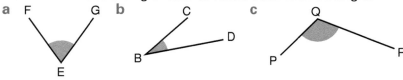

Review 2 Use three letters to name the marked angles in Review **1**.

Review 3 Use three letters to name each marked angle.

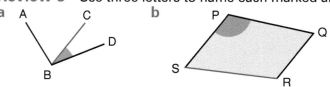

Review 4 Name the red line segments in Review **3**.

Parallel and perpendicular lines

Remember
Parallel lines are always the same distance apart (equidistant).
Perpendicular lines intersect at right angles.

Parallel lines never meet.
We write AB//CD. We read // as 'is parallel to'.

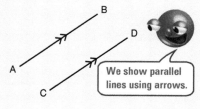

We show parallel
lines using arrows.

We write AB⊥CD.
The symbol ⊥ is
read as 'is
perpendicular to'.

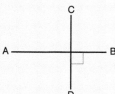

This symbol is used to
show the lines are at
right angles.

Exercise 2

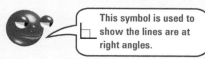

1 Look at this photograph.
Find as many parallel and perpendicular lines as possible.

2 Which line is parallel to
   a AB   **b** AD?

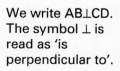

3 **a** Which line is perpendicular to SP?
   b Which lines are perpendicular to PQ?
   c Which line is parallel to QR?

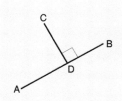

4 Which of // or ⊥ goes in the gaps?
   a DE ___ HI   **b** DE ___ EF   **c** GF ___ DE
   d HG ___ GF   **e** EF ___ DE

5 This is a drawing of a cuboid.
   a Which edges are parallel to AB?
   b Which edges are perpendicular to DH?

There is more about
drawing 3-D shapes
on page 320.

Review 1
a Name three parallel lines in this shape.
b Name the perpendicular lines.

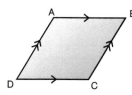

Review 2   This a drawing of a box.
a Which edges are parallel to CB?
b Which edges are perpendicular to EF?

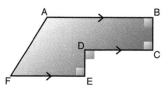
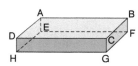

Constructing lines and bisectors

Parallel and perpendicular lines can be drawn with a **set square**. One angle of a set square is a right angle.

If you do not have a set square, you can make a paper one as shown in the following practical.

 ## Practical

You will need paper and a stapler.
To make your own set square follow these steps.

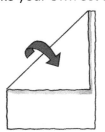 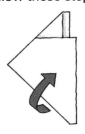 

| Fold one corner of a piece of paper over to make a straight edge. | Fold the paper again so that half of the straight edge is *exactly* on top of the other half. | Tidy up the edges. Staple the paper through all thicknesses. Make sure the edges don't move as you staple. |

These diagrams show how to draw two parallel lines using a ruler and set square.

Set
Square

| Ruler | Ruler | Ruler | Ruler |
| Put the set square on the ruler. | Draw a line. | Slide the set square along the ruler. | Draw another line. |

Discussion

The diagrams below show how to draw a line through a point C that is parallel to the line AB.

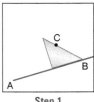   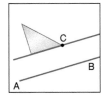

| Step 1 | Step 2 | Step 3 | Step 4 |

Discuss these steps. (**Hint**: in Step 3, the set square slides along the ruler.)

We can use a set square and ruler to draw a line segment perpendicular to another line segment.

M————N

We use compasses and a ruler to construct the perpendicular bisector of a line segment BC.

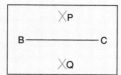

Open the compasses to a little more than half the length of BC.
With compass point first on B and then on C, draw arcs to meet at P and Q.

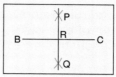

Draw the line through P and Q.
R is the point that bisects BC.

Bisect means cut in half. R is the mid-point of BC.

The following diagrams show the construction of the bisector of the angle P.

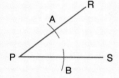

Open out the compasses to a length less than PR or PS.
With compass point on P, draw arcs as shown.

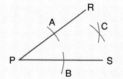

With compass point on first A, then B, draw arcs to meet at C.

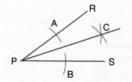

Draw the line from P through C. This line, PC, is the bisector of angle P.

It is very important to draw accurately.

Exercise 3

1 Draw a line segment GH.
Using your set square and ruler, draw a line that is parallel to GH.

2 a Draw a line segment PQ.
  b Mark a point, S, above the line segment.
  c Use your set square and ruler to draw a line through S that is parallel to PQ.

3 a Draw a line segment, CD.
  b Mark a point, E, above the line.
  c Use your set square to draw a line through E that is perpendicular to CD.

4 Repeat question **3** but mark the point, E, below the line segment CD.

5 a Draw a line segment, AB.
  b Use your compass to draw the perpendicular bisector of AB.

6 Practise the construction 'bisecting an angle' as follows:
Use your ruler to draw angles of different sizes.
Use your compass to bisect the angles.
Check the accuracy of your construction by measuring with a protractor.

7 Draw a triangle, ABC.
Use your compass and ruler to draw the
perpendicular bisector of each side.
What do you notice about the three lines
you have drawn?

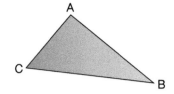

*8 **a** Trace this diagram into your book.
  b Using the set square, draw the line through A which is at
    right angles to XY.
  c Name the point where this line meets YZ as B.
  d Bisect angle AYZ.
  e Name the point where this bisector meets AB as C.
  f Measure the length YC to the nearest mm.

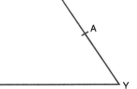

*9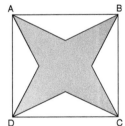

Make an accurate drawing of this diagram as follows.

Draw a line DC, 6 cm long.
Construct the square ABCD.
Draw in the diagonals AC and BD.
Bisect the angles between the diagonals and the sides.
Shade as shown.

Review 1
a Draw a line segment, MN.
b Mark a point, P, anywhere on the line segment.
c Use your set square and ruler to draw a line
  through P that is perpendicular to MN.

Review 2   Repeat Review **1** but mark P above the line segment.

Review 3
a Draw a line segment, KL.
b Use your set square and ruler to draw a line that is parallel to KL.

Review 4
a Draw a line segment, JK.
b Use your compass and ruler to draw the perpendicular bisector of JK.

*Review 5
a Trace this diagram.
b Bisect the angle B.
c Bisect the line BC. Label the mid-point as P.
d Through P, draw the line that is parallel to BA.
e Label as Q, the point where the lines constructed in **b** and **d** meet.
f Measure the length of QP.

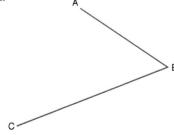

Practical

1 Make a logo for your class or school using parallel and/or perpendicular lines and shading.

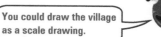

You could draw the village as a scale drawing.

2 A new village is to be built.
Design a street plan for the village using parallel and perpendicular lines.
Give your village a name.

Angles

T

Practical

You will need acetate sheets.

This practical can also be done using a dynamic geometry software package.

1 Draw a line segment AB.

Draw another line segment CD on a separate sheet of acetate. Lay it on top of the first piece of acetate so the lines intersect.

Rotate the line segment CD about the point of intersection. As you rotate you will get diagrams similar to the ones below.
What can you find out about the size of the marked angles in each diagram?
Which ones are equal?
Which ones add to 180°?
Which ones add to 360°?

2 Draw three lines that intersect at the same point.
Name the angles a, b, c, d, e and f.
Which angles are equal?
Which angles add to 180°?

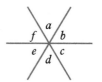

a and *b* are called **vertically opposite angles**.
They are opposite each other when two lines intersect.
Vertically opposite angles are equal.

$a = b$
$c = d$

Example

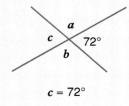

$c = 72°$

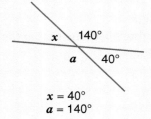

$x = 40°$
$a = 140°$

a, *b*, *c* and *d* are called **angles at point**.
Together, *a*, *b*, *c* and *d* make a complete turn.
Angles at a point add to 360°.

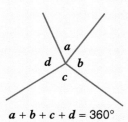

$a + b + c + d = 360°$

Worked Example
Find the size of angle *x*.

a

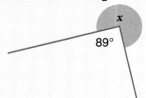

b

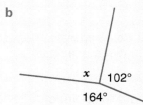

Answer

a We can form an equation.
   $x + 89° = 360°$
   $x = 360° - 89°$
   $x = \mathbf{271°}$

b We can form an equation.
   $x + 102° + 164° = 360°$
   $x = 360° - 102° - 164°$
   $x = \mathbf{94°}$

Angles on a straight line add to 180°.

A pair of angles that add to 180° are called
supplementary angles.

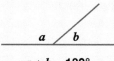

$a + b = \mathbf{180°}$

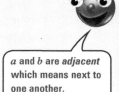

a and *b* are *adjacent*
which means next to
one another.

Worked Example
Find the size of angle *a*.

a

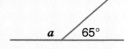

b

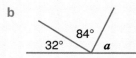

Answer

a $a + 65° = 180°$
   $a = 180° - 65°$
   $= \mathbf{115°}$

b $a + 32° + 84° = 180°$
   $a = 180° - 84° - 32°$
   $= \mathbf{64°}$

Exercise 4

1 Find the size of the angles marked with letters.

a 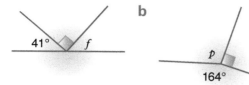 b c 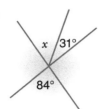 d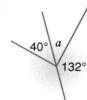

e 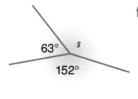 f 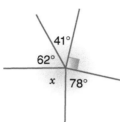 g 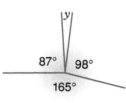 h

2 Find the value of *a*, *b* and *c*.

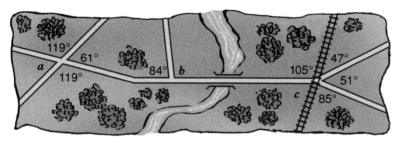

*3 Calculate the value of *x*. You will need to solve an equation.

a  b 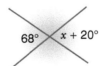 c   d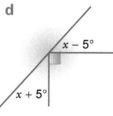

e 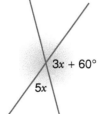 f 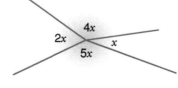 g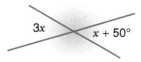

 There is more about solving equations on page 197.

T | **Review 1**

67°	47°	35°	57°	70°	113°		110°	151°	138°		151°

N

122° 70° 64° 70° 35° 151° 113° 47° 151° 65°

Use a copy of this box. Calculate the shaded angles. Put the letter that is beside each diagram above its answer.

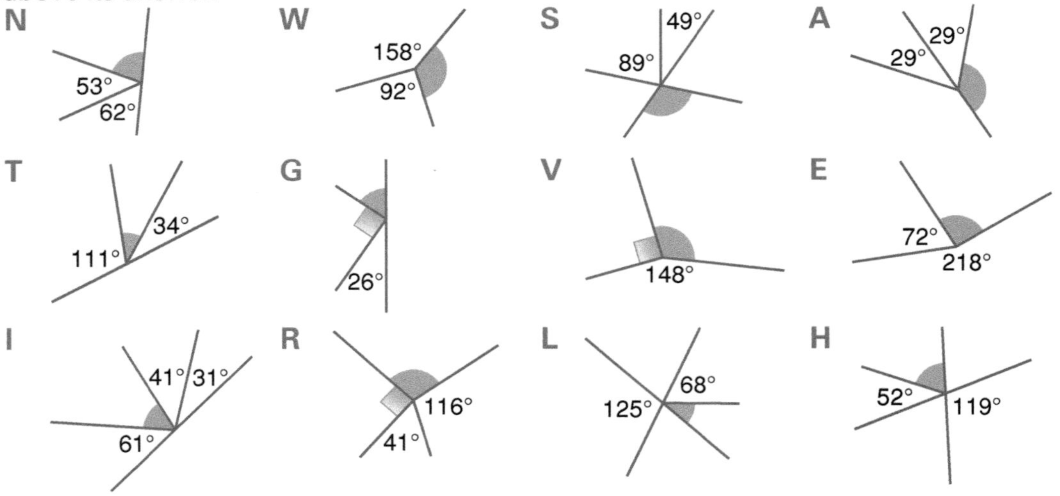

N
53° 62°

W
158° 92°

S
49° 89°

A
29° 29°

T
34° 111°

G
26°

V
148°

E
72° 218°

I
41° 31° 61°

R
116° 41°

L
68° 125°

H
52° 119°

*Review 2   Calculate the value of *x*.

a
3x 86° x + 42°

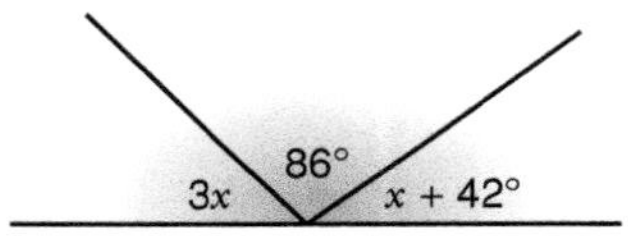

b
115° x + 36°
2x 24°
2x + 10°

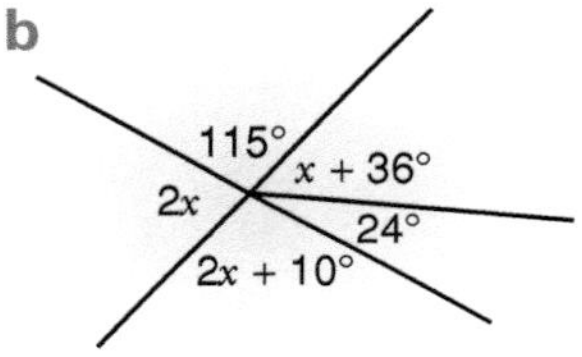

Angles made with parallel lines

T

⭐ **Practical**

You will need a geometry software package or some acetate sheets.

Draw two intersecting lines.
Draw a third line which is parallel to one of them.
Name the angles as shown.
Which angles are equal?

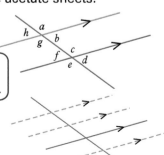

If you use acetate, draw the third line on a separate sheet.

Move the third line you drew so that it always remains parallel.

What happens to the sizes of *a*, *b*, *c*, *d*, *e*, *f*, *g* and *h*?

In each of these diagrams we can translate angle *a* onto angle *b*.

angle *a* = angle *b*.

Angle *a* and angle *b* are called **corresponding angles**.

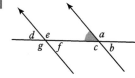

In each of these diagrams we can rotate angle *x* onto angle *y*.

angle *x* = angle *y*

Angle *x* and angle *y* are called **alternate angles**.

Worked Example
Find the value of the angles marked with letters.

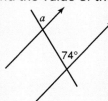 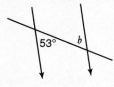 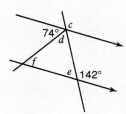

Answer
a Angle *a* and 74° are corresponding angles. They are equal. ***a* = 74°**.
b Angle *b* and 53° are alternate angles. They are equal. ***b* = 53°**.
c *c* = **142°**   corresponding angles
   74° + *d* = *c*   vertically opposite angles
   74° + *d* = 142°
       d = 142° − 74°
       ***d* = 68°**
   e + 142° = 180°   angles on a straight line
       e = 180° − 142°
       ***e* = 38°**
       ***f* = 74°**   alternate angles

1 i Name the angle that is alternate to the marked angle
   ii Name the angle that is corresponding to the marked angle.

a   b   c   d

2 Name all the pairs of corresponding angles in each of the following diagrams.
   The diagrams show a section of trellis and a roof.

a   b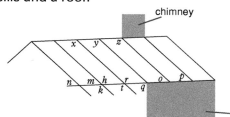

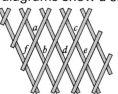

3 Find the size of the shaded angles. Give a reason.

a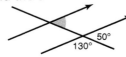
50°
130°

b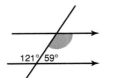
140°
40°

c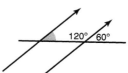
120° 60°

d
121° 59°

e
71° 109°

4 Name all the angles that are equal to
a *e*
b *b*
c *k*
d *d*

c d
b a e

m j
l k

i f
h g

5 Find the size of *x*.

a
68° 87°
x

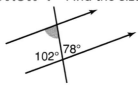
You'll need to use some other angle properties to work these out.

b
x
72°

c
x
61°

*6 Find the size of the angles marked *a, b, c, d, e, f, g*.

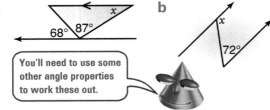
112° e
d 40°
b
a c
f g
96°

*7 Vanessa drew this triangle diagram. The lines BF and CE are parallel. AE and BD are parallel and so are DF and CA.
Find the size of *x, y, z, a, b, c*.

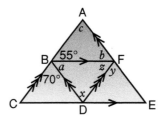
A
c
B 55° b F
a z y
70°
C D E
x

*8 Find the value of *n*.

a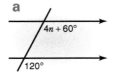
4n + 60°
120°

b
3n + 25°
68°

c
5n – 10°
3n + 38°

Review 1 Find the size of the marked angles. Give a reason.

a
102° 78°

b
82°
98°

c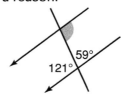
59°
121°

Review 2 For each of the following diagrams, write down the letters of all the angles which are equal to angle *a*.

a

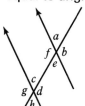

b

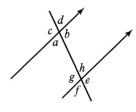

c

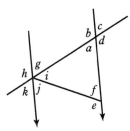

Review 3 Find the size of the angles marked as *a, b, c, d, e, f*.

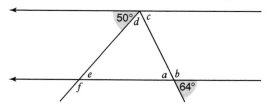

*Review 4** The arrows show the path of a billiard ball from X to H.

XY and ZH are parallel paths.

Find the values of *a, b, c, d* and *e*.

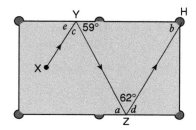

Angles in triangles

Remember

The angles *a, b* and *c* are called **interior angles** of a triangle. They are the angles inside the triangle.

The sum of the interior angles of a triangle is 180°.

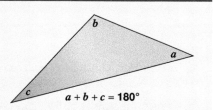

$a + b + c = 180°$

Discussion

Diana **showed** that the angles of a triangle add to 180° like this.

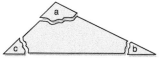

Prove means it is true for every case.

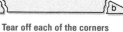

Tear off each of the corners   Put the torn off corners side by side

How could Diana use this to prove that $a + b + c = 180°$? **Discuss.**

Hint *x, c* and *z* are adjacent angles on a straight line.

You could use a dynamic geometry package.

Diana wanted to **prove** that the three interior angles of a triangle add to 180°.

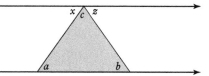

We know that

$x = a$ alternate angles, parallel lines, are equal

$z = b$ alternate angles, parallel lines, are equal

Angle a is called an **exterior** angle. It is outside the triangle.

The exterior angle of a triangle is equal to the sum of the two opposite interior angles.

$a = b + c$

Example $y = 57° + 59°$
$= \mathbf{116°}$

Discussion

Marcus and Abbie wanted to **prove** that the exterior angle of a triangle is equal to the sum of the two opposite interior angles. Marcus drew this diagram and wrote

$a + b + c = \mathbf{180°}$ interior angles of a △ add to 180°
$d + c = \mathbf{180°}$ adjacent angles on a straight line add to 180°

So a plus b must equal d or $a + b = d.$

Is Marcus correct? **Discuss.**

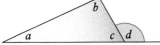

Abbie drew this diagram. The line parallel to CA cuts angle d into two angles, x and y.
How could Abbie use this diagram to prove that $a + b = d$? **Discuss.**

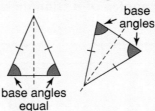

Remember
An **isosceles triangle** has two equal sides and two equal angles.
It has one axis of symmetry.

The unequal side is called the base.
The equal angles are at each end of the base.
They are called the **base angles**.

base
angles

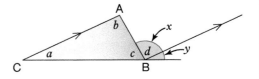

We mark equal
sides with a dash.

Worked Example
Find the value of n.

base angles
equal

Answer
Before we can find n, we must find the value of the unmarked angle.

unmarked angle $= 360° - 272°$ angles at a point add to 360°
$= 88°$
$n + n + 88° = 180°$
$2n + 88° = 180°$
$2n = 180° - 88°$
$2n = 92°$
$n = \mathbf{46°}$

Exercise 6

1 Find the value of x.

a
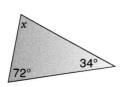
55°
71°
x

b
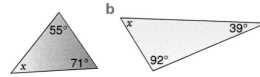
x
39°
92°

c
57°
x

d
50°
58°
x

e
x
72°
34°

f
324°
x
83°

g
286°
53°
x

h
314°
72°
x

2 Find the size of a.

a
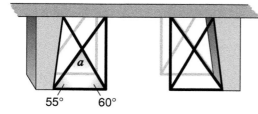
a
55°   60°

b
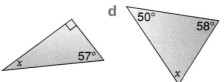
a
52°

c
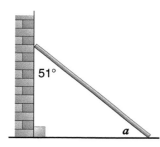
51°
a

d
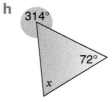
32°
a

3 Find the value of e.

a
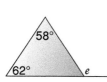
58°
62° e

b
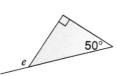
50°
e

c

112°
60° e

d
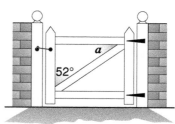
53°
120° e

4 Find the size of a.

a

56°
a a

b

a
100° a

c

88°
a a

d

a

e

a
72°

f
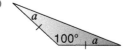
82°
a

g
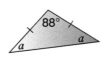
a
274°

h

a
296°

5 Find the value of x and y.

a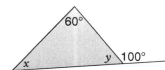

b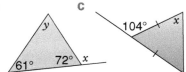

c

d

6 Use a copy of this diagram.

a The angle between Caley Drive and Raven Road is 105°. Which of the two angles between Caley Drive and Raven Road will this be? Write this angle on your diagram. The angle between Caley Drive and Gresham Avenue is 60°. Which angle is this? Write this angle on your diagram.

b Calculate the angles at the three corners of Bell Park.

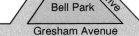

7 This tile is made from eight identical triangles.

a Find the size of angle x.

b Find the size of angle y.

8 Explain why a triangle can never have a reflex angle within it.

*9 Find the size of the angles marked with letters.

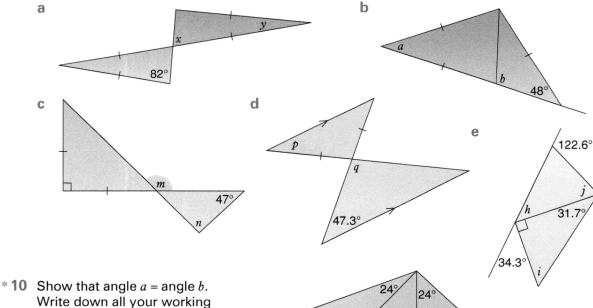

*10 Show that angle a = angle b.
Write down all your working
and reasons.

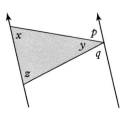

*11 Use a copy of this.
Fill in the gaps.

$p + q + y = \underline{\hspace{1cm}}$ adjacent angles on a straight line

$p = \underline{\hspace{1cm}}$ alternate angles, parallel lines are equal

$q = \underline{\hspace{1cm}}$ alternate angles, parallel lines are equal

How can you finish this to prove that $x + y + z = 180$?

Review 1 Find the value of x.

a

b
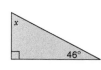

c

d

e
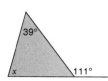

f

g

h

T

Review 2 Use a copy of this crossnumber.
Calculate the value of each unknown.
Using these values, fill in the crossnumber.

Across	**Down**
1. c	1. h
2. d	3. e
5. g	4. f
8. b	6. a
9. i	7. j

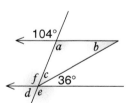

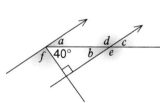

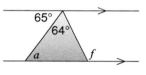

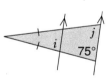

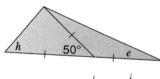

* **Review 3** Find the size of the angles marked with letters.

a

b

c

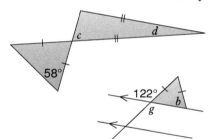

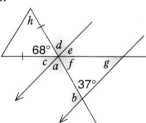

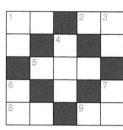

T

* **Review 4** Use a copy of this.
Fill in the gaps.

$x + y + z =$ _____ interior angles of a △ add to $180°$
$a + z =$ _____ adjacent angles on a straight line
How can you finish this to prove $a = x + y$?

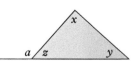

  Puzzle

1 Draw two equilateral triangles by drawing just five lines.

2 Draw two equilateral triangles by drawing just four lines.

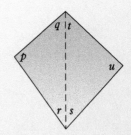

Investigation

*Angles in a quadrilateral

Martina drew this quadrilateral.
She divided it into two triangles as shown.

How could she use this diagram to prove the sum of the
interior angles of any quadrilateral is 360°? **Investigate**.

Summary of key points

line
infinite length

P————————Q
line segment
finite length

Two straight lines on the same **plane** must either **intersect** or be **parallel**.

We name this angle

   1 using the letter at the vertex, $\angle Q$

or **2** using three letters, the middle letter
   being the vertex, $\angle PQR$ or $\angle RQP$
   or   PÔR or RÔP.

If there is more than one angle at the vertex always use three letters to name an
angle.

B   **Parallel lines** never meet.
We show parallel lines with arrows.

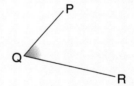

We show lines are **perpendicular** using the symbol

Continued ...

Parallel and perpendicular lines can be drawn using a **ruler and set square**.

Example   These diagrams show how to draw parallel lines.

Set
Square

| Ruler | Ruler | Ruler | Ruler |

Put the set square
on the ruler.

Draw a line.

Slide the set square
along the ruler.

Draw another line.

We use compasses and a ruler to construct the **perpendicular bisector** of a line segment, BC.

Open the compasses to a little more than half the length of BC. With compass point first on B and then on C, draw arcs to meet at P and Q.

Draw the line through P and Q. R is the Point which bisects BC.

 Angles

$a = b$
**Vertically opposite
angles are equal**

$x + y + z = 180°$
**Angles on a straight
line add to 180°**

$c + d + e = 360°$
**Angles at a point
add to 360°**

 Angles made with parallel lines

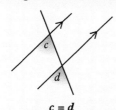

$c = d$
**corresponding angles
are equal**

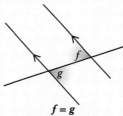

$f = g$
**alternate angles
are equal**

 The interior angles of a triangle add to 180°

Example $m + 57° + 64° = 180°$

$m = 180° - 57° - 64°$

$= 59°$

Interior angles are
inside the shape.

**The exterior angle of a triangle is equal to
the sum of the two opposite interior angles.**

Example $f = 47° + 58°$

$= 105°$

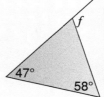

Exterior angles are
outside the shape.

Test yourself

1 Use a single letter to name the marked angles.

2 Use three letters to name the marked angles in question **1**.

3 a Name the parallel lines in this shape.
  b Name the perpendicular lines.

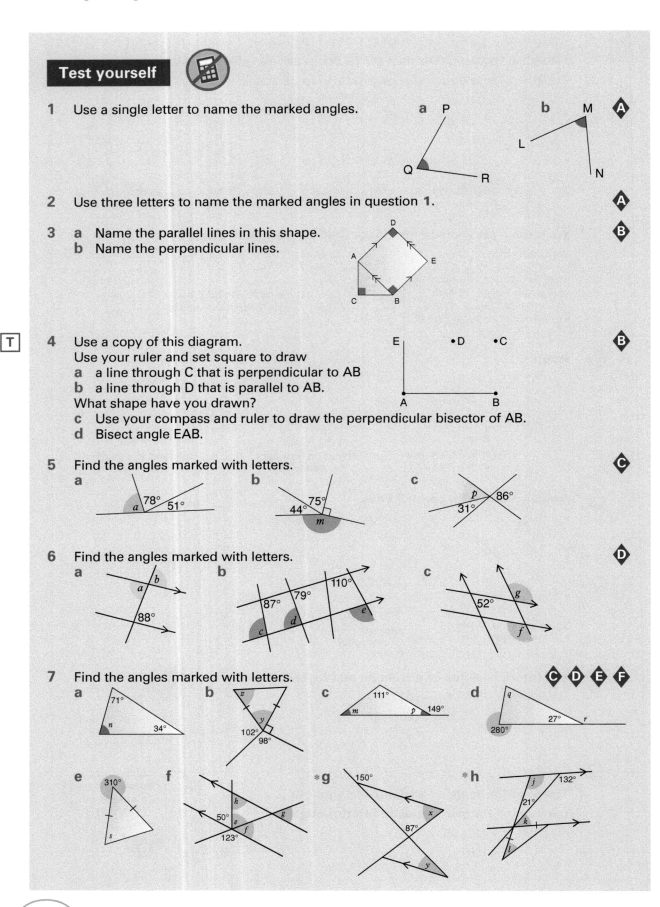

T 4 Use a copy of this diagram.
Use your ruler and set square to draw
a a line through C that is perpendicular to AB
b a line through D that is parallel to AB.
What shape have you drawn?
c Use your compass and ruler to draw the perpendicular bisector of AB.
d Bisect angle EAB.

5 Find the angles marked with letters.

6 Find the angles marked with letters.

7 Find the angles marked with letters.

12 Shape. Construction

You need to know

✓ 2-D shapes — triangles   page 271
  — quadrilaterals
  — polygons

✓ symmetry   page 273

✓ 3-D shapes   page 272

✓ faces, edges, vertices   page 272

✓ nets   page 272

Key vocabulary

adjacent, base (of solid), centre, circle, concave, convex, diagonal, edge, face, horizontal, identical, irregular, isometric, opposite

polygon: pentagon, hexagon, octagon

quadrilateral: arrowhead, delta, kite, parallelogram, rectangle, rhombus, square, trapezium

regular, shape, side

solid (3-D): cube, cuboid, cylinder, hemisphere, prism, pyramid, sphere, square-based pyramid, tetrahedron, three-dimensional (3-D)

triangle: equilateral, isosceles, right-angled, scalene

tessellate, two-dimensional (2-D), vertex, vertical, vertices

 ## Three of a kind

Triangles is a game for two students.

You will need a pen each.
Put six dots on a piece of paper so that no three are in a straight line.

Take turns to join two dots with a line.
The first player forced to make a triangle, loses.

Note If the dots are the vertices of a regular hexagon this game is called Sim.

> Each of the vertices of your triangle must be on a dot.

Naming triangles

A **triangle is named** using the capital letters of the vertices.
We start with one of the letters and go round in order clockwise
or anticlockwise.
This triangle could be named as △PQR or △QPR or △PRQ.
How else could it be named?

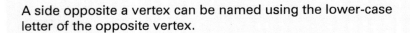

A side opposite a vertex can be named using the lower-case
letter of the opposite vertex.

Example   The side opposite ∠Q is named as *q*.
            The side opposite ∠P is named as *p*.

Exercise 1

1   Name the red triangle in each.

a
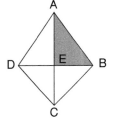

b
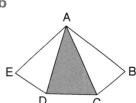

c
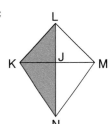

d
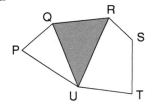

2   Use a single lower-case letter to name the side opposite the marked angle.

a
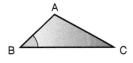

b
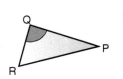

c

d
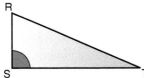

Review 1   Name the purple triangle in each.

a
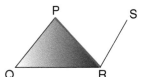

b
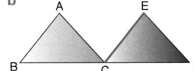

c

Review 2   Name the red side of each triangle in Review **1** using a single lower-case letter.

Describing and sketching 2-D shapes

Discussion

How might Benjamin and Simone have got their shapes?
Are any other shapes possible? **Discuss**.

Exercise 2

1 Imagine a square with both diagonals drawn on. Remove one of the small triangles.
   What shape is left?

2 Imagine a regular hexagon. Join adjacent mid-points of all of the sides of the hexagon.
   What shape is formed by the new lines?

   > Adjacent means next to one another.

3 Imagine a rectangle with one corner cut off.
   Describe the shape left.

4 Imagine an equilateral triangle. Fold it along one of its lines of symmetry. What angles are
   in the folded shape?
   Explain your answer.

5 Imagine a square. Cut off all the corners. What shape have you got left? Is this the only
   possible shape?

6 Imagine folding a piece of square paper in half and then half again to get a smaller square.
   One of the corners of this smaller square is the centre of the larger square of paper. Cut a
   small triangle off this corner. If the paper was opened out, what shape must the hole be?
   Give a reason.

7 Imagine two identical parallelograms put together along sides of equal length. What shape
   is made? Is this the only possible shape?

8 Imagine two identical rhombuses put together along one side.
   What shape is made? Is this the only possible shape?

9 Imagine two identical isosceles triangles put together along
   sides of equal length. What shape is made? Is this the only
   possible shape?

*10 Imagine two identical equilateral triangles put together along one edge.
  a What shape is made? Explain why.
  b If you add a third equilateral triangle, what shape is formed?
     Sketch this shape. Is this the only possible shape?
  c Add a fourth equilateral triangle. What shape is formed?
     Is this the only possible shape? Make some sketches.

Review 1 Imagine a regular pentagon. Cut it into two pieces.
What shapes have you got? Are these the only possible shapes?

Review 2 Imagine a semicircle cut out of paper. Fold it in half along its line of symmetry.
Fold it in half again and then in half again.
How many degrees is the angle at the vertex of the shape now?

Review 3 Imagine two identical scalene triangles put together along sides of equal length.
What shape is formed? Is this the only possible shape?
Explain your answer.

Properties of triangles, quadrilaterals and polygons

Investigation

Properties

You will need a geometry software package.

1 Draw two equal line segments from a common point.
   Draw two parallel lines to form a rhombus.
   Draw the two diagonals.
   Use the cursor to drag one of the vertices.
   What happens to the sides? Does the rhombus stay a rhombus?
   What happens to the diagonals and angles?

2 Repeat **1** except draw a parallelogram by drawing two unequal line segments from a common point.

A **polygon** is a 2-dimensional shape made from line segments which enclose a region.
2-D is short for **two-dimensional**. 2-D shapes have length and width but no height.
We sometimes call 2-D shapes **plane** shapes. A plane is a flat surface.

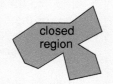

closed region

Discussion

You will need 3 × 3 pinboard or square dotty paper.

Discuss if it is possible to construct these on the pinboard.

- a trapezium with one line of symmetry
- a rhombus which is not a square

Once you have decided, check if you are correct by trying.

Investigation

T

Properties of triangles, quadrilaterals and polygons

A You will need two large copies of each of these:
equilateral triangle, isosceles triangle, square, rectangle,
parallelogram, rhombus, kite, isosceles trapezium,
trapezium, arrowhead.

> An isosceles trapezium
> has two equal sides.

1 A square has four lines of symmetry.
Check this by folding your square.
It should fold in half along each line shown.

How many lines of symmetry have each of these got?
Check by folding.
rectangle, parallelogram, isosceles trapezium, trapezium, kite, rhombus,
arrowhead, equilateral triangle, isosceles triangle

2 A parallelogram has rotation symmetry of order 2.
As you turn it one complete turn it fits onto itself in
two different positions

Turn each of your other shapes.
What is the order of rotation symmetry of each?

Use a copy of this table. Fill it in.

Shape	Number of lines of symmetry	Order of rotation symmetry
Equilateral △		
Isosceles △		
Parallelogram		
Rectangle		
Square		
Rhombus		
Kite		
Isosceles trapezium		
Trapezium		
Arrowhead		

3 A rhombus has rotation symmetry of order 2 and
2 lines of symmetry.

How can these symmetry properties be used to
show the following for a rhombus?

all the sides are of equal length
the diagonals cross at right angles
the diagonals bisect each other
the diagonals bisect the angles

> Remember rotation
> symmetry of order 2 means
> it fits onto itself twice as
> you turn it 360°.

Use the symmetry properties of the other shapes to fill in this table.

Quadrilateral	Diagonals equal	Diagonals cross at right angles	Diagonals bisect each other	Diagonal bisects the angles	Sides
Parallelogram	✗	✗	✓	✗	opposite sides equal
Rectangle					
Square					
Rhombus	✗	✓	✓	✓	4 equal
Kite					
Isosceles trapezium					
Trapezium					
Arrowhead					

B You will need a 3 × 3 pinboard and some square dotty paper.

1   There are eight *distinct* triangles you can make on a 3 × 3 pinboard.

Make all eight and draw them on square dotty paper.
Write down all the properties you know about each.

Example   Richard made this triangle.
He wrote

> These are properties of an isosceles triangle.

This is an isosceles triangle.
It has 2 equal sides and its base
angles are equal. It has one line of symmetry.

is the same triangle as

and and .

Put the eight triangles into groups.
For example, you could put all the isosceles triangles in one group.
Is there more than one way to group the triangles?

*2   There are 16 *distinct* quadrilaterals you can make on a 3 × 3 pinboard.
Make all 16 and draw them on square dotty paper.

Put them into groups according to their properties.
Some quadrilaterals will be in more than one group.

Example   Tessa made this trapezium.
She put it into the following groups.

quadrilaterals with a right angle
quadrilaterals with one pair of parallel sides
quadrilaterals with no lines of symmetry

Properties of triangles

Right-angled	**Isoceles**	**Equilateral**	**Scalene**
one angle is a right angle	2 equal sides base angles equal 1 line of symmetry	3 equal sides 3 equal angles 3 lines of symmetry	no 2 sides are equal no 2 angles are equal

Properties of quadrilaterals
A quadrilateral has 4 sides.

These are the **special quadrilaterals**.

Square
4 equal sides
4 right angles
4 lines of symmetry
Diagonals bisect at right angles

Rectangle
2 pairs of opposite sides equal
4 right angles
2 lines of symmetry
Diagonals bisect each other

Parallelogram
Opposite sides equal
and parallel
Diagonals bisect each other

Rhombus
A parallelogram with
4 equal sides
2 lines of symmetry
Opposite angles equal
Diagonals bisect at right angles

Trapezium
1 pair of opposite sides parallel

Isoceles trapezium
1 pair of opposite sides parallel
Non-parallel sides equal length
Diagonals equal length
1 line of symmetry

Kite
2 pairs of adjacent sides
of equal length
1 line of symmetry
1 pair of equal angles
Diagonals cross at right angles

Arrowhead or Delta
2 pairs of adjacent
sides equal
1 reflex angle
1 line of symmetry
1 pair of equal angles
Diagonals cross at right
angles outside the shape

A shape which has no reflex angles is a
convex shape.

Convex shapes

A shape which has one or more reflex
angles is a **concave** shape.

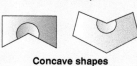

Concave shapes

We can use the properties of 2-D shapes to
find unknown angles.

Worked Example
Find *a* and *b*.

Answer
This shape is a rhombus.
The diagonals are lines of symmetry.
By symmetry, $a = 124°$.

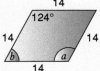

This diagonal cuts angle *a* and the 124° angle in half.
The blue triangle is isoceles.

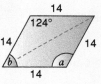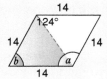

$62° + 62° + b = 180°$ angle sum of △ = 180°
$124° + b = 180°$
$b = 180° - 124°$
$= \mathbf{56°}$

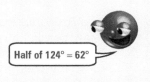
Half of 124° = 62°

1 Write true or false for each of these.
  a A rhombus has no equal angles.
  b An arrowhead has two pairs of adjacent sides equal.
  c An arrowhead is a concave shape.
  d A rectangle and a rhombus both have two lines of symmetry.
  e A square has only two lines of symmetry.
  f A parallelogram has opposite sides and opposite angles equal.
  g A square and a rhombus have diagonals that bisect at right angles.
  h A kite and an isosceles trapezium have diagonals that are equal.
  i The diagonals of a parallelogram cross at right angles.
  j The diagonals of a square, arrowhead and rhombus cross at right angles.

2 What shape am I?
  a I am a quadrilateral.
    I have four equal sides.
    I have only two lines of symmetry.

  b I am a quadrilateral.
    I have two pairs of adjacent sides equal.
    I am a concave shape.

  c I am a quadrilateral.
    My diagonals cross but do not bisect at right angles.
    I have just two lines of symmetry.

  d I am a quadrilateral.
    My diagonals bisect each other but not at right angles.
    I have no lines of symmetry.

  e I am a quadrilateral.
    None of my angles is reflex.
    My diagonals cross at right angles.
    I have just one line of symmetry

3 Name all the special triangles and quadrilaterals which have
  a Two axes of symmetry
  b Four axes of symmetry
  c no axes of symmetry
  d all angles equal
  e all sides equal
  f no rotation symmetry
  g rotation symmetry of order 2
  h diagonals that bisect each other
  i diagonals that bisect at right angles
  j diagonals that bisect the angles.

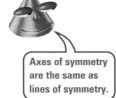

Axes of symmetry are the same as lines of symmetry.

4 a A quadrilateral with just one pair of parallel sides is a
      A parallelogram   B kite   C trapezium   D rhombus.
  b A quadrilateral with just one line of symmetry is a
      A parallelogram   B kite   C square   D rhombus.
  c A quadrilateral with equal diagonals that are also perpendicular is a
      A rhombus   B rectangle   C kite   D square.
  d A quadrilateral with parallel sides and equal diagonals is a
      A rhombus   B rectangle   C kite   D parallelogram.

5 Find the angles named with letters.
Use the properties of the shapes to help.

Remember the angle properties you learnt in Chapter 11.

a

125° *a*

b

132° *b*
48° *c*

Remember the dashes on the sides mean the sides are equal.

c

d

d

10 10 60°
f *e*
15 15

e

20 20
g
20

f

12
130°
12 12
i *h*
12

*g

15 *k*
58°
15
15
j
15

*h

l
82°

6 Explain why a triangle can never have a reflex angle but a quadrilateral can.

7 Explain why it is always possible to make an isosceles triangle from two identical right-angled triangles.

8 Morgana made this shape with three identical yellow and three identical blue tiles.
Each tile is in the shape of a rhombus.
Find the size of the angles in the yellow and blue tiles.

65°

9 Use a larger copy of this table.
This trapezium is symmetrical.
It has two pairs of equal angles and one pair of parallel sides.
It is filled in on the table as shown.

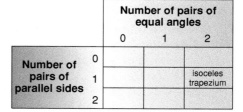

		Number of pairs of equal angles		
		0	1	2
Number of pairs of parallel sides	0			
	1			isoceles trapezium
	2			

i Fill these in on the table.
  a square   **b** rectangle   **c** kite
  d non-symmetrical trapezium
  e rhombus   **f** arrowhead (delta)

ii Some of the boxes are still empty. Is it possible to draw quadrilaterals for these? If not, explain why. Use the properties you know about quadrilaterals in your explanation.

A box could have more than one shape in it.

*10 Find the value of *x*. You will need to write an equation first.

a

x + 60°
2*x* − 10°

b

3*x* − 53° 2*x* + 10°

c

8
2*x* + 40°
8
8
x − 10°
8

Review 1 Name all the special quadrilaterals which have
a 1 axis of symmetry   **b** rotational symmetry of order 4   **c** equal diagonals.

Review 2 Which special quadrilateral am I?
a My diagonals intersect at right angles.
  I have four equal sides and no right angles.
b My diagonals bisect each other but not at right angles.
  My diagonals are not lines of symmetry.

Review 3 Find the angles marked with letters.

a

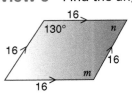

b

This shape is made
with three identical
rhombuses.

*__Review 4__ Is it possible to draw a quadrilateral with two pairs of parallel sides and one pair of
equal sides? If so, draw it. If not, explain why not.

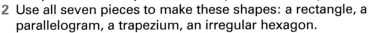

Investigation

Making shapes

You will need three equal sized squares of acetate sheet or tracing paper.

1 What shapes can you make by overlapping two
squares? **Investigate**.

Example   Rosemary made a quadrilateral and
a pentagon.

Which of these shapes cannot be made?
Explain why not.
  rectangle, decagon, hexagon, kite, octagon,
  isosceles triangle, trapezium, rhombus

2 What shapes can you make by overlapping three squares?
Investigate.

Put coloured outlines on
your squares to help.

3 Hamish put a vertex of one square at the centre of
the other square.
The area of overlap is one quarter of the area of
the square.
He rotated one square, keeping the vertex in the middle of the other square.
Is the area of overlap **always** a quarter of the area of one square? **Investigate**.

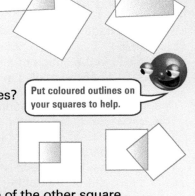

Puzzle

*

You will need squared paper or a set of tangram pieces.

This diagram shows how you can make a set of 7 tangram
pieces from squared paper.
1 Make a triangle using
   a 2 pieces   **b** 3 pieces   **c** 4 pieces   **d** 5 pieces.
2 Use all seven pieces to make these shapes: a rectangle, a
   parallelogram, a trapezium, an irregular hexagon.
3 Use all seven pieces to make these digits.
   Try and make some other digits.

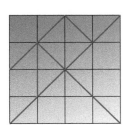

Practical

You will need LOGO.

1 A regular hexagon can be divided into six equilateral triangles. From this we can work out that each red shaded angle is 120°. Explain how.

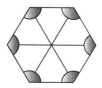

These instructions will draw a regular hexagon using Logo.

```
     forward  right turn  60°
        ↓        ↓      ↓
repeat 6 [fd  100  rt  60]
```

Explain why the angle turned through must be 60°. Use this diagram to help.

2 Draw a regular pentagon on the screen. Part of the instructions are
    repeat 5 [fd 100 rt ___].

[Fill in the angle.]

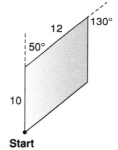

3 Predict what shape these instructions will draw, then check to see if you are right.
    repeat 3 [fd 100 rt 120]

4 Draw a regular octagon and a regular decagon on the screen.

5 These instructions draw this parallelogram.
    repeat 2 (fd 10, rt 50, fd 12, rt 130)
Draw some other parallelograms.

Write the instructions to draw a kite with two angles of 100° and another of 110°.

12   130°
50°

*6 These instructions draw five octagons arranged in a circle.
    repeat 5 [rt 72 repeat 8 [fd 50 rt 45]]
Draw some other designs.

10

Start

Investigation

Polygons and right angles

Is it possible to have two right angles in a triangle?

Is it possible to have just two right angles in a quadrilateral? Is it possible to have just three?
Is it possible to have 4?

What is the greatest number of right angles it is possible to have in a 5-sided polygon?

What if the polygon had six sides?
What if the polygon had seven sides?
What if ...

Investigate to find the maximum possible number of right angles in polygons with different numbers of sides.

Investigation

Polyiamonds

You will need some triangle dotty paper.
Polyiamonds are shapes made from equilateral triangles.
An equilateral triangle can be drawn using triangle dotty paper.

1 Here are two polyiamonds drawn by joining four equilateral triangles.
How many *different* polyiamonds can you make by joining four *identical* equilateral triangles?
Draw them on dotty paper.

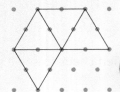 is the same shape as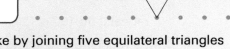

These shape are reflections of each other.

2 How many *different* polyiamonds can you make by joining five equilateral triangles together?

*3 There are 12 different ways of joining six identical equilateral triangles together.
Draw them all on dotty paper. Cut them out.

These 12 different shapes can be put together to make a parallelogram.
Investigate how.

Tessellations

A shape will **tessellate** if it can be used to completely fill a space with no overlapping and no gaps.
The shape may be translated, reflected or rotated to fill the space.

Examples

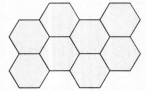

These hexagons tessellate.

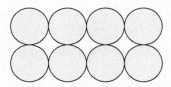

These circles do not tessellate.

Discussion

How would you describe this tiling pattern to someone over the phone? **Discuss**.
The pattern has been made with these two tiles.

How could you construct these tiles? **Discuss**.
You could use a computer tiling software package to make this pattern.

T

Investigation

Tessellations

1

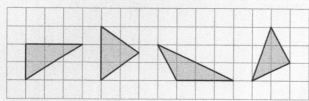

Choose one of the triangles shown.
On squared paper, tessellate this triangle to tile an area.
Colour your design, using at most three colours.

Do all triangles tessellate? **Investigate**. Explain why or why not using alternate and corresponding angles.
You could use computer tiling software.

2 a A polyiamond is a shape made from equilateral triangles.
Choose one of the polyiamond outlines shown below or draw another. On triangle dotty paper, tessellate this polyiamond to tile an area. Colour your tessellation, using at most four colours.

   b Do all polyiamonds tessellate?

3 Use computer tiling software to explore if other shapes will tessellate.

4 Design a poster, using the tessellation of a shape.

Constructing triangles and quadrilaterals

We can **construct a triangle** using a **ruler and protractor** if we are given
two sides and the included angle
or two angles and the side between them.

These diagrams show how to construct ABC. They are not full size.

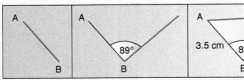

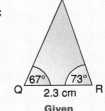

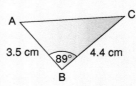

Draw AB, 3.5 cm long.

Draw an angle of 89 at B. Use your protractor.

Draw BC, 4.4 cm long. Join A to C.

Given
Two sides and the angle between them (called SAS)

Given
Two angles and the side between them (called ASA)

If the three sides of a triangle are given we can construct the triangle using a ruler and compasses.

Example This shows how to construct triangle *xyz*.

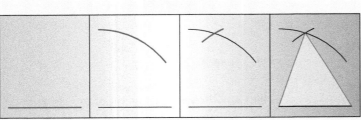

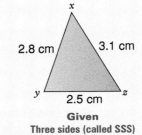

Given
Three sides (called SSS)

Draw a line 2.5 cm long.

Open the compasses out to 2.8 cm. With the point of the compasses on one end of the line, draw an arc.

Open the compasses out to 3.1 cm. With the point of the compasses on the other end of the line, draw an arc to cross the first arc.

Complete the triangle.

Be careful to keep the compasses open at the length you want.

To construct an accurate drawing you need to measure and draw carefully.

Discussion

How could you construct ABCD and PQRS? **Discuss.**

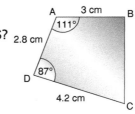

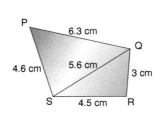

Exercise 4   **In this exercise, measure and draw very carefully.**

1 Use your ruler and compasses to draw triangles with sides of these lengths. Measure the angle between the first two sides given.
   a   3 cm, 4 cm, 5 cm   b   5 cm, 5 cm, 4 cm   c   8 cm, 4 cm, 6·6 cm   d   9 cm, 6 cm, 7 cm

2 Use your ruler and protractor or ruler and compasses to construct the quadrilaterals that are sketched. Measure the coloured lengths and angle.

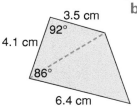
a   3.5 cm   92°   4.1 cm   86°   6.4 cm

b   3.1 cm   5.2 cm   7.5 cm   4.8 cm   8.2 cm

c   3.2 cm   115°   3.6 cm

3 Construct a rhombus with sides of 4 cm and an angle equal to 75°.

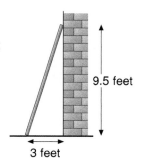
See page 288 for how to construct parallel lines.

4 Linda drew these sketches of her sister's toys. Make an accurate drawing of each.
   Note you will need a ruler, protractor and compasses.

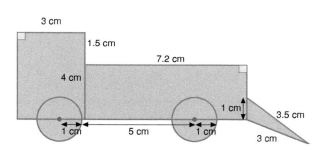
a   3 cm   1.5 cm   7.2 cm   4 cm   1 cm   3.5 cm   1 cm   5 cm   1 cm   3 cm

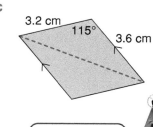
b   4 cm   77°   4.2 cm   4 cm   9 cm   100°   1 cm   3 cm   2 cm

*5 A 10 m high tree casts a 4 m shadow. Using a scale of 1 cm to represent 1 m, construct the triangle shown in the diagram.
   a   Use the protractor to measure the angle the sun's rays make with the ground.
   b   Write the scale you used as a ratio.

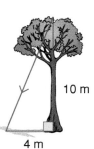
10 m   4 m

*6 A 10 foot ladder is leaning against a wall.
   The ladder reaches 9·5 feet up the wall and the foot of the ladder is 3 feet from the wall. Construct a triangle to represent this. Use a suitable scale. Use a protractor to measure the angle the ladder makes with the wall.

9.5 feet

3 feet

319

Review 1 Use your ruler and protractor to construct the quadrilateral that is sketched. Measure the length of the dashed red line on your diagram.

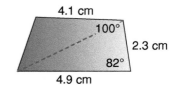

Review 2 Construct a parallelogram with sides of 4·5 cm and 3 cm and an angle of 68°.

Review 3 Wasim made a card with a picture of a yacht. This is a sketch of a yacht. Use your ruler and protractor to make an accurate drawing of the yacht. Use a scale of 1 cm represents 1 m.

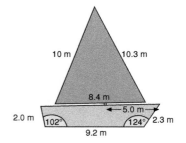

Describing and sketching 3-D shapes

Remember

3-D stands for **three-dimensional**.

3-D shapes have length, width and height.

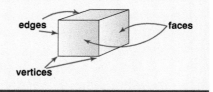

 Practical

You will need some Multilink or centimetre cubes and a partner.
Sit back to back with your partner.
Without letting your partner see, make a model using at least six cubes.

Example Beth and Tony made these.

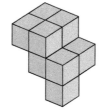 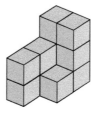

Take turns to tell your partner how to make the model.

Exercise 5

1 Imagine painting a dot on each face of three cubes.
a Imagine glueing the three cubes together in a row.
How many dots would be showing altogether?
b Imagine glueing the three cubes together in an L-shape.
How many dots would be showing altogether?

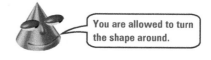

You are allowed to turn the shape around.

2 Imagine a pyramid with a square base.
Cut a slice off the top.
a Describe the new face that is created.
b How many faces, edges and vertices does the new solid have?

This is the base

3 Imagine two identical square-based pyramids.
Put them together, matching the bases.
a Describe the new solid.
b How many faces, edges, vertices does the new solid have?

4 a Imagine you are standing in front of a large square-based pyramid.
A genie gives you a magic potion and you shrink to the size of a mouse.
i If you walk around the pyramid what is the maximum number of faces you can see at one time?
ii The genie now gives you wings to fly. You fly above the pyramid.
What is the maximum number of faces you can see at one time?

b Imagine you are standing in front of a large cube that comes up to your waist.
Is it possible for you to stand so that you can see just
i 1 face　ii 2 faces　iii 3 faces　iv 4 faces?
Sketch what you would see each time.

*5 Imagine a jewellery box. Imagine any two opposite sides painted yellow and the rest of the box painted blue.
a At how many edges do a yellow and blue face meet?
b At how many edges do two blue faces meet?

Review 1 Imagine you are standing in front of a large tetrahedron as tall as a house.
a Is it possible to stand so you can see just
i 1 face　ii 2 faces　iii 3 faces?
b If you flew above the tetrahedron, what is the maximum number of faces you can see at one time?

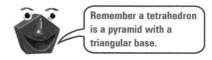

Remember a tetrahedron is a pyramid with a triangular base.

Review 2 How could you cut a cube to create an hexagonal face?

* **Review 3** Imagine two cubes. Paint two opposite faces of each cube red. Paint the other four faces on each cube blue.
a Put the two cubes together with two red faces next to each other as shown.
i How many edges does the new shape have?
ii How many faces does the new shape have?
iii At how many edges of the new shape do a red and a blue face meet?
iv How many faces of the new shape are red?
b Put the two cubes together with two red faces *joined*.
Repeat questions i–iv of a.

Investigation

Cubes

You will need isometric paper and Multilink or centimetre cubes.

1 This shows you how to draw a cube on isometric paper.

Isometric paper is triangle dotty paper.

Simon wanted to know how many different ways four cubes could be put together. He drew these diagrams on isometric paper.

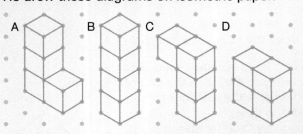

Make sure the isometric paper is the right way up. If it is the dots will make vertical lines.

A and C are the same shape.
Simon has drawn just three different ways of putting four cubes together. How many different ways are there of putting four cubes together? **Investigate**. Draw them on isometric paper.

*2 How many different ways are there of putting five cubes together? **Investigate**. Draw them on isometric paper.

3 **You will need** eight cubes.
Make a larger cube with the eight small cubes.
a How many ways can you split your larger cube into two *identical* parts? **Investigate**.
Draw each on isometric paper.
b How many ways can you split your larger cube into two *different* parts? **Investigate**.
Draw each on isometric paper.

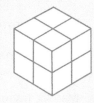

Nets

A 2-D shape that can be folded to make a 3-D shape is called a **net**.

Example   This net folds to make the shape shown.

Many nets are made up of squares, rectangles and triangles.
We can construct nets using a ruler, set square, protractor and compasses.

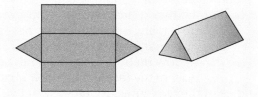

Discussion

Will both of the nets below fold to make this prism? **Discuss**.

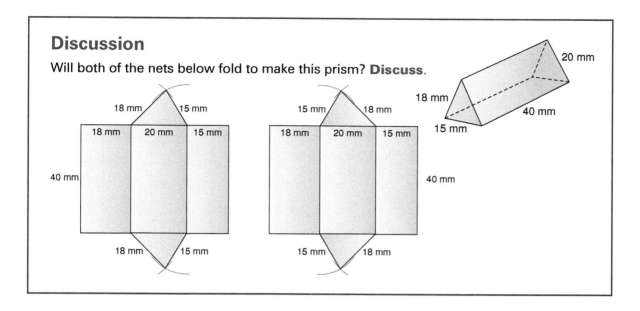

Exercise 6

T

1 a Jasmine drew this net for a cube.
   She wanted her cube to have a red and a
   blue strip along the top edges like this.
   Use a copy of the net.
   Colour a red strip on the other
   edge that should be red.
  b Jasmine drew a different net for a cube.
   When the net is folded, which colour will be
   opposite the green face?
  c Copy the net.
   Put a dot on the two corners that will meet
   the corner with the dot on it.

2 Imagine folding these nets.
   Which ones will fold to make cuboids?

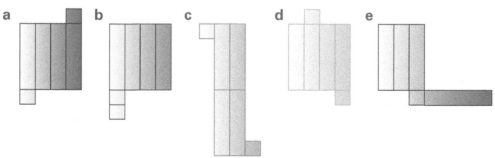

  a   **b**   **c**   **d**   **e**

3 Use a ruler and set square to draw nets for these cuboids.
   a length 5 cm, width 4 cm, height 2 cm   **b** length 4 cm, width 2 cm, height 3 cm
   c length 6 cm, width 5 cm, height 2 cm   **d** length 4 cm, width 3 cm, height 1 cm
   Fold your nets to make the cuboids.

Remember to put
tabs on.

4 **Sketch** the nets for these shapes.

a b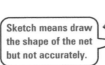

Sketch means draw the shape of the net but not accurately.

T

5 Joanne started the net below for a regular tetrahedron.
Use two copies of this.
Use a ruler and protractor to finish the net in two different ways. Make your drawings accurate.

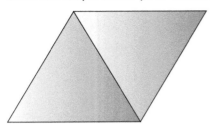
3.5 cm

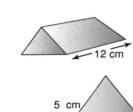
6 cm
60°
6 cm
6 cm

6 Robyn made this pyramid-shaped box for an Easter egg.
It has a square base.
Draw the net for this box accurately.

7 David made some chocolates for his aunt's birthday. He designed a box for them with triangular ends.
This sketch shows the triangular end.
Construct this triangle. Using your triangle as part of the net for the box, finish the net accurately.
Fold it to make the box.

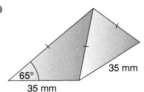
12 cm

5 cm
55°
6 cm

This shape is a *prism*.
It has two identical ends.

8 Draw a net for each of these.

a
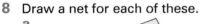
48°   63°   4 cm
5 cm
triangular prism

b
65°
35 mm
35 mm
square-based pyramid

9 Jason wanted to make a cardboard stand for two of his model cars.
He sketched this diagram.
Use a ruler and set square to draw the net for Jason's stand.
Fold your net to make the shape.

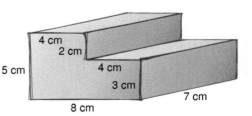
4 cm
2 cm
4 cm
5 cm
3 cm
8 cm
7 cm

10 Use a ruler, compasses and set square to construct a net for each of these.

a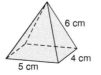
6 cm
4 cm
5 cm

b
6 cm
4 cm
4 cm

c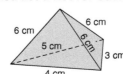
6 cm
6 cm
6 cm
5 cm
3 cm
4 cm

11 Make a net for each of the prisms sketched below.
Fold your nets to make these prisms.

a

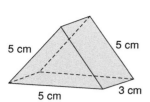

5 cm 5 cm
5 cm 3 cm

b

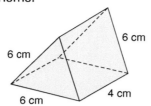

6 cm 6 cm
6 cm 4 cm

Remember to put tabs on your net.

Review 1 Ralph wanted to make this cube. He drew this net.

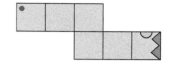

Copy the net.
a Draw the other half of the diamonds and circle on the net.
b Put a dot on the other two corners that will meet the corner with the red dot in it.

Review 2 Use a ruler and set square to draw nets for these cuboids.
a length 8 cm, width 3 cm, height 2 cm
b length 50 mm, width 40 mm, height 25 mm
Fold your nets to make the cuboids.

Review 3 Asha made a box for her little sister's front tooth.
It had a triangular end like the one shown.
Draw a net for this box.

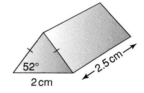

52°
2 cm
2.5 cm

Review 4 Use a ruler, compasses and set square to construct the net for each of these.

a

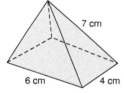

7 cm
6 cm 4 cm

b

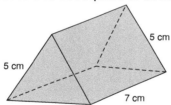

5 cm
5 cm
4 cm 7 cm

? **Puzzle**

Here are four pictures of the same cube.

Which shapes are opposite each other?

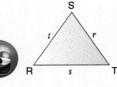

*Investigation

Least wastage

Jayne's class are going to sell sweets.
They have decided to pack the sweets into open boxes which are cubes with 10 cm sides.
First, they have to make the nets from sheets of cardboard 1 m by 1 m.
Andrew suggested they should use the net that would waste the smallest amount of cardboard.

Investigate to find this net. Assume the net is put together using sticky tape and no tabs.

Summary of key points

 A **triangle is named** using the capital letters at the vertices.

This triangle could be named as △STR or △SRT or △RST or ...
The side opposite each vertex is named with the lower-case letter of the vertex.

> Start with one letter and go round in order.
> ↻ or ↺

 A **polygon** is a 2-D shape made from line segments enclosing a region. 2-D is short for **two-dimensional**.

Properties of triangles A triangle is a 3-sided polygon.

Right-angled	**Isoceles**	**Equilateral**	**Scalene**
one angle is a right angle	2 equal sides base angles equal 1 line of symmetry	3 equal sides 3 equal angles 3 lines of symmetry	no 2 sides are equal no 2 angles are equal

Properties of Quadrilaterals A quadrilateral is a 4-sided polygon.

  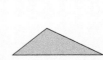

Square	**Rectangle**	**Parallelogram**	**Rhombus**
4 equal sides 4 right angles 4 lines of symmetry Diagonals bisect at right angles	2 pairs of opposite sides equal 4 right angles 2 lines of symmetry Diagonals bisect each other	Opposite sides equal and parallel Diagonals bisect each other	A parallelogram with 4 equal sides 2 lines of symmetry Opposite angles equal Diagonals bisect at right angles

Trapezium	**Isoceles trapezium**	**Kite**	**Arrowhead or Delta**
1 pair of opposite sides parallel	1 pair of opposite sides parallel Non-parallel sides equal length Diagonals equal length 1 line of symmetry	2 pairs of adjacent sides of equal length 1 line of symmetry 1 pair of equal angles Diagonals cross at right angles	2 pairs of adjacent sides equal 1 reflex angle 1 line of symmetry 1 pair of equal angles Diagonals cross at right angles outside the shape

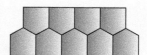

 A shape will **tessellate** if it can be used to completely fill a space with no overlapping and no gaps.

 We can **construct triangles and quadrilaterals** using a set square and ruler or compasses and ruler.

Examples

To construct this triangle
1. Draw PR 2·6 cm long
2. Draw an angle of 85° at R
3. Draw RQ 2·8 cm long
4. Join P to Q.

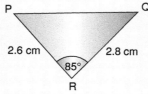

To construct this triangle
1. Draw AB 2·8 cm long.
2. Open compasses to 2·5 cm and with point on A draw an arc.
3. Draw an arc from B, 2 cm long.
4. Complete the triangle.

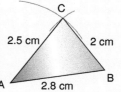

 3-D shapes

3-D stands for **three-dimensional**. 3-D shapes have length, width and height. 3-D shapes can be drawn on triangle dotty paper.

 A 2-D shape that can be folded to make a 3-D shape is called a **net**.

Example The net folds to make a tetrahedron.

Test yourself

1 Name these triangles, using three letters.

a 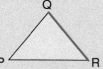 b 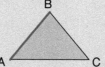 c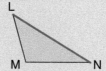

2 Using a lower-case letter, name the red side of each triangle in question **1**.

3 Write true or false for these.
  a A kite has one line of symmetry and two equal angles.
  b A rhombus has diagonals which bisect at right angles and two lines of symmetry.
  c A parallelogram has two lines of symmetry and diagonals that bisect each other.

4 Write down all the special quadrilaterals
  a with two lines of symmetry
  b which are concave
  c with diagonals that bisect at right angles.

5 Find the value of the angles marked with letters. **B**

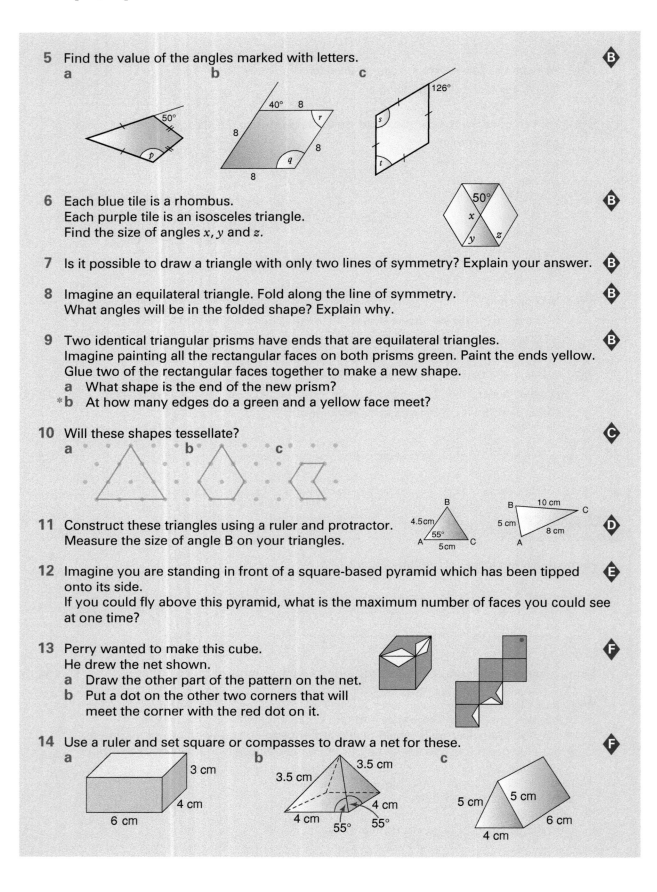

a

b

c

6 Each blue tile is a rhombus. **B**
Each purple tile is an isosceles triangle.
Find the size of angles x, y and z.

7 Is it possible to draw a triangle with only two lines of symmetry? Explain your answer. **B**

8 Imagine an equilateral triangle. Fold along the line of symmetry. **B**
What angles will be in the folded shape? Explain why.

9 Two identical triangular prisms have ends that are equilateral triangles. **B**
Imagine painting all the rectangular faces on both prisms green. Paint the ends yellow.
Glue two of the rectangular faces together to make a new shape.
a What shape is the end of the new prism?
*b At how many edges do a green and a yellow face meet?

10 Will these shapes tessellate? **C**
a   **b**   **c**

11 Construct these triangles using a ruler and protractor. **D**
Measure the size of angle B on your triangles.

12 Imagine you are standing in front of a square-based pyramid which has been tipped **E**
onto its side.
If you could fly above this pyramid, what is the maximum number of faces you could see
at one time?

13 Perry wanted to make this cube. **F**
He drew the net shown.
a Draw the other part of the pattern on the net.
b Put a dot on the other two corners that will
meet the corner with the red dot on it.

14 Use a ruler and set square or compasses to draw a net for these. **F**
a   **b**   **c**

13 Coordinates and Transformations

You need to know

✓ coordinates   page 273

✓ 2-D shapes   page 271

✓ symmetry   page 273

✓ transformations   page 273

Key vocabulary

axes, axis, axis of symmetry, centre of enlargement, centre of rotation, congruent, coordinates, direction, enlarge, enlargement, grid, image, line of symmetry, line symmetry, mirror line, object, order of rotation, origin, position, quadrant, reflect, reflection, reflection symmetry, rotate, rotation, rotation symmetry, scale factor, symmetrical, tessellation, transformation, translate, translation, *x*-axis, *x*-coordinate, *y*-axis, *y*-coordinate

 That sinking feeling

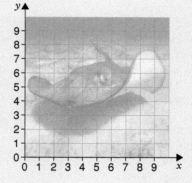

Sink the Ships — a game for 2 players
You will need a grid for each player.
To play

● Each player puts these ships on the grid.
    1 submarine (1 dot)
    2 destroyers (2 dots each)
    1 cruiser (3 dots)
    1 battleship (4 dots)
The ships must not cross.

Example

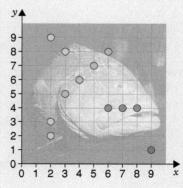

● Players take turns to name a coordinate.
If one of the other player's dots is at this coordinate
this player must say 'hit' and cross out that dot.

● Once all the dots of a ship have been 'hit', call out
'ship sunk'.

● The winner is the first person to sink all of the other
player's ships.

Note You could use a grid with positive and negative
      numbers instead.

Coordinates

We use **coordinates** to give the position of a point on a grid.
The coordinates of Q are (⁻3, 2). We always give the *x* coordinate first.

The coordinates of the **origin** are (0, 0).

The *x*- and *y*-axes make four quadrants as shown.
Q is in the second quadrant.

We name the quadrants anticlockwise. ↰

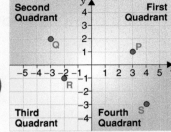

Exercise 1

1 **a** Which quadrant is each of these in?
     i (4, 3)   **ii** (⁻3, ⁻4)   **iii** (⁻3, 2)   **iv** (2, ⁻1)
   b Which of these points are on the *x*-axis?
     (0, 4)   (⁻3, 0)   (0, ⁻2)   (4, 0)
   c Which of these points are on the *y*-axis?
     (0, 4)   (⁻3, 0)   (0, ⁻2)   (5, 0)

2 Draw *x*- and *y*-axes on squared paper. Number the *x*-axis from ⁻8 to 8 and the *y*-axis from ⁻6 to 10.
   a These are the coordinates of the vertices of a shape. Plot them on your grid. Join them in order.
     (⁻1, 2), (5, 2), (6, 0), (2, ⁻1), (⁻2, 0), (⁻1, 2)
   b How many lines of symmetry does this shape have?
   c The points (⁻4, 0), (1, 0), (1, ⁻3) are three vertices of a rectangle.
     Write down the coordinates of the fourth vertex.
   d The points (⁻2, 2) and (3, 2) are two vertices of a rectangle.
     What might the coordinates of the other two vertices be?

3 Draw and label axes on squared paper. Choose a suitable scale so that you can plot these points.
   A (⁻4, ⁻1)   B (4, 2)   C (0, ⁻2)   D (1, 1)
   E (4, ⁻1)   F (1, ⁻1)   G (1, 2)   H (⁻2, 1)
   a Which of the points could you join to make a
     i square   **ii** parallelogram   **iii** trapezium?
   b Which of the points could you join to make
     i a right-angled triangle   **ii** an isosceles triangle without a right angle?

4 Draw *x*- and *y*-axes on squared paper. Number both axes from ⁻4 to 10.
   a Plot the points (⁻2, 1), (1, 2), (⁻1, 3).
   b What fourth point could you plot to make
     i a parallelogram   **ii** a kite   **iii** an arrowhead?
     Is it possible to make a rectangle? Explain your answer.

*5 **a** What are the mid-points of the lines joining
     i (2, 4) to (6, 4)   **ii** (4, 2) to (4, 8)   **iii** (4, 2) to (10, 4)?
   b Try to write a rule for finding the mid-point of the line joining (x_1, y_1) to (x_2, y_2).

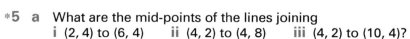

You could plot the points.

Review Draw and label axes on squared paper. Have the *x*-axis from ⁻7 to 5 and the *y*-axis from ⁻10 to 5.

a Plot the points (⁻2, 3), (⁻1, ⁻2), (⁻6, ⁻3).
What fourth point could you plot to make a square?
Is it possible to make a rectangle that is not a square? Explain why or why not.

b Plot the points (⁻4, ⁻3), (2, ⁻1) (⁻2, 1).
What fourth point will make
  i a kite
  ii a parallelogram?

Practical

1 Make up a game that uses coordinates.
Write instructions for your game.
Play your game to check it works.

2 **You will need** a graphical calculator.

  a Use the plot and line on your graphical calculator to draw these.

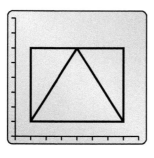 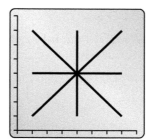 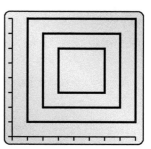

  *** b** Fabian Peters put his initials on the screen.
Draw your initials on the screen.

  c Plot the points (4, 2) and (7, 2).
These two points are the ends of one side of a trapezium.
Plot two other points to finish the trapezium.
Plot two different trapezia with the points (4, 2) and (7, 2) as the ends of one side.
Make other shapes which have the points (4, 2) and (7, 2) as the ends of one side.
What if the points are (4, 2) and (5, 6) instead of (4, 2) and (7, 2)?

  *** d** Draw this outline of a cube on your screen.

Trapezia is the plural of trapezium.

Reflection

Discussion

As Jenny walked into a jeweller's she saw these reflections of three clocks in a mirror. What time was each of these clocks actually showing?

● The word AMBULANCE is to be written across the front of an ambulance. The driver of a car in front of the ambulance can see AMBULANCE in the rear vision mirror. How should the word AMBULANCE be written on the front of the ambulance?

When we **reflect** an **object** in an **axis of symmetry** or **mirror line** we get an **image**.
ABCDE has been reflected in the mirror line to get the image A'B'C'D'E'.
We say ABCDE **maps** onto A'B'C'D'E'.

To map A'B'C'D'E' onto ABCDE we use the **inverse** reflection. The inverse reflection is the reflection in the same mirror line.

The image is **congruent** (same shape and size) to the original.

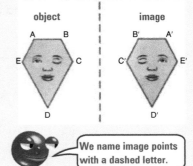

We name image points with a dashed letter.

Practical

You will need a geometry software package like Geometer's Sketchpad.

Draw a line to be a mirror line on your screen.
Draw a shape on one side of the line.
From one vertex of your shape, draw a line perpendicular to the mirror line.
Extend this line to a point an equal distance on the other side of the mirror line. This will be the image point.
Repeat this for the other vertices.
Join the points to give the image.

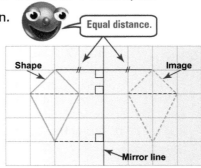

Equal distance.

Use the cursor to drag the vertices of the original shape.
What happens to the image?
What happens if one of the vertices of the original shape is on the mirror line?
What happens if the original shape crosses the mirror line?

The line joining any point to its image is perpendicular to the mirror line.

Example   AA′ and CC′ is perpendicular to the mirror line.

The image is the same distance behind the mirror line as the object is in front.
Points **on** the mirror line do not change their position when reflected.

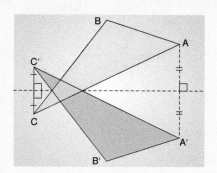

Exercise 2

1 ABC is reflected in the dashed line.
  a Which points will not change their position? Why?
  b Draw the reflection of ABC. Label it A′B′C′.
  c A′B′C′ is reflected in the dashed line.
    What happens?

You can use tracing paper if you like.

T

2 Use a copy of this.
Reflect these shapes in the dashed line.
Write down the name of the quadrilateral you get.

  a   **b**   **c**   **d**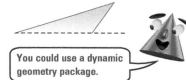

You could use a dynamic geometry package.

T

3 Use a copy of this.
Reflect these shapes in **m**.

  a 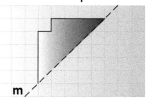   **b**    **c**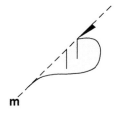

T

4 Use a copy of this.
Reflect these words in the mirror line.

  a    **b**    **c**    **d**

5 Explain why A'B'C'D' is not the reflection of ABCD in the mirror line **m**.

6

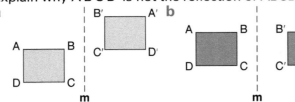

    a Some letters look the same after reflection in a vertical mirror line.
        What do these letters have in common?
    b Some letters look the same after reflection in a horizontal mirror line.
        What do these letters have in common?
    c Some letters look the same in both a vertical and a horizontal mirror line.
        What do these letters have in common? Which letters are they?

7

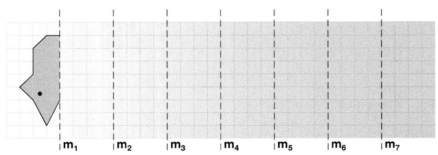

Use a copy of this.
    a Reflect the shape in m_1.
    b Reflect the image you got in **a** in m_2.
    c Continue reflecting the images in m_3, m_4, m_5, m_6, m_7.
        Describe what happens.

8 Use a copy of this.
    a Reflect the shape in m_1.
    b Reflect the image you got in **a** in m_2.

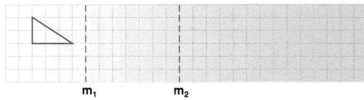

    c Repeat **a** and **b** for this diagram.

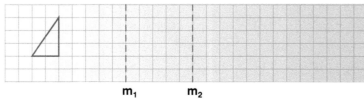

    d What *single* transformation is equivalent to reflection in two parallel lines?
        A translation   **B** rotation   **C** reflection

9 Rick is making patterns by reflecting a shape in two perpendicular lines.

  a Use a copy of the diagrams below. For each
   1 reflect the shape in the *x*-axis
   2 reflect the image you got in **1** in the *y*-axis.
  b What *single* transformation is equivalent to **1** *and* **2** in **a**?
   A reflection   **B** rotation of 180°   **C** rotation of 90°

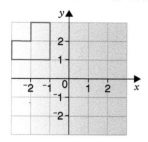

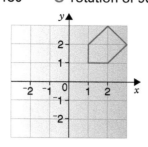

10 a Use the diagrams from question **9**. For each
   1 reflect the shape in the *y*-axis.
   2 reflect the image you got in **1** in the line *y* = 1.
  b What *single* transformation is equivalent to **1** *and* **2** in **a**?
   A reflection   **B** rotation of 180°   **C** rotation of 90°
  c What single transformation is equivalent to reflection in any two perpendicular lines?

Review 1 Use a copy of these.
Reflect each in the dotted line.

a

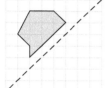

b

c

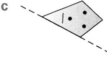

Review 2 Use a copy of this.
  a Reflect this shape in the line *x* = ⁻1 and then reflect the image in the *x*-axis.
  b Use another copy. Reflect the shape in the *x*-axis and then reflect the image in the line *x* = ⁻1.
  c Do you get the same result for **a** and **b**?
  d What *single* transformation is the same as **a**?
   A translation   **B** rotation of 180°   **C** reflection

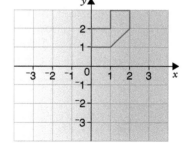

Practical

You will need a dynamic geometry software package.

Draw a triangle.
Reflect it in one of its sides.
Name the shape that the image and object together make.
What if you reflect the triangle you drew in one of its other sides?
Try and make these shapes by reflecting a triangle in one of its sides.
Which ones can you make?

mirror line

The image and object together make a kite.

  rectangle   square   kite   parallelogram
  arrowhead   rhombus   trapezium
  ∗ For the shapes you cannot make, explain why not.

Rotation

Jordan went to the fair.
He went on two rides where he was **rotated**.
A **rotation** is a movement round a centre point.

$\frac{1}{4}$ turn = 90° $\frac{1}{2}$ turn = 180° $\frac{3}{4}$ turn = 270°

In both of the diagrams the blue shape has
been **rotated** 90° anticlockwise about O.
A maps to its image point A'.
O is called the **centre of rotation**.
The centre of rotation can be inside
the shape or outside the shape.

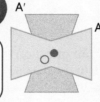

The centre of
rotation stays in
the same place as
you rotate a shape.

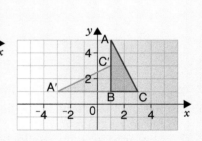

The image is **congruent** to the original.

To rotate a shape you need to know **the centre of rotation** and the **angle of rotation**.
If the direction is not given, the angle of rotation is always in an **anticlockwise** direction.

Worked Example
Rotate ABC 90° about (1, 1).

Answer
We can use tracing paper.

1 Trace ABC onto tracing paper.
2 Put a pin or sharp pencil through
   the tracing paper at (1, 1).
3 Turn the tracing paper 90° anticlockwise.
4 Press hard on the tracing paper to give the outline of A'B'C'.
5 Remove the tracing paper. Draw A'B'C'.

Exercise 3

T

1 a

b

c

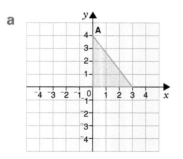

Use a copy of these.
  a Rotate the shape in **a** through 180° about the point (2, 0).
  b Rotate the shape in **b** through 90° about the point (1, 1).
  c Rotate the shape in **c** through 45° about the point (0, 0).
  d Write down the coordinates of point A' after the rotation in
     i part **a**   and   **ii** part **b**.

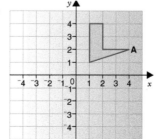

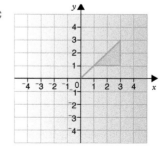

You can use tracing
paper.

2 Use another copy of question **1**.
  a Rotate the shape in **a** through 270° using (⁻1, ⁻1) as the centre of rotation.
  b Rotate the shape in **b** through 180° using (1, 0) as the centre of rotation.
 *__c__ Rotate the shape in **c** through 135° using (0, 0) as the centre of rotation.
 *__d__ Write down the coordinates of A′ after the rotation in
    i part **a**   and   **ii** part **b**.

3 A rotation through 90° about (0, 0) maps **A** onto **A**′.
  Which of these maps **A**′ onto **A**? There are two correct answers.
  a rotation through 180° about (0, 0)
  b rotation through 90° about (0, 0)
  c rotation through 270° about (0, 0)
  d rotation through 90° clockwise about (0, 0)

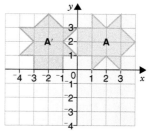

4 A rotation through 270° about (1, 2) maps B onto B′.
  Which of these maps B′ onto B?
  There are two correct answers.
  a rotation through 90° about (0, 0)
  b rotation through 90° about (1, 2)
  c rotation through 270° clockwise about (1, 2)
  d rotation through 180° about (1, 2)

5 Copy this using just one of the words in red in each gap.
  The inverse of a rotation is either an **equal/unequal** _____ rotation about **the same/
  a different** _____ point in the **same/opposite** _____ direction, **or** a rotation about
  the same/a different _____ point in the same direction so that the angle of the original
  rotation and the angle of the inverse rotation add to 360°.

6 Use another copy of question **1**.
  a Rotate the shape in **a** 90° about the origin. Then rotate the image 180° about the origin.
    What single rotation is equivalent to these two rotations?
  b Rotate the shape in **b** 180° about (1, 1). Then rotate the image 90° about (1, 1).
    What single rotation is equivalent to these two rotations?
  c Rotate the shape in **c** 90° about (0, 0). Then rotate the image 45° about (0, 0).
    What single rotation is equivalent to these two rotations?
  d Which of these statements is true?
    A There is no single rotation that is equivalent to two rotations about the same point.
    B Two rotations about the same point in the same direction is equivalent to a single
      rotation about the same point in the same direction by the sum of the angles of the
      two rotations.
    C Two rotations about the same point in the same direction is equivalent to a single
      rotation about the same point in the same direction by 360° — the sum of the angles
      of the two rotations.

*__7__ Use three copies of this triangle.
  a Rotate the triangle 180° about A.
    What shape is formed by the triangle *and* its image?
    Mark any equal sides using a dash.
    Mark any equal angles using ● or ×.
    Explain why they are equal.
  b Repeat **a** but rotate the triangle about B.
  c Repeat **a** but rotate the triangle about C.

*__8__ **a** ABCD is rotated about (0, 0) through $x°$ and then $x − 20°$.
    Write an expression for the equivalent single rotation.
  b The equivalent single rotation is 145°. Write and solve an equation to find x.

*__9__ Repeat question **8** but ABCD is rotated through $x + 40°$ then $2x − 30°$.

T **Review 1**

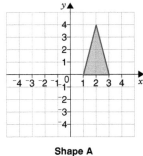

Shape A

Shape B

a Rotate **shape A** through 90° about the point (1, ⁻1).
b Rotate the **shape B** through 180° about the point (0, 3).
c Which two of these gives the inverse rotation of the rotation in **a**?
    A a rotation of 90° about the point (⁻1, 1)
    B a rotation of 270° about the point (1, ⁻1)
    C a rotation of 90° clockwise about the point (⁻1, 1)
    D a rotation of 270°clockwise about the point (1, ⁻1)

Review 2 Use another copy of Review **1 shape A**.
a Rotate the shape through 90° about (1, 0).
b Rotate the image you got in **a** through 180° about (1, 0).
c What single rotation is equivalent to these two rotations?

Review 3
a PQRS is rotated about (0, 0) through $2x + 25°$ and then $x - 15°$.
    Write an expression for the equivalent single rotation.
b The equivalent single rotation is 175°. Write and solve an equation to find x.

⭐ **Practical**

T

A You will need a copy of the triangle.

Samara made this tessellation by rotating a scalene triangle 180° about the mid-point of one of its sides.

Use a copy of Samara's triangle.
Try to make a different tessellation by rotating the triangle about the mid-point of one of the other sides.

B You will need a geometry software package.
Mark drew a trapezium on the screen.
He used rotations and reflections to make this tessellation.

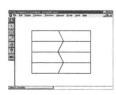

Draw one of these shapes on the screen.
Use rotations, reflections and/or translations to tessellate your shape.

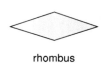  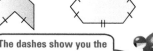

rhombus   kite

The dashes show you the sides of equal length.

Translation

When we **translate** a shape, every point moves the same distance in the same direction.

Example   A has moved the same distance and in the same direction to A', as B has moved to B'.

The original shape is not turned during translation. It has the same **orientation**.

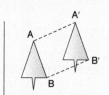

Example   ABCDE has been translated to A'B'C'D'E'. The translations are described under each diagram.

The shape and its image are congruent.

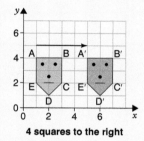

4 squares to the right

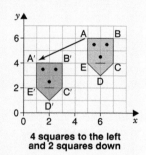

4 squares to the left and 2 squares down

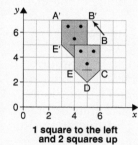

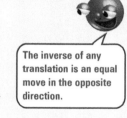

1 square to the left and 2 squares up

The inverse of any translation is an equal move in the opposite direction.

Exercise 4

T

1 Use a copy of these.
Draw the image after these translations.

a
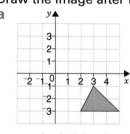

2 units left and 3 units up

b
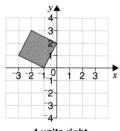

4 units right and 3 units down

c
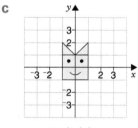

1 unit right and 2 units down

2 ABCD has been translated 5 units to the right and 3 units down to A'B'C'D'.
   a What translation would map A'B'C'D' to ABCD?
   b What translation is the inverse of these?
      i a translation 5 units to the left
      ii a translation 4 units to the right
      iii a translation 3 units up
      iv a translation 7 units down
      v a translation 4 units to the right and 6 units down
      vi a translation 3 units to the left and 4 units up

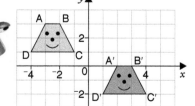

The *inverse translation* undoes the translation.

3 A shape has been translated 3 units to the right and 5 units down.
The image is then translated 1 unit to the left.
Which of these translations would have given the same result?
A 4 units to the right and 5 units down
B 2 units to the right and 4 units down
C 2 units to the right and 5 units down

4 Which two of these are equivalent translations?
A 3 units up, 2 units down, 3 units right
B 5 units down, 3 units right
C 1 unit up, 3 units right

5 Plot these points on a grid. Join them to make a triangle.
  A (⁻1, 2)   B (2, 0)   C (⁻2, ⁻3)
  a Translate ABC 3 units right and 4 units down. Write down the coordinates of the image A′B′C′.
  b Translate the image you got in **a** 1 unit left and 2 units up. Write down the coordinates of the new image, A″B″C″.
  c What single translation is equivalent to the two translations in **a** and **b**?
  d What single translation is equivalent to these two translations?
    i   3 units right and 6 units up then 4 units left and 3 units down
    ii 4 units left and 2 units down then 1 unit left and 5 units up
    iii 2 units right and 1 unit down then 3 units right and 2 units down

Review 1   PQRS has been translated 3 units to the left and 2 units up to P′Q′R′S′.
a Write down the translation that would map P′Q′R′S′ onto PQRS.
b What translation is the inverse of 3 units left and 7 units up?

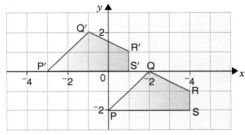

Review 2   Plot these points on a grid. Join them to make a square.
  A (⁻1, ⁻1)   B (⁻1, 3)   C (3, ⁻1)   D (3, 3)
a Translate ABCD 2 units right and 1 unit up.
Write down the coordinates of the image A′B′C′D′.
Translate A′B′C′D′ 5 units left and 2 units up.
Write down the coordinates of the new image A″B″C″D″.
b What single translation is equivalent to the two translations in **a**?

Combinations of transformations

Discussion

Brendon was experimenting with combinations of transformations. To map **A** onto **B** he rotated **A** 90° about (1, 0) then reflected the image in the x-axis to get **B**.

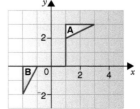

What other two transformations could Brendon have used to map **A** onto **B**? Choose from the list. **Discuss**.
  Rotate **A** 270° clockwise about (1, 0) followed by reflection in the x-axis.
  Rotate **A** 180° about (0, 0) followed by reflection in the line y = ⁻1.
  Reflect **A** in the line x = 1 followed by a rotation of 90° about (1, 0).

Exercise 5

1 Which of the transformations below will map
   a P to R   **b** R to S   **c** Q to R   **d** P to S?
   A reflection in the line $y = -\frac{1}{2}$
   B rotation of 180° about (0, 0)
   C rotation of 180° about (4, 0)
   D translation 8 squares right

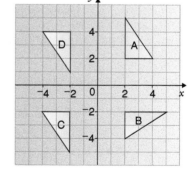

2 Look at the diagram in question **1**.
   Which combination of transformations below will
   map
   a P to R   **b** P to Q   **c** Q to R   **d** Q to S
   e R to S   **f** P to S   **g** Q to P?
   Try them out.
   A rotation 180° about (4, 2) followed by translation 4 units down
   B reflection in the line $y = 1$ followed by translation 3 units down
   C rotation 180° about (0, 0) followed by reflection in the line $y = \frac{1}{2}$
   D reflection in the x-axis followed by reflection in the y-axis
   E reflection in the line $x = {}^-1$ followed by reflection in the line $x = 3$
   F reflection in the y-axis followed by translation of 1 unit down
   G reflection in the x-axis followed by translation 8 units right and 1 unit down.

3 A($^-$2, 3), B(1, 3), C($^-$2, $^-$1)
   Plot A, B and C to make a triangle ABC.
   Write down the coordinates of A, B and C after these transformations.
   rotate ABC 90° about ($^-$2, $^-$1), **then**
   reflect the image A′B′C′ in the line $x = {}^-1$, **then**
   translate the new image A″B″C″ 4 units up and 3 units left.

∗4 Use the diagram in question **1** to find a different combination of transformations that will
   map
   a P to R   **b** Q to R   **c** P to S.

∗5 Abbie drew these triangles by transforming triangle A.
   a Find a single transformation that will map
      i C to D   **ii** A to C.
   b Find a combination of two transformations that will map
      i A to D   **ii** B to C   **iii** A to B.

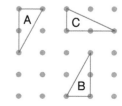

∗6 You will need square dotty paper for this question.
   a Jennasee drew triangle A on a piece of 5 by 5 dotty paper.
      She drew it in two other positions. She then found a
      transformation or combination of transformations she could have
      used to draw them.
      What transformation or combination of transformations could
      Jennasee have used to map
      i A to B   **ii** A to C?
   b Draw a triangle like Jennasee's on a piece of 5 by 5 dotty paper.
      Call it A.
      Draw it in two other positions. Call them B and C.
      What transformation or combination of transformations will map
      i A to B   **ii** A to C?

Name any dots you use as
the centre of rotation or lines
you use as mirror lines.

Review 1 Which of the transformations below will map
a J to M   b L to K   c J to L   d K to J?

    A rotation of 180° about (0, 0)
    B translation 6 squares right
    C rotation of 180° about (3, 0)
    D reflection in the line $x = ^-1$

Review 2 Use the diagram in Review **1**.
Which combination of transformations below will map
a M to L   b K to J   c J to L   d M to K?

    A rotation 180° about (0, 0), followed by reflection in
        the line $x = 4$
    B rotation 180° about (1, 1) followed by translation 2 units left and 2 units down
    C translation 2 units right followed by reflection in the x-axis
    D rotation 180° about (2, 0) followed by translation 2 units right

* **Review 3** Find a different combination of transformations that will map
a J to L   b M to K.

Symmetry

A shape has **reflection symmetry** or **line symmetry** if one or more
lines can be found so that one half of the shape reflects to the other half.

Cassie drew this shape for the cover of her science folder.
It has three lines of symmetry.

A shape has rotation symmetry if it fits onto itself **more than once**
during a full turn.
The **order of rotation symmetry** is the number of times a shape fits
onto itself in a 360° turn.
Cassie's shape has rotation symmetry of order 3.

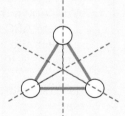

Shapes with rotation
symmetry of *order 1* do *not*
have rotation symmetry.

Worked Example
Describe the reflection and rotation symmetry of these shapes.
a

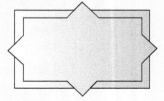

b

Answer
a This shape has **two lines of symmetry**.
   It fits onto itself twice as it is turned through 360°.
   It has **rotation symmetry of order 2**.
b This shape has **no lines of symmetry**.
   It fits onto itself four times as it is turned through 360°.
   It has **rotation symmetry of order 4**.

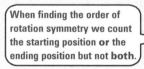

When finding the order of
rotation symmetry we count
the starting position **or** the
ending position but not **both**.

Exercise 6

1   Describe the reflection and rotation symmetry of these.

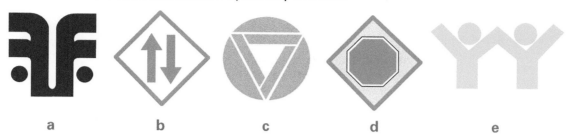

       a   b   c   d   e

2   Imagine this is how a piece of paper looks after it has been folded twice, each time along a line of symmetry.
    What shape might the original piece of paper have been?
    Is there more than one answer?

3   Show how you can put these shapes together to make a single shape with reflection symmetry *and* rotation symmetry.

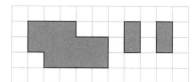

4   a   Rochelle put these two shapes together, edge to edge. Her new shape had just 1 line of symmetry. Show two ways Rochelle might have done this.
    b   Show a way of putting the two shapes together edge to edge to get a new shape with
        i   rotation symmetry of order 2 but no lines of symmetry (four ways)
        ii two lines of symmetry *and* rotation symmetry of order 2.

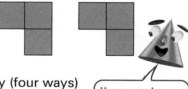

Use squared paper to help with questions 4 and 5.

5   Show how you can put these shapes together to make a single shape with rotation symmetry of the order given and no lines of symmetry.

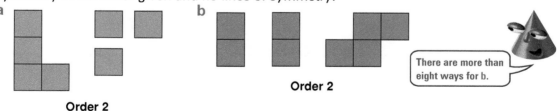

    a   b

                                            Order 2

    Order 2

There are more than eight ways for b.

6   This shape has rotation symmetry of order 6.
    As it is rotated through 60° angles about the centre point, it maps onto itself in six different positions in a full turn.
    a   Explain how you know that the yellow shape in the middle is a regular hexagon.
    b   If a pen was attached to point A, what would it draw as the whole shape is rotated 360° about its centre?

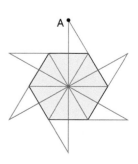

T *7

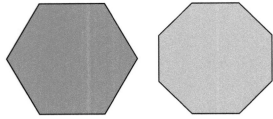

Use a larger copy of the hexagon and octagon.
On each, draw patterns which have both rotation and reflection symmetry.
Describe the symmetry of each pattern.

Review 1 Describe the reflection and rotation symmetry of these.

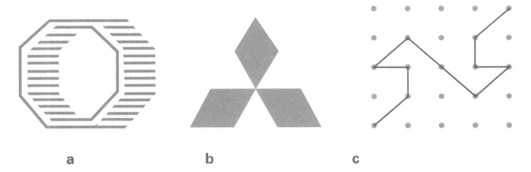

a   b   c

Review 2 This shape has rotation symmetry of order 8.
As it is rotated through 45° angles about the centre point, it maps onto itself in eight different positions in a full turn.
a Explain how you know that the blue shape in the middle is a regular octagon.
b If a pen is attached to point A, what would it draw as the whole shape is rotated 360° about its centre?

Review 3 Show how you can put these shapes together to make a single shape with reflection and rotation symmetry.

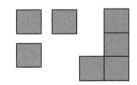

 Practical

You will need LOGO or a similar ICT package.
Make a shape on your screen which has rotation symmetry of order
a 2   b 4   c 5   d 6   e 8   f 9

Example Melanie made this shape by rotating a triangle through 45° angles.
Her shape has rotation symmetry of order 8.

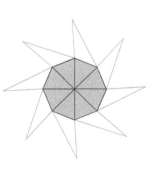

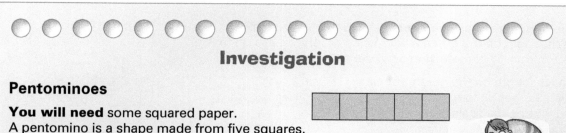

Investigation

Pentominoes

You will need some squared paper.
A pentomino is a shape made from five squares.
There are 12 pentominoes altogether.
Try and draw them all.
Investigate which pentominoes have line
symmetry and which have rotation symmetry.
Which ones tessellate? **Investigate**.
Make a poster or booklet to show your results.

? Puzzle

What goes in the centre square?

Enlargement

Discussion

Jake had some photos enlarged.
Think of some other everyday examples where
the word 'enlarged' is used. **Discuss**.

The brown tree has been **enlarged** to the green tree.
Each length on the green tree is twice as long as it is on the
brown tree. The **scale factor** of the enlargement is 2.

If each length on the green tree had
been 3 times as long as it is on the
brown tree the scale factor would be 3.

The shape and its
image are not
congruent.

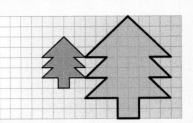

Exercise 7

1 Shape A has been enlarged to shape B.
  What is the scale factor of the enlargement?

a

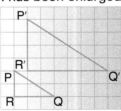

b

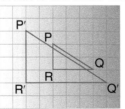

c

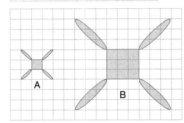

d

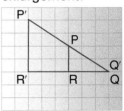

2 PQR has been enlarged to P'Q'R'. What is the scale factor of the enlargement?

a

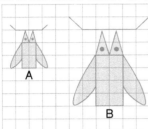

b

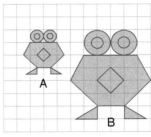

c

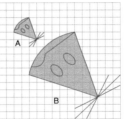

Review Each shape on the right has been enlarged to the shape on the left.
Give the scale factor for each of these enlargements.

a

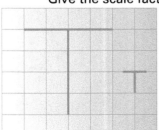

b

c

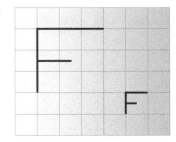

ABC has been mapped to A'B'C' under an **enlargement**, scale factor 2.
Each side of A'B'C' is twice as long as the corresponding side of ABC.

The distance from O to A' is **twice** the distance from O to A.
The distance from O to B' is **twice** the distance from O to B.

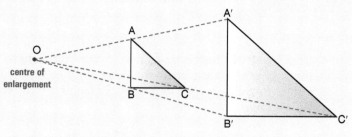

centre of enlargement

The distance from O to C' is **twice** the distance from O to C.
The **centre of enlargement** is O.

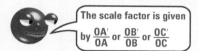

The scale factor is given by $\frac{OA'}{OA}$ or $\frac{OB'}{OB}$ or $\frac{OC'}{OC}$

Discussion

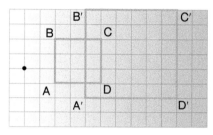

In both of these diagrams ABCD has been enlarged to A'B'C'D', by scale factor 2.
Why is A'B'C'D' in a different position in each diagram? **Discuss**.

To **draw an enlargement** we need to know both the scale factor **and** the centre of enlargement.

We need the scale factor to make the enlargement the right size.
We need the centre of enlargement so we know where to put the image.

Example   The smaller 'kite' has been enlarged by scale factor 2 and centre of enlargement P, to get the larger kite. This was done by following these steps.

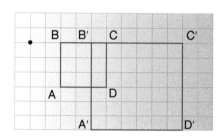

1 Draw a line from the centre of enlargement, P, to some of the points on the smaller kite.
2 Extend each line drawn in 1 to make it twice as long (×2).
3 Draw the larger kite so that each length on it is twice as long as it was on the smaller kite.
4 Label the image points A', B', C', D', E' and F'.

You can do this by measuring or counting squares.

Exercise 8 **You will need** some squared paper.

1 Draw five sets of x- and y-axes on squared paper with x- and y-values from 0 to 16.
Copy each of these shapes onto one of the sets or axes.
Enlarge each by the scale factor given.
Use the origin as your centre of enlargement for each.
Write down the coordinates of A'.

Remember the origin is the point (0, 0).

a
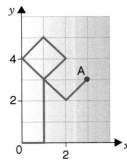
Scale factor 2

b
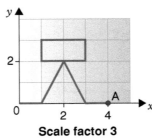
Scale factor 3

c
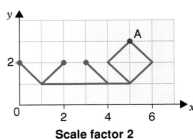
Scale factor 2

d
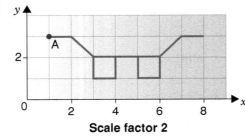
Scale factor 2

e
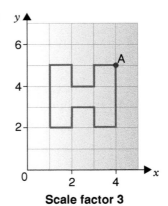
Scale factor 3

2 a Draw this shape on squared paper.
Use the × as the centre of enlargement.
Enlarge the shape by a scale factor of 2.
What has happened to the lengths of the purple lines?
What has happened to the shaded angles?
Measure them if you need to.
b Repeat **a** for scale factor 3.
Try other centres of enlargement.

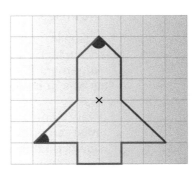

c Copy and finish this sentence. Use these words to fill in the gaps: angles, lengths.
When a shape is enlarged the _____ stay the same size but the _____ change.

T

3

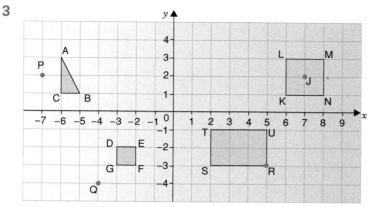

Use a copy of this diagram.

a Using P as the centre of enlargement, enlarge triangle ABC by a scale factor of 3.
Write down the coordinates of the vertices of the image triangle.

b Enlarge square DEFG by a scale factor of 4, centre of enlargement Q. Write down the coordinates of D′, E′, F′ and G′.

c With R as centre of enlargement, enlarge rectangle RSTU by a scale factor of 3.
What are the coordinates of the vertices of the image rectangle?

d KLMN maps to K′L′M′N′ under an enlargement, centre J, scale factor 2.
Draw the square K′L′M′N′.
Give the coordinates of K′, L′, M′ and N′.

*e Draw a shape of your choice.
Choose a centre of enlargement. Choose a scale factor.
Draw the enlargement of your shape.

*4 P(⁻1, 3), Q(4, 3), R(3, 0)
Plot P, Q and R to give triangle PQR.
What are the coordinates of P, Q and R after an enlargement, scale factor 2, centre of enlargement (2, 2)?

Review 1 Draw a set of axes with **x- and y-values** from 0 to 14.
Copy this diagram onto the axes.
With centre of enlargement (0, 0), enlarge this shape by a scale factor of 2.
Write down the coordinates of A′.

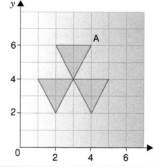

T

Practical

⭐

You will need a dynamic geometry software package like Geometer's Sketchpad.

Make a shape of your own on the screen.
Choose a point as the centre of enlargement.
Choose a scale factor.
Enlarge your shape on the screen using the **Transform** menu.
Which of these stay the same and which change when you enlarge a shape:
    length   angles   orientation   area?

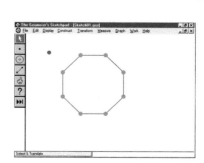

Summary of key points

 A We use **coordinates** to give the position of a point on a grid.

The coordinates of A are ($^-$2, 1). We always give the x-coordinate first.

The coordinates of the **origin** are (0, 0).

The x- and y-axes make four quadrants as shown.

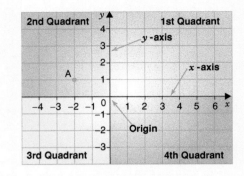

 B **Reflection**

ABC has been reflected in the mirror line to get the image A'B'C'.

The line joining a point to its image is prependicular to the mirror line.

A'B'C' is the same distance behind the mirror line as ABC is in front. The shape and its image are congruent.

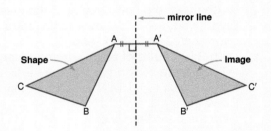

The **inverse of a reflection** is a reflection in the same mirror line. If a reflection maps A onto A' then it also maps A' onto A.

Reflection in two parallel lines is equivalent to a translation.

Reflection in two perpendicular lines is equivalent to a 180° rotation.

 C To **rotate** a shape we need to know the **centre of rotation** and the **angle of rotation**.

The centre of rotation is a fixed point. It can be inside or outside the shape being rotated.

Example   ABCD has been rotated 270° using (0, 1) as the centre of rotation.

An angle of rotation given as 90° means 90° in an **anticlockwise** direction.

The **inverse of a rotation** is either an equal rotation about the same point in the opposite direction **or** a rotation about the same point in the same direction so the sum of the angle of rotation and its inverse add to 360°.

Two rotations about the same centre are equivalent to a single rotation.

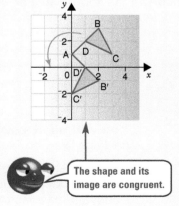

The shape and its image are congruent.

Continued ...

D To **translate** a shape we slide it without turning. Every point on the shape moves the same distance in the same direction. The shape and its image are congruent.

Example This shape has been translated 4 units right and 3 units up.

The **inverse of a translation** is an equal move in the opposite direction.

Two translations are equivalent to a single translation.

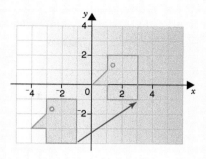

 E A shape has **reflection symmetry** if one half of the shape can be reflected in a line to the other half. The line is a **line of symmetry**.

A shape has **rotation symmetry** if it fits onto itself **more than once** during a complete turn.

The **order of rotation symmetry** is the number of times a shape fits exactly onto itself during one complete turn.

If a shape has rotation symmetry of order 1, we say it does not have rotation symmetry.

This shape has 3 lines of symmetry and rotation symmetry of order 3.

 F To **enlarge** a shape we need to know the **scale factor** and the **centre of enlargement**.

Example PQR has been enlarged by scale factor **2** and centre of enlargement X.

Each point on the image, P'Q'R, is **2** times as far from X as the same point on PQR.

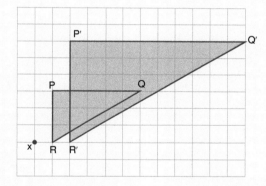

Test yourself

 1 Use a copy of this grid.
   a These are the coordinates of the vertices of a shape. Plot them on your grid and join them in the order given.
   (⁻1, 3), (1, 1), (⁻1, ⁻1), (⁻3, 1), (⁻1, 3)
   b How many lines of symmetry does the shape have?
   c The points (1, 4) and (4, 2) are two of the four vertices of a rectangle.
   What might the coordinates of the other two vertices be? Is it possible to make a square? Explain why or why not.
   d Which quadrant does the point (⁻3, 4) lie in?

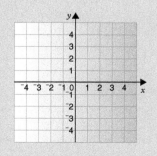

T   **2**   Use a copy of this.   **B**
Shade in two more squares to make a shape which has the dashed lines as lines of symmetry.
You may use a mirror or tracing paper to help you.

a 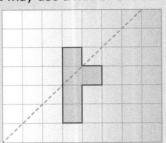   **b**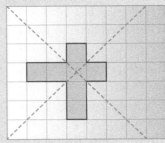

You can use a mirror if you want to.

T   **3**   Use a copy of this diagram.   **B**
    a   Reflect the shape in the *y*-axis.
    b   Reflect the image you got in **a** in the *x*-axis.
    c   What single transformation is equivalent to
       a and **b**?
    d   Would you get the same result if you had
       reflected the shape in the *x*-axis and then
       reflected the image in the *y*-axis?

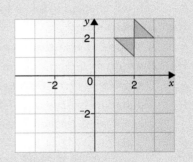

T   **4**   Use a copy of these shapes.   **C**
The centre of rotation and angle of rotation are given.
Draw the image shapes.

    a   (0, 0), 90°   **b**   (1, 1), 180°   **c**   (0, 0), 45°

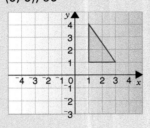   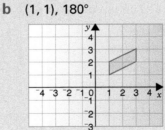   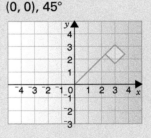

5   A rotation through 90° about (2, 1) maps P onto P′.   **C**
Which of these maps P′ onto P? There are two correct answers.
    A   a rotation through 90° about (2, 1)
    B   a rotation through 270° about (2, 1)
    C   a rotation through 90° about (1, 2)
    D   a rotation 90° clockwise about (2, 1)

You can use tracing paper.

T   **6**   Use a copy of question **4** shape **a**.   **C**
Cameron rotated the shape 180° about the point (1, 1). Then he rotated the image
90° about (1, 1).
What single rotation is equivalent to these two rotations?

T

7 Use a copy of these.
Draw the image after these translations.
a 3 units down

b 2 units left 3 units up

c 4 units right 3 units down

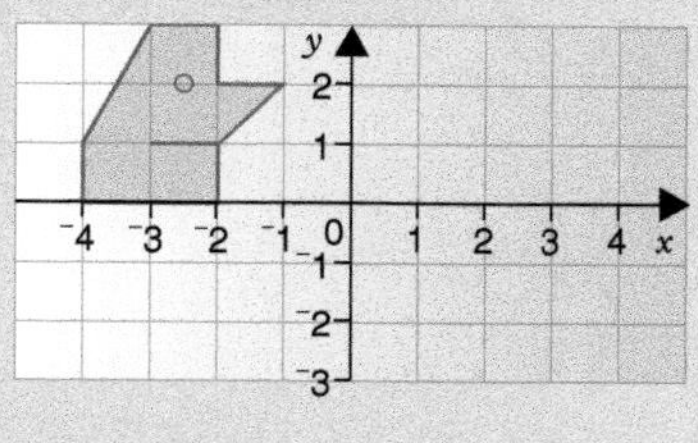

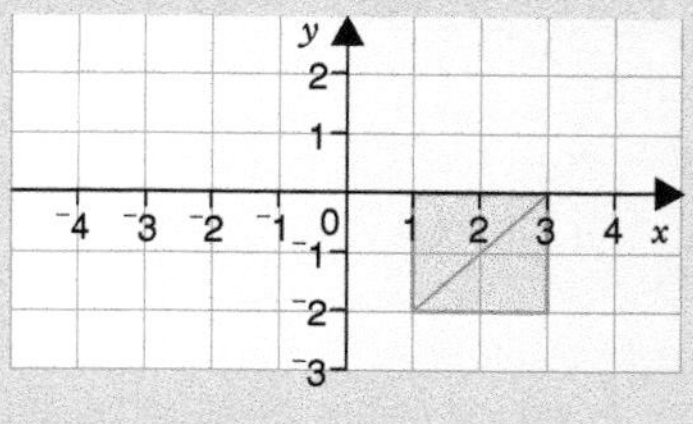

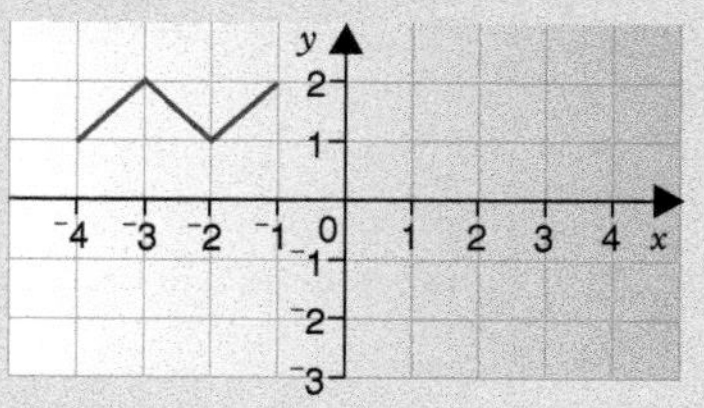

8 What translation is the inverse of each of these?
a 5 units right and 3 units down
b 4 units left and 6 units up

9 A($^-$2, 2)   B(0, 0)   C($^-$2, $^-$2)   D($^-$4, 0)
a Raymond plotted these points on a grid. He joined them to make a square.
He translated his shape 4 units right and 2 squares down.
Write down the coordinates of the image A′B′C′D′.
He then translated A′B′C′D′ 1 unit right and 5 units up.
Write down the coordinates of the new image, A″B″C″D″.
b What single translation is equivalent to the two translations?

10 i Which of the transformations below will map
a K to L   **b** L to N   **c** M to L?
 A translation 6 units right
 B translation 6 units left
 C rotation 180° about (1, 0)
 D rotation 180° about the origin
 E reflection in the line $y = ^-1$ 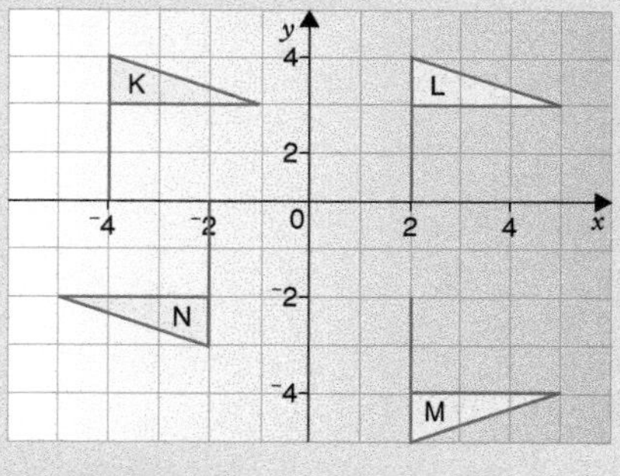
ii Which of the combinations of transformations
below will map
a L to N   **b** L to K   **c** K to M   **d** N to K?
 A reflection in the y-axis followed by reflection in the line $x = ^-3$
 B rotation 180° about ($^-$4, 0) followed by translation 2 units right
 C reflection in the x-axis followed by reflection in the y-axis
 D reflection in the x-axis followed by translation 6 units right and 2 units down

11 Vladimir found these Russian designs in a book.
Describe the reflection and rotation symmetry of each.
a   **b** 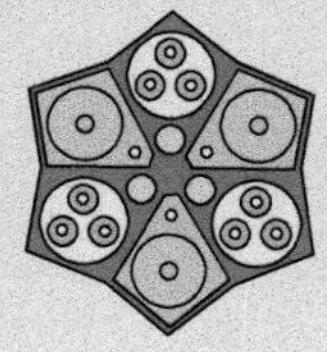   **c**

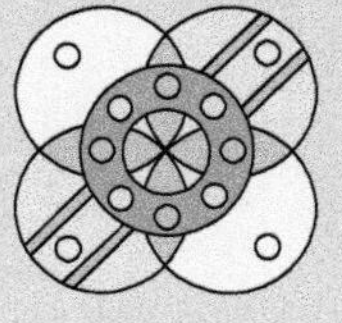

12 Join these shapes together to make a single shape
with both reflection and rotation symmetry.

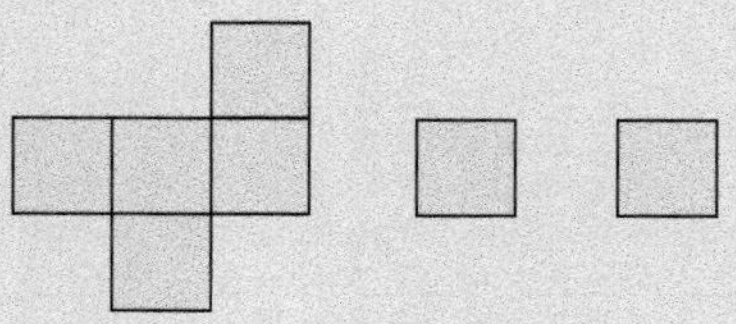

T

13 a Use a copy of this.
Enlarge this 'fish' by a scale factor of 2,
centre of enlargement (0, 0).
Write down the coordinates of A'.
b Use another copy of the fish.
Enlarge it by a scale factor of 2, centre of
enlargement (3, 2).
Write down the coordinates of A'.

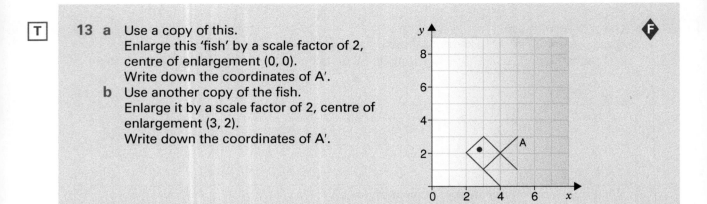

14 Measures

You need to know

✓ measures — time
  — metric measures
  — metric and imperial

  page 274

✓ reading scales

  page 274

✓ perimeter and area

  page 275

Key vocabulary

area: hectare, square centimetre, square kilometre, square metre, square millimetre
capacity: centilitre, gallon, litre, millilitre, pint
length: centimetre, foot, kilometre, metre, mile, millimetre, yard
mass: gram, kilogram, pound, tonne
time: century, day, decade, hour, millennium, minute, month, second, week, year
volume: cubic centimetre (cm^3), cubic metre (m^3), cubic millimetre (mm^3)
depth, distance, height, high, perimeter, surface area, width

Topsy turvy

● In the following 'story' most of the measurements are in the wrong places. Rewrite the story so the measurements are in the correct places.

On the way to the football I saw a labrador dog that weighed 5 kg, a car that was 3 cm long, a cat with a 20 cm tail, a baby that weighed 40 kg and a house that was 2 m tall. At the football I bought a pie that was 3 m thick from a man who was 6 m tall.

Write another 'story' in which you mix all, or most, of the measurements up.
Ask someone else to rewrite your story.

● Two systems for measuring lengths, weights and capacities are the **metric** system and the **imperial** system.
Think of some metric units.
Think of some imperial units which are still used in Britain.
Which system is easier to use? Why?

Metric measurements

Fiona needs 2·5 centimetre long nails to make a bookcase. At the store, all the measurements are in millimetres. She needs to change 2·5 centimetres to millimetres.
We often need to **change one metric unit into another**.

Remember

Length	1 km = 1000 mm	1 m = 100 cm	1 m = 1000 mm	1 cm = 10 mm
Mass	1 kg = 1000 g	1 tonne = 1000 kg		
Capacity	1 ℓ = 1000 mℓ	1 ℓ = 100 cℓ	1 cℓ = 10 mℓ	

We can use this chart to help us convert between one metric unit and another.

Worked Example
Michael's bed is 190 cm long.
How long is this in metres?

Answer
190 cm = (190 ÷ 10 ÷ 10) m
= **1·9 m**

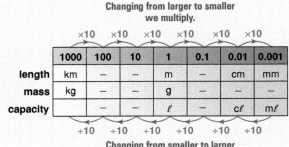

Changing from larger to smaller
we multiply.

	×10	×10	×10	×10	×10	×10

	1000	100	10	1	0.1	0.01	0.001
length	km	–	–	m	–	cm	mm
mass	kg	–	–	g	–	–	–
capacity		–	–	ℓ	–	cℓ	mℓ

	÷10	÷10	÷10	÷10	÷10	÷10

Changing from smaller to larger
we divide.

We use **tonnes** to measure very heavy objects.

Worked Example
Lump, the elephant, weighs 4520 kg.
How many tonnes is this?

Answer
4520 kg = (4520 ÷ 1000) tonnes
= **4·52 tonnes**

If you need to remind yourself how to multiply and divide by 10, 100 and 1000 go to page 19.

Exercise 1

Except question 2 and Review 2.

T

1 Use a copy of this cross number. If decimal points are needed give them a full space.

Across
Write
1 1450 cm in m
4 37 mm in cm
6 2·8 cm in mm
7 0·048 ℓ in mℓ
10 0·707 kg in g
12 564 cm in m
14 7·35 m in cm
15 31·42 ℓ in cℓ
16 9090 g in kg
18 0·026 km in m
19 4100 kg in tonnes
20 3·4 cm in mm
22 0·523 kg in g
23 7070 mm in m

Down
Write
1 0·12 ℓ in mℓ
2 0·048 m in mm
3 5·2 cm in mm
4 0·39 tonnes in kg
5 0·074 kg in g
8 8·05 kg in g
9 3460 kg in tonnes
10 7400 cℓ in ℓ
11 7·7 ℓ in mℓ
12 510 mm in cm
13 6·2 cm in mm
15 3·425 tonnes in kg
16 9870 mℓ in ℓ
17 0·932 kg in g
21 4·3 m in cm
22 5·6 cm in mm

2 Change
a 500 minutes to hours and minutes
b 3·8 hours to hours and minutes
c 426 minutes to hours
d 803 days to years (no leap years)
e 1262 months to years and months
f 962 months to decades, years and months
g 427 weeks to years and weeks
h 3000 minutes to days and hours

T

Review 1

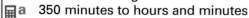

							E			
0·04	8·6	0·352	8·6	0·086	8·6	4	0·4	8·6	8	0·04

	E		E									
3·52	0·4	0·04	0·4	0·086	0·075	0·86	8	8·6	0·086	8·6	0·75	0·86	0·8

Use a copy of this box.
What goes in the gaps?

E 400 mm = 0.4 m   M 4000 g = ___ kg   N 80 mm = ___ cm   I 86 cm = ___ m
T 75 g = ___ kg   D 3520 mℓ = ___ ℓ   B 750 m = ___ km   C 80 cℓ = ___ ℓ
R 86 g = ___ kg   H 35·2 cℓ = ___ ℓ   S 40 kg = ___ tonnes A 8600 mℓ = ___ ℓ

Review 2 Change

a 350 minutes to hours and minutes
b 842 days to years and days (no leap years)
c 1741 months to years and months
d 8460 minutes to days and hours.

T

? Puzzle

Use a copy of this.
Cut out the nine squares.
Fit the nine squares together so that
the edges which touch show
measures which are equal to one
another.

4 cm	400 mm / 5600 mℓ	2·4 ℓ
3 kg	504 cm	18000 mℓ
4·56 km / 4·56 m	18 ℓ / 1·8 km	18 m / 5·04 m
	40 mm	2400 mℓ
8·64 kg / 1800 m	5·6 ℓ / 8640 g	1800 cm / 3000 g
40 cm	456 cm	4560 m

More metric units

Capacity

1 litre of water would fill a container with a volume of 1000 cm³.
1 millilitre of water would fill a container with a volume of 1 cm³.
1000 litres of water would fill a container with a volume of 1 m³.

1ℓ = 1000 cm³   1 mℓ = 1 cm³   1000 ℓ = 1 m³

Worked Example
A petrol tank holds 5000 cℓ of petrol
a How many ℓ does the tank hold?
b What is the volume of the tank in cm³?

Answer
a 5000 cℓ = (5000 ÷ 100) cℓ   b 50 ℓ = 50 × 1000 cm³
        = **50 ℓ**   = **50 000 cm³**

Area

Hectares are used to measure large areas.
1 hectare (ha) = 10 000 m²

Examples   The area of a school's grounds might be about 5 hectares.
             The area of a farm might be about 80 hectares.

Exercise 2

1 Write
   a 3 ℓ in cm³   b 5 ℓ in cm³   c 5 mℓ in cm³   d 9 mℓ in cm³
   e 5000 ℓ in m³   f 7000 ℓ in m³   g 4 m³ in ℓ   h 3 m³ in ℓ
   i 4000 cm³ in ℓ   j 8·6 mℓ in cm³   k 9·2 mℓ in cm³   l 4·6 ℓ in cm³
   m 8·3 ℓ in cm³   n 0·82 mℓ in cm³   o 896 ℓ in m³   p 437 ℓ in m³
   q 4·6 m³ in ℓ   r 0·82 m³ in ℓ   s 43 cm³ in mℓ   t 8420 cm³ in ℓ
   u 523 cm³ in ℓ   v 84 cm³ in ℓ   w 43 ℓ in m³.

2 How many m² in each of these parks?
   Hyde Park 255 ha   Kensington Gardens 111 ha   Regents Park 197 ha   Kew Gardens 120 ha

3 How many hectares is
   a 50 000 m²   b 70 000 m²   c 120 000 m²   d 90 000 m²?

4 A tank holds 5682 ℓ.
   How many m³ is this?

5 A large tin of tomato sauce holds 260 mℓ of sauce.
   a How many ℓ of sauce is in this tin?
   b What volume of sauce, in cm³, is in the tin?

6 a How many ℓ of coke is there in a 1200 mℓ bottle?
   b How many cm³ are there in the bottle?

Review 1 Write
a 4 ℓ in cm^3 b 6000 ℓ in m^3 c 4·2 ℓ in cm^3 d 562 ℓ in m^3
e 5·2 m^3 in ℓ f 842 cm^3 in mℓ g 5832 cm^3 in ℓ.

Review 2 Four friends share a 1 litre bottle of drink.
a How many mℓ does each friend have?
b How many cm^3 does each have?

Review 3
a The area of a park is 80 000 m^2. How many hectares is this?
b 1·2 ha of the park is a children's playground. How many m^2 is this?

Solving measures problems

Mohammed trains for 2 hours and 20 minutes each week. He trains for the same time each day. His trainer asked him how long he trains each day?

Mohammed needs to divide 2 hours and 20 minutes by the number of days in a week, 7. It will be easier if he converts 2 hours and 20 minutes to minutes.

2 hours and 20 minutes = 120 + 20 minutes
$\qquad\qquad\qquad\quad$ = 140 minutes
$\frac{140}{7} = 20$

Mohammed trains for **20 minutes each day**.

When solving measures problems

1 Write down what you know from the problem.
2 Work out what you have to do to find the answer.
3 Make sure the measures are in the same units, or in the units you want for the answer.

Worked Example
A rectangular park is 150 m by 200 m. Find the area of the park in hectares.

Answer
Area = length × width
$\qquad$ = 150 × 200
$\qquad$ = 30 000 m^2
30 000 m^2 = $\frac{30\,000}{10\,000}$ ha
$\qquad\qquad$ = **3 ha**

200 m

Park   150 m

1. We know the length and width.
2. We have to find the area.
3. We need the answer in ha.

Exercise 3   **Only use a calculator if you need to.**

1 A 1·2 kg box of 'Wake Up' cereal is enough for 25 servings.
   How many grams is each of these servings?

2 On a map 1 cm represents 5 km.
   A lane on the map is 1·2 cm long.
   What is the length of the lane in kilometres?

Link to ratio and proportion page 152.

3 Patty typed eight pages in 1 hour and 40 minutes. How many minutes on average did it take to type each page?

4 Khalid made 450 cℓ of curry sauce.
How many 75 mℓ servings will he get from this?

5 Ash Wednesday is the first day of Lent, coming $6\frac{1}{2}$ weeks before
Easter Sunday. In the year 2000 Easter Sunday was on the
23 April.
What date was Ash Wednesday in the year 2000?

6 Paul and Asmat are stacking boxes in a cupboard which
is 1·88 m high. Each box is 24 cm high.
How many layers will fit in the cupboard?

Round sensibly.

7 A rectangular playing field is 200 m by 350 m.
How many hectares is this?

8 A box weighs 800 g. The box is filled with packets of crisps.
Each packet weighs 85 g. The full box weighs a total of 4·88 kg.
How many packets of crisps are in the box?

9 How many decades are there in a millennium?

Work out how long it is to the
nearest hour to the next millennium.

10 A recipe for a fruit punch has
    1·4 ℓ of orange juice,
    750 mℓ of pineapple juice and
    3 ℓ of lemonade.
a How many litres of punch does this recipe make?
b How many cm^3 does it make?
c How many 150 mℓ glasses could be filled with the punch?

11 Mandarins cost £1·80 per kilogram.
One mandarin weighs 150 g.
Find the cost of one mandarin.

12 On 5 May 1961 Alan Shepard became the first American in Space. 291 days later John
Glenn became the first American to orbit Earth. On what date did he do this?

13 Gavin has a small square garden of side 150 cm.
a What is its area in m^2?
b Gavin wants to fertilise it. The instructions say to sprinkle 800 g of fertiliser per square
    metre.
    How many kg of fertiliser does Gavin need?

14 A fish tank holds 3000 cℓ of water.
a How many ℓ of water does the fish tank hold?
b What is the volume of the tank in cm^3?
c What would be the volume, in m^3, of 20 of these fish tanks?

15 Roydon School grounds are rectangular in shape.
One side is 150 m long. The area of the grounds is 3 ha.
What is the length of the other side?

16 The mass of an unloaded lorry was 14·2 tonnes. When the lorry was loaded with
16 containers the laden mass was 22·68 tonnes.
What was the mass of each container, in kg?

17 Each day Manisha bikes to school. She leaves at 8:15 a.m. and arrives at about 8:35 a.m.
Each Tuesday, she posts a letter on the way to school and doesn't arrive until 8:45 a.m.
About how long does Manisha spend getting to school
a each week   b each term of 16 weeks
c each year of 39 weeks?

Give the times in
sensible units.

18 A box of fruit punch holds 1250 m*ℓ*.
Zeta likes her fruit punch strong. She adds between 3 and
3·5 parts lemonade to the fruit punch.
About how many litres of drink can she make with one
bottle of fruit punch?

1250m*ℓ*
**FRUIT
PUNCH**
Add 4 parts
lemonade

19 Ribbons are 40 cm long.
Jake joins some ribbons together to make a streamer.
They overlap each other by 4 cm.
a How long is a streamer made from
i 2 ribbons   ii 5 ribbons   iii 10 ribbons?
b Nasreen made a streamer that was 220 cm long.
How many ribbons did she use?

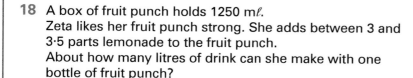

4 cm   4 cm

40 cm   40 cm

20 A scale on a map says '2 cm to 1 km'. This means every 2 cm you measure on the map
represents 1 km in real life.
a How far is a distance in real life if on the map it is i 4 cm ii 8 cm iii 5 cm iv 3·5 cm?
b How far on the map would these real-life distances be? i 4 km   ii 10 km   iii 4·5 km?

21 About how long would it take to spend one thousand pounds at a penny a minute?
A 7 days   B 70 days   C 7 years   D 70 years

22 What time of what day of what year will it be
a 3000 minutes after the start of 2005
b 1000 hours after the start of 2010?

∗23 It takes 23 hr 56 min 4 sec for the Earth to make one complete rotation on its axis. It takes
24 hr 37 min 23 sec for Mars to make one complete rotation on its axis.
How much longer does it take for Mars to make one rotation?

∗24 a 'Present Extras' make fancy bows that each use 68·3 cm of ribbon.
On Thursday 200 000 of these bows were made.
How many km of ribbon were used?
b What is the maximum number of these bows that can be made from 50 m of ribbon?
c To reduce costs, Present Extras decides to reduce the amount of ribbon needed to
make a bow to 63 cm.
How many more bows can be made from 50 m of ribbon?

∗25 The scouts at a jamboree are putting on an outdoor concert.
360 chairs are to be put outside. Each chair is a square, 55 cm wide.
No more than 12 chairs must be put next to each other in a row.
There must be 0·6 m between rows.
Down the middle of the area, there is to be an aisle 1·5 m wide.
What is the minimum space needed to set out the chairs?

Draw a diagram.

Review 1 Tyrone is a tour guide for 'Lakes Tours'. Each tour takes 4 days. Last year he
guided 20 tours. How many weeks and days did this take?

Review 2

a The weight of an unloaded 4 wheel drive is 1·956 tonnes. Find the loaded weight, in kg, of this 4 wheel drive once the Weston family and their luggage are in.
   Their luggage weighs a total of 124 kg. The weights of the family are: Mr. Weston – 68 kg, Rashni – 56 kg, Michael – 62 kg, Mrs. Weston – 54 kg.

b What is the weight of the loaded 4 wheel drive in tonnes?

Review 3 A large tin of tomato sauce contains 500 cℓ.
How many 250 mℓ bottles can be filled from this tin?

Review 4 How many books each weighing 760 g can be sent in this service lift?

Maximum
Load
60 kg

Review 5 The first woman to orbit Earth was Valentina Tereshkova in the Vostok 6.
She was launched on June 16th, 1963 at 9:30 a.m. (Greenwich Mean Time). She returned after 2 days, 22 hours and 46 minutes. When did Valentina land back on Earth?

Review 6 Susan has made 8 litres of grape juice. To store it she has bottles of two different sizes, 1·25 ℓ and 750 mℓ.

To store the juice, bottles should be filled right to the top. If an equal number of large and small bottles are used, how many bottles are filled?

Review 7 What will the time and date be 12 682 minutes after the start of 2005?

Puzzle

1 Each link of this chain is 24 mm from outside edge to outside edge.
   Each link is made from brass 5 mm thick.
   What is the length of the 6-link chain?

24mm

2 Jimmy has four containers which, when full, will hold 2 ℓ, 3 ℓ, 4 ℓ and 5 ℓ of wine. The 3 ℓ and 5 ℓ containers are full. By pouring wine from one container to another, show how to get 2 ℓ of wine into each of the containers in as few moves as possible.

3 Three winemakers had between them 21 barrels. 7 were empty (E), 7 were half-full of wine (H) and 7 were full of wine (F).
   These barrels were divided between the winemakers so that each had the same number of barrels and the same amount of wine. There are two ways in which this could be done. Find them.

4 Josh, Becky and Ryan have to cross a river in a rowboat which can carry at most 140 kg. Josh weighs 100 kg, Becky weighs 50 kg and Ryan weighs 75 kg. They all have rucksacks. Josh's weighs 40 kg, Becky's weighs 15 kg and Ryan's weighs 25 kg. How do they cross the river if none of them will leave their rucksacks with any of the others?

5 Shalome is given nine £1 coins, one of which is fake. This fake coin looks the same as the others but is lighter. Shalome is allowed to keep all the coins if she can find the fake one by just 2 weighings on a balance like this. She is not allowed to use any weights during the weighing. How could she find the fake one in two weighings?

* 6 In three minutes time it will be three times as many minutes to 9 o' clock as it was past 8 o'clock nine minutes ago. What is the time now?

* 7 This clock shows the correct time at 1:00 p.m. on May 5th. The clock loses 20 minutes a day. When will it next show the correct time?

* 8 When Angie asked her mother what year she was born, her mother replied 'The sum of the last two digits is the same as the number of months in a year and the product of all four digits is the number of hours in twelve days'. When was Angie's mother born?

* ## Practical

Choose a room in your house. Make a scale drawing of the room and furniture using the scale 5 cm represents 1 m.
Rearrange the furniture to a layout that you like.

Metric and imperial equivalents

Discussion

Hien is helping on her family farm.
Shalome is helping at a holiday camp.
Alex is helping at his football club.
Megan is helping to care for her baby brother.
Barry is helping to plan a party.
Bonny is helping to care for the animals at the RSPCA.
Tim is helping out at his mother's car sales business.

What imperial measures will Hien, Shalome, Alex, Megan, Barry, Bonny and Tim be likely to use? **Discuss**.

Learn these approximate **metric and imperial equivalents**.

length	mass	capacity
8 km ≈ 5 miles	1 kg ≈ 2·2 lb	4·5 ℓ ≈ 1 gallon
1 m ≈ 1 yard	30 g ≈ 1 oz	1 ℓ ≈ 1·75 pints or just
1 m ≈ 3 feet		less than 2 pints
2·5 cm ≈ 1 inch		

Worked Example
A recipe for fudge cake uses 8 oz of butter.
a About how many grams is this?
b How many kg is this?

Answer
a 1 oz ≈ 30 g
   8 oz = 8 × 30 g
       = **240 g**
b 240 g = 240 ÷ 1000 kg
         = **0·24 kg**

Worked Example
A measuring jug can hold 2·5 litres.
What is the approximate capacity of this jug in pints?

Answer
1 litre is about 1·75 pints.
2·5 litres is about 2·5 × 1·75 pints or 4·375 pints.
The approximate capacity of the jug is 4·4 pints.

We could also give the answer as 'about $4\frac{1}{2}$ pints' or 'nearly $4\frac{1}{2}$ pints'.

Worked Example
Dee walked 28 km. About how many miles did Dee walk?

Answer
Firstly, find how many lots of 8 km Dee walked. She walked $\frac{28}{8}$ = 3·5 lots of 8 km.

Since there are 5 miles in each lot of 8 km, we now multiply by 5 to find the total number of miles.
5 × 3·5 = 17·5 miles
Dee walked about **$17\frac{1}{2}$ miles**

Exercise 4	**Only use a calculator if you need to.**

1 'Delicious Drinks' packages its juice in the following sizes. Find the approximate equivalent in pints.
   a 10 litres   b 2 litres   c 5 litres   d 0·5 litre   e 20 litres

2 Potatoes had these amounts written on the bags. Find the approximate equivalent in lbs.
   a 5 kg   b 2 kg   c 10 kg   d 3000 g   e 4000 g

3 Distances on a map were given in km.
   Josefa wanted to convert them to miles. Approximately how many miles are there in the following?
   a 40 km   b 80 km   c 48 km   d 36 km   e 92 km   f 115 km

4 Packages to send overseas were weighed. Find the approximate metric equivalent for these.
   a 4·4 lb   b 20 oz   c 5 oz   d 13·2 lb   e 20 lb   f 27 lb

5 A recipe for raspberry jam uses 6 lb of sugar. About how many kg of sugar is used?

6 Water containers have the following capacities. Find the approximate metric equivalent of each.
   a 2 gallons   b 60 gallons   c 17·5 pints   d 7 pints   e 22 pints

7 Argene is visiting England. In her country, distances are measured in km. Convert these distances to km.

    a 20 miles   **b** 200 miles   **c** 50 miles   **d** 12 miles   **e** 52 miles

8 Measurements for scenery for the school play were taken in inches and feet. Amanda needed these measurements in cm. Convert the following for her.

    a 6 inches   **b** 2 inches   **c** 18 inches   **d** 3 feet   **e** 9 feet

9 Jenufa's family own a 15 foot caravan.

    a Find the approximate length of this caravan, in metres.

    b How many caravans like this could be stored end to end in a shed 20 m long?

10 A book measures 25 cm by 18 cm.
Find the approximate dimensions of this book, in inches.

11 At the supermarket, Ruski bought the quantity of fruit shown.

    a Find the approximate total mass (in lbs) of this fruit.

    b Apples are £0·89 per pound. How much would 4 kg of apples cost?

    c About how many ounces is 250 g?

MEMO

Apples	5kg
Pears	250g
Peaches	500g
Bananas	250g
Oranges	1kg

12 A washing machine holds about 27 ℓ of water.
About how many gallons of water does this washing machine hold?

13 Highlife paint can be bought in 2 ℓ and 5 ℓ tins.
About how many more pints does the larger tin hold?

Review This is part of a letter Marcia sent to her Swiss penfried Yurg.
Yurg converted each of the measures given in this story to the unit in the brackets.
What answers should he get?

Melanie took all of her pets to the school 'pet day'. Melanie lives 30 miles (km) from her school so she left home very early. The goldfish tank had 4 gallons (litres) of water so it was heavy. The dog was on a 4 metre (feet) long lead. The cat was entered in the large cat section because it weighed 11 lb (kg). The cat won the 'longest tail' section. Its tail is 15 inches (cm) long.

Reading scales

There are ten divisions between 1 and 2.
Each division on this scale is 1 m ÷ 10 = $\frac{1}{10}$ m = 0·1 m.

The pointer is at 1·4 m.

There are five small divisions between 8 and 9.
Each division is 1 cm ÷ 5 = $\frac{1}{5}$ cm = 0·2 cm.

The pointer is at 8·6 cm.

There are ten small divisions between 0·1 and 0·2.
Each division on this scale is 0·1 kg ÷ 10 = 0·01 or $\frac{1}{100}$ kg.

The pointer is at 0·14 kg.

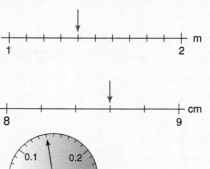

Sometimes we have to estimate the reading on a scale.
The pointer is about halfway between 5 ℓ and 6 ℓ or at about 5$\frac{1}{2}$ ℓ.

Exercise 5

1 Write down the readings given by A, B and C.

a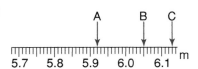

b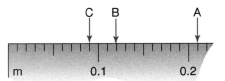

c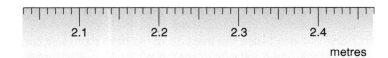

d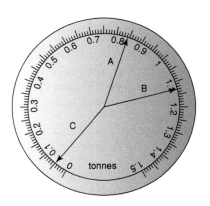

2 Use a copy of this.

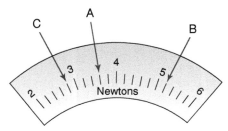

Use arrows to show these. Label each arrow.
a 2·3 m   b 2·35 m   c 2·42 m   d 2·17 m   e 2·08 m   f 2·26 m

3 Maria only had a broken ruler to measure lines. Find the length of these.

a 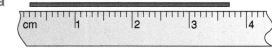   b

c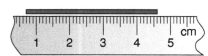

4 Estimate the readings for A, B and C on these scales

a 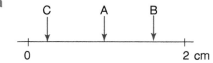   b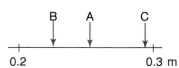

5 Use of a copy these scales.
Use an arrow to show the number given beside the scale.

a 2.9   b 4.7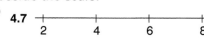

6 What is the volume to the nearest 10 mℓ?

a    b    c

7 How long is the string to
a the nearest 10 cm
b the nearest 100 cm
c the nearest cm
d the nearest mm?

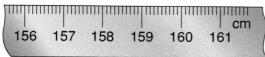

8 Use this scale to convert these to °F.
a 40 °C   b 100 °C   c 0 °C

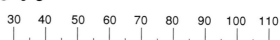

9 Use the scale in the previous question to convert these to °C.
a 50 °F   b 200 °F   c 140 °F

Review 1 What measurement is given by the pointers?

a 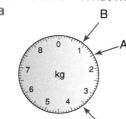   b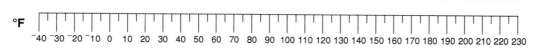

Review 2 Estimate the readings for A, B and C.

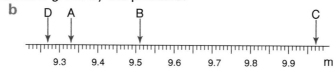

Review 3 Use a copy of the scale in Review **2**.
Use an arrow to show 1·6 m.

T

Review 4 Use this scale to convert these to kilograms.
a 84 lb   **b** 130 lb   **c** 74 lb

Review 5 Use the scale to convert these to pounds.
a 65 kg   **b** 30 kg   **c** 25 kg

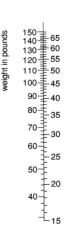

Units, measuring instruments and accuracy

Sarah measured the length of the school grounds.
She says they are 200 m long.
We do not know what she used to measure with or how accurately she measured.
She could have measured to the nearest centimetre, nearest metre or nearest 10 metres.

All measurements are approximate. The **measuring instrument** and **degree of accuracy** we choose depend on how accurate we need the measurement to be.

Discussion

● Suggest an imperial or metric unit you could use to measure these. **Discuss.**
　　weight of a nail   width of a pencil
　　distance from Mars to Jupiter   mass of a car
　　amount of petrol in a petrol tanker time to travel to the moon
　　time to grow a water cress plant   thickness of a hair
　　diameter of a 20p coin
　　Suggest a measuring instrument and a way to measure each. **Discuss.**

● Melanie is making scones. Should she measure the flour to the
　nearest g, nearest 5 g, nearest 10 g or nearest 50 g? **Discuss.**

SCONES
375 g flour
3 tsp baking powder
1/4 tsp salt
1 cup milk
25 g butter

● Rajiv takes 12·5 minutes to walk to his piano lesson.
　Do you think this time is correct to the nearest second,
　nearest 30 seconds or nearest minute? **Discuss.**

● The distance to Chipping Campden is given in miles.
　Do you think the distance on this road sign is correct to the
　nearest mile, nearest $\frac{1}{2}$ mile or nearest $\frac{1}{4}$ mile? **Discuss.**
　Use your answer to decide
　　the shortest and longest distances to Chipping Campden
　　the shortest and longest possible distance between Chipping Campden
　　and Snowshill.

$2\frac{1}{2}$ Chipping Campden   Snowshill $2\frac{3}{4}$

● Do you think the volume given on this bottle will be correct to the
　nearest mℓ, nearest 10 mℓ, nearest 50 mℓ or nearest 100 mℓ? **Discuss.**
　Why is it important that goods have the mass or volume shown on the
　label? **Discuss.**

Olive Oil
800 ml

Exercise 6

1

kg	g	km	m	cm	mm	ℓ	mℓ	hours	min	sec

Which of the units in the box would you use to measure each of the following?

a width of a netball court   b mass of a packet of biscuits
c capacity of a cup   d time to fly from London to New York
e length of an ant   f distance to the moon
g capacity of a fridge   h capacity of a tablespoon
i mass of a chair   j time to eat an ice cream

2 Choose a sensible degree of accuracy for these.
Choose from the box.

a the diameter of an eyeball
b the mass of a cricket ball
c the length of an airport runway
d the distance a train travels in an hour
e the mass of a small box of chocolates
f the height of a chair
g the mass of a bicycle
h a dose of cough medicine
i the mass of a train
j the capacity of a car's petrol tank
k the time to walk to school
l the time to run 1500 m

> nearest g
> nearest 100g
> nearest kg
> nearest tonne
> nearest mℓ
> nearest 100mℓ
> nearest ℓ
> nearest mm
> nearest cm
> nearest m
> nearest 10m
> nearest 500m
> nearest km
> nearest second
> nearest 30 sec
> nearest minute

3 Do you think these measurements are given to
    A the nearest cm   B the nearest m   C the nearest 10 m   D the nearest 100 m?

a Juliet says her room is 4 m long.
b Forrest says he lives 850 m from the supermarket.
c Humphrey says his thumb is 5 cm long.

4 Do you think these measurements are given to
    A the nearest g   B the nearest 10 g   C the nearest 500 g   D the nearest kg?

a Rebecca says she weighs 48·5 kg.
b Martin says he needs 550 g of sugar to make fudge.
c Olivia says her school bag weighs 18 kg.

5 Do you think these measurements are given to
    A the nearest $\frac{1}{2}$ ℓ   B the nearest ℓ   C the nearest 10 ℓ   D the nearest 100 ℓ?

a Manisha says her vase holds 3·5 ℓ.
b Todd says his water tank holds 640 ℓ.
c Nina says her family uses 55 500 ℓ of water each year.

6 Do you think these measurements are given to
    A the nearest second   B the nearest 30 seconds
    C the nearest minute   D the nearest hour?

a Paul says it takes him 25 minutes to cycle to work.
b Alex says it takes her 7 seconds to write her full name.
c Damion says he has spent 103 hours practising for his piano exam.

7 Write down two things that you would measure to these degrees of accuracy.
  a to the nearest m   **b** to the nearest ℓ   **c** to the nearest hour
  d to the nearest mm   **e** to the nearest g   **f** to the nearest minute
  g to the nearest kg   **h** to the nearest mℓ   **i** to the nearest second

Review 1

kg	g	mg	km	m	cm	mm	ℓ	mℓ	min	sec

Which of the units in the box would you use to measure each of the following?
  a mass of a school bag   **b** length of a Boeing 747
  c time to run 5 km   **d** mass of a calculator
  e distance around the Equator   **f** time for 8 heartbeats
  g amount of water used in a shower

Review 2   To what degree of accuracy do you think it would be sensible to give these?
Choose from the box in question **2**.
  a the length of an arm   **b** the mass of a lorry
  c the time to make breakfast   **d** the amount of water in a large jug
  e the width of a tooth

 Practical

1 Choose the degree of accuracy needed to measure some lengths around your
school.
Decide whether you need to use a trundle wheel, metre tape, ruler with cm
and/or mm divisions or a micrometer screw gauge.
Measure these lengths as accurately as possible.

2 Design an experiment, such as measuring reaction time, to
measure small intervals of time.
Choose the degree of accuracy you will use.

3 The pendulum clock was invented in the 17th century. Before this,
time was sometimes measured using the clocks described below.

The Candle Clock
Marks were made at equal distances down the side of a candle.
How close do you think these marks might be if it took 5 minutes to
burn the length of candle between two marks?

The Water Clock
A container, such as that shown, was filled with water.
The water dripped out of a hole at the bottom.
Marks were made on the inside of the container. Equal intervals
of time were read off by watching the level drop from one mark
to the next.
Do you think the marks should be equally spaced?

Sinking Bowl Clock
A small bowl, with a hole in the bottom, was put in a large container of water.
The time it took for the bowl to sink measured a particular interval of time.

Design and make a clock to measure an interval of time, maybe 2 minutes,
5 minutes or 15 minutes.
You may like to make a candle clock, a water clock or a sinking bowl clock.
You may like to make a different sort of clock.

Make your clock as accurate as possible.

Estimating measurements

Estimation — a game for a group

You will need a tape measure and ruler.

To play
- Choose someone to start.
- This player chooses a length or a height and a unit of length for this measurement.

  Examples   height of my desk in centimetres
         or length of Dale's hair in millimetres
         or width of the room in metres

- Each player writes down an estimate for this.
- The length or height is measured.
- The player with the closest estimate gets 5 points.
- Take turns choosing the length or height and unit for this.
- The winner is the person with the most points at the end of the game.

Note You could play this game again but choose a mass and a unit of mass or choose a volume and a unit of volume.

Discussion

Jonathon was asked to estimate the time it would take him to walk 5 miles.
He knows that it takes him about 20 minutes to walk the mile to school.
He said '5 miles will take about 5 times as long as it takes me to walk to school. It would take about 100 minutes or 1 hour and 40 minutes'.

Jonathon used what he already knew to help estimate the answer.
We say Jonathon used what he knew as a **benchmark**.
Other possible benchmarks are
    the height of a door is about 2 m
    the capacity of a glass is about 250 mℓ
    the area of a piece of A4 paper is about 600 cm^2
    the mass of a glass of water is about 300 g.

What else could you use as a benchmark when estimating? **Discuss.**

Use benchmarks to estimate these. **Discuss.**
    the height of the tallest tree you can see
    the capacity of a sink
    the area of your school grounds
    the area of the cover of this book
    the mass of this book

When we **estimate a measurement** it is a good idea to give a **range** for it.

Gareth was asked to estimate the width of his teacher's desk to see if it would fit through the door.
He said 'Its bigger than 0·8 m but less than 1 m'.

He wrote this as   0·8 m < width of desk < 1 m.
This gives the range for the estimate.

Remember < means is less than.

Exercise 7

1   Estimate the length of each of these lines.
Measure each line to the nearest 0·1 cm (nearest mm).

a ——————————   b ————————————

c ——————————   d ————————————

e ——————————————————————

2   What should go in the gaps?
Choose from the numbers given in the box below.
Use each number just once.

30   4000
1500
12   9
4   150
45   500

a	the temperature on a cold day	____ °C
b	mass of an elephant	____ kg
c	time to run 100 metres	____ sec
d	length of a bus	____ m
e	amount of cola in a large bottle	____ mℓ
f	length of a cat's tail	____ cm
g	weight of a boy	____ kg
h	length of a calculator	____ mm
i	mass of a box of chocolates	____ g

3   Choose the best range.

a   A   1 m < width of classroom door < 2 m
    B   0·5 m < width of classroom door < 1 m
    C   1·0 m < width of classroom door < 1·1 m

b   A   0·1 ℓ < capacity of a cup < 0·3 ℓ
    B   0 mℓ < capacity of a cup < 500 mℓ
    C   0·5 ℓ < capacity of a cup < 1 litre

c   A   500 g < mass of a small apple < 1 kg
    B   50 g < mass of a small apple < 100 g
    C   1 kg < mass of a small apple < 2 kg

d   A   1 m < length of bed < 2 m
    B   1 m < length of bed < 5 m
    C   1·5 m < length of bed < 2·5 m

4 Write A, B or C for each item.

i

Item	Mass between		
	A	B	C
	10 g and 500 g	0·5 kg and 1·5 kg	1·5 kg and 5 kg
a a pen			
b a cup			
c a large cat			
d a ruler			
e litre of milk			
f a pumpkin			

ii

Item	Capacity between		
	A	B	C
	0 mℓ and 100 mℓ	100 mℓ and 1 ℓ	1 ℓ and 10 ℓ
a a basin			
b a bucket			
c a shampoo bottle			
d a can of cola			
e a teaspoon			
f a perfume bottle			

5 Using metric units, write down an approximate measurement for each of these.
   Give a range for each.
   a   the height of your classroom
   b   the mass of a classroom chair
   c   the volume of water in a full bucket
   d   the mass of an orange
   e   the length of an adult arm from shoulder to wrist
   f   the volume of coffee in a full coffee mug

Remember to use a benchmark to help.

6 The man in this picture is 1·8 m tall.
   Estimate the height of these.
   a   the child
   b   the garage
   c   the tree
   d   the light

Review 1   Write A, B, or C for each of the following to show which box would give your estimate for its length.

Item	Capacity between		
	A	B	C
	0 m and 1 m	1 m and 2·5 m	2·5 m and 5 m
a a book			
b a telephone			
c a tissue box			
d a family car			
e a room			
f a mountain bike			

Review 2 Using metric units, write down an approximate measurement for each of these. Give a range for each.

a mass of a light bulb

b length of a floor-length curtain

c capacity of a kitchen sink

d mass of a small car

e length of a long-jump pit

f capacity of a vacuum flask

Review 3 The man in this picture is 1·8 m tall. Estimate the length of the house in this picture.

 Practical

1 **You will need** a tape measure and chalk.
Mark a point on the ground with chalk.
Estimate a distance of 1 m from this mark and put another mark.
Measure the actual distance.
How good was your estimate?

Repeat for other distances.

Estimate the length of some things such as the length of a netball court, width of a window, length of the school gymnasium, length of a paper clip, width of a book, ... Give a range for your estimate.
Measure the actual lengths.
You could draw a table like this one.

> You could estimate by pacing.

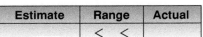

Estimate	Range	Actual
	< <	

2 **You will need a** trundle wheel or long measuring tape.
Choose two objects in the school grounds. Estimate the distance between them. Give a range for your estimate.
Use a measuring tape or trundle wheel to measure the actual distance.
Was it within your range?

3 **You will need** a stopwatch.
Work in pairs to estimate intervals of time such as 1 minute, 5 minutes and so on.
Before you begin, decide how you are going to do this.

Estimate the time to run 200 m. Give a range for your estimate.
Check by timing someone to run 200 m.

4 'Invent' a person who is about your age. Give your person a name.
Estimate sensible measurements for height, weight, arm length, foot length, waist measurement, ...
Make a poster showing your person.

Perimeter and area of a triangle, parallelogram and trapezium

Remember
Perimeter is the distance around the outside of a shape.
Perimeter is measured in km, m, cm or mm.

Area is the amount of surface a shape covers. Area is measured in square millimetres (mm²),
square centimetres (cm²), square metres (m²) or square kilometres (km²).

Discussion

- **You will need** a sheet of paper and some scissors.
  Draw a triangle on the sheet of paper.
  Draw a rectangle around the triangle, as shown below. (One side of the triangle
  becomes a side of the rectangle.)

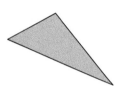

Cut out the two pieces of the rectangle that are outside the triangle. (Pieces A and B).

Try and fit these two pieces exactly onto the triangle.
Repeat this with other triangles.

Do you think this statement is true? **Discuss.**
  'The area of a triangle is always half the area of a rectangle'.

- A parallelogram can be made from a rectangle.

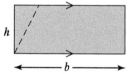

Cut a triangle off one side   Add it to the opposite side of
of the rectangle.   the rectangle.

Is the area of the parallelogram the same as the area of the rectangle? What is the
formula for the area of the parallelogram? **Discuss.**
Can all parallelograms be made this way? **Discuss.**

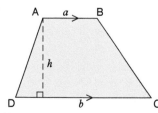

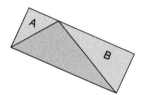

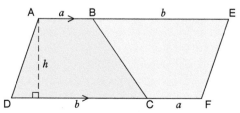
a + b is the sum of the parallel sides.

Trapeziums ABCD and BEFC are congruent (same shape and size).
AEFD is a parallelogram. What is its area? **Discuss.**

Use this to show that the area of ABCD is $\frac{1}{2}(a + b) \times h$.
Is this true for all trapezia?

The **area of any triangle** is given by A = $\frac{1}{2}bh$ where b is the base of the triangle and h is the perpendicular height.

 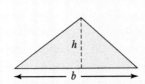 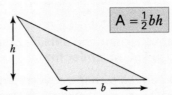

$A = \frac{1}{2}bh$

The **area of a parallelogram** is given by **A = bh** where b is the base of the parallelogram and h is the perpendicular height.

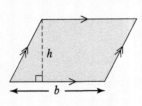 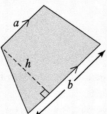

$A = bh$

The **area of a trapezium** is given by **A = $\frac{1}{2}(a + b)h$** where a and b are the parallel sides of the trapezium and h is the perpendicular distance between them.

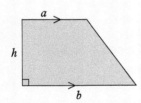 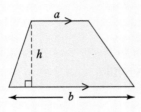

$A = \frac{1}{2}(a+b)h$

Examples

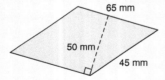 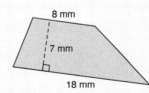

A = bh
= 65 × 50
= 3250 mm^2

Remember to put the units with the answers.

A = $\frac{1}{2}(a + b) \times h$
= $\frac{1}{2}(8 + 18) \times 7$
= 91 mm^2

Worked Example
This window hanging has a triangle inside a circle.
The circle has a diameter of 12 cm.
Calculate the area of the triangle.

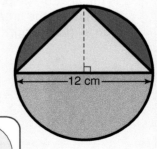

Answer
The height of the triangle is the radius of the circle.
The height of the triangle is 6 cm.

Area of triangle = $\frac{1}{2}bh$
= $\frac{1}{2} \times 12 \times 6$
= **36 cm^2**

Remember the radius is half of the diameter.

Exercise 8   **Only use a calculator if you need to.**

1 i Calculate the area of these triangles.

a
6.4 m   4 m
5 m

b
3 cm
5.8 cm   5 cm

c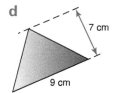
16 mm
20 mm

d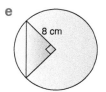
7 cm
9 cm

e
8 cm

ii Calculate the perimeter of the triangles in **a** and **b**.

2 The 'Bermuda Triangle' is a part of the Atlantic Ocean in which more than 50 ships and 20 planes have mysteriously disappeared.

Find the area of the 'Bermuda Triangle' if the base is 1560 km and the height is 1320 km.

3 Find the area of these parallelograms.

a **b** **c** 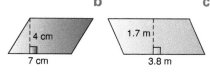 **d**   **e**

a: 4 cm, 7 cm
b: 1.7 m, 3.8 m
c: 2 mm, 2.3 mm, 5 mm
d: 1.2 m, 0.8 m, 1.6 m
e: 4.2 m, 2.8 m, 3.6 m

4 The area of a parallelogram is 20 cm². The base is 16 cm.
What is the height of this parallelogram?

5 a What is the area of this flower bed?
b Mary wants to fertilise the flower bed.
One box of fertilizer is needed for each 10 m².
How many boxes does she need?
c What fraction of a box will she have left?

6 m
8 m

6 The area of this triangular glass window hanging is 24 cm².
What might the lengths of *a* and *b* be?

a
b

7 Even without measurements written on this diagram, we know that the area of the shaded figures are equal. How do we know?

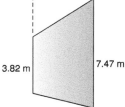

8 i Find the area of these trapeziums.

a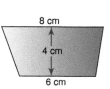
8 cm
4 cm
6 cm

b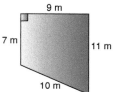
9 m
7 m   11 m
10 m

c
4 m
3.82 m   7.47 m

d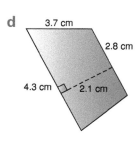
3.7 cm
2.8 cm
4.3 cm   2.1 cm

ii Find the perimeter of the trapezium in **b**.

9 This shows the cross-section of a swimming pool. What is the area of this cross section?

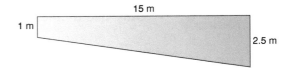

15 m
1 m
2.5 m

***10** Charlotte made this banner to support her favourite team. The flag is a parallelogram.
It has an area of 4·2 m^2.
How far apart are the poles?

Manchester United
1.2 m
pole → ← pole

***11** This diagram shows a park of area 25 000 m^2.
Find h.

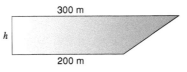

300 m
h
200 m

***12** The area of a trapezium is 10 cm^2.
What might be the value of h, a and b?

b
h
a

***13** Ruth drew this shape on square dotty paper.
She drew a rectangle around it.
She found the area as follows.

Area of green shape = area of rectangle − area A − area B − area C
$$= 4 \times 6 - \tfrac{1}{2} \times 4 \times 4 - \tfrac{1}{2} \times 2 \times 3 - \tfrac{1}{2} \times 1 \times 6$$
$$= 24 - 8 - 3 - 3$$
$$= \textbf{10 units}^2$$

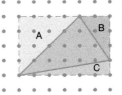

A B
C

Use Ruth's method to find the areas of these.

a   **b**   **c**   **d**   ***e**

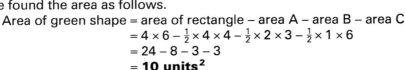

***14** Ayler found the area of this shape by cutting it into pieces.
He found the areas of C, D and E and added them.
a Copy and finish Ayler's working.

Area of red shape = area C + area E + area D
$$= \tfrac{1}{2} \times 2 \times 2 + \tfrac{1}{2} \times \underline{\hspace{1cm}} + \tfrac{1}{2} \times \underline{\hspace{1cm}}$$
$$= \underline{\hspace{1cm}} + \underline{\hspace{1cm}} + \underline{\hspace{1cm}}$$
$$= \underline{\hspace{1cm}} \text{ units}^2?$$

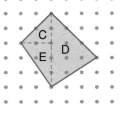

C
D
E

b Use Ayler's method to find the areas of these.
i   **ii**

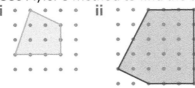

c Which of the shapes in question **13** can you find using this method?

*15 You will need some square dotty paper for this question.
Draw a triangle or quadrilateral on your square dotty paper.
Make sure each vertex is on a dot.
Find the area of your shape using either Ruth's method (question 13) or Ayler's method (question 14).
Repeat for other shapes.

Review 1 Triangles of pastry like the shapes shown, are used to make savouries.
What area of pastry is used for each?

a

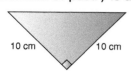

10 cm   10 cm

b

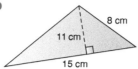

11 cm   8 cm
15 cm

Review 2
i Find the area of these shapes.

a

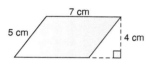

7 cm
5 cm   4 cm

b

1.8 m
2.5 m
2.9 m   2.4 m
3.8 m

c

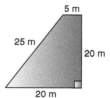

5 m
25 m   20 m
20 m

ii Find the perimeter of the shapes in i.

*Review 3 This diagram shows an end wall of a shed.
The area of this wall is 26·4 m².
What is the length of the wall?

2.4 m
2 m

*Review 4 Find the area of these shapes.

a   b

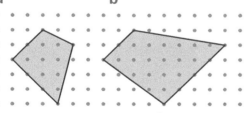

Practical

You will need some 1 cm squared paper.
On the squared paper draw a triangle with an area of 10 cm².
What if the area is 6 cm²?
What if the area is 7·5 cm²?

Perimeter and area of shapes made from rectangles

Example   To find the perimeter of this shape we need to find the missing lengths.

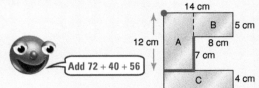

This dot shows the starting point.

red side = 16 − 5 − 4 = 7 cm
Perimeter = 14 + 5 + 8 + 7 + 8 + 4 + 14 + 16
= **76 cm**

To find the **area** of this shape divide it into rectangles.
One way is shown.

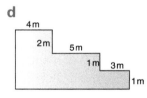

purple length = 14 − 8 = 6 cm
Area of A = 12 × 6 = 72 cm²
Area of B = 8 × 5 = 40 cm²
Area of C = 14 × 4 = 56 cm²
Total area of shape = **168 cm²**

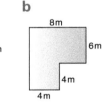

Add 72 + 40 + 56

What other ways could you use to find the area of the shape?

Exercise 9

1   Find the perimeter of these shapes.

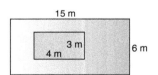

a   b   c   d

2   Find the area of the shapes in question **1**.

3   In each of these rectangular lawns a flower bed has been cut out.
Find the area of the remaining lawn.

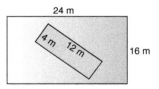

a   b   c

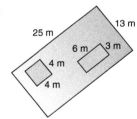

4   Sarah's lawn was 10 m by 12 m. She cut a rectangular vegetable patch, 2 m by 5 m, into the lawn.
Find the area of lawn left.

5   This shows a path through a courtyard.
Find the area of the path (grey shaded).

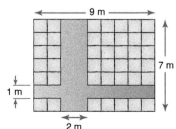

6 A floor 4·5 m by 6 m is covered with square tiles of side 25 cm.
  a Find the number of tiles needed.
  b If each tile costs £2·25, find the cost of tiling the floor.

7 Johnny uses this shape to make a pattern.
What is the perimeter of the pattern?

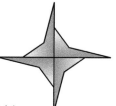

*8 The perimeter of a rectangular lawn is 84 m. The length is twice the width.
What is the length of the shortest side?

*9 Rachel makes a patio with two sizes of square
paving stones.
  a What is the length of the large paving stone?
  b What is the length of the small paving stone?

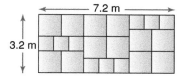

* 10 The dimensions of a box are 1·5 cm by 2·5 cm by 3·2 cm.
What is the largest number of these boxes that can be put in a
tray with a base of 9 cm by 22 cm?

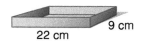

*11 This is the plan of an open-plan living
area of a home. All measurements are
in cm.
This living area is to be carpeted with
30 cm square carpet tiles. How many
will be needed?

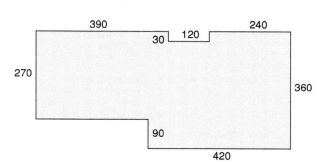

Review 1 Find the perimeter and area of the green shapes.

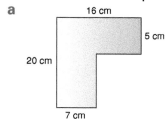
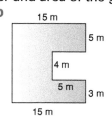
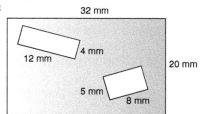

*Review 2 Sunita paved an area with slabs.
What is the area of one slab?

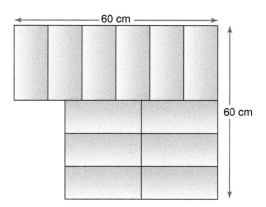

Practical

You will need LOGO or a dynamic geometry package.
Draw a square.
Now draw a square that is one quarter of the area of the first square.

How do you know the area of the second square is one quarter of the area of the first square? Explain.

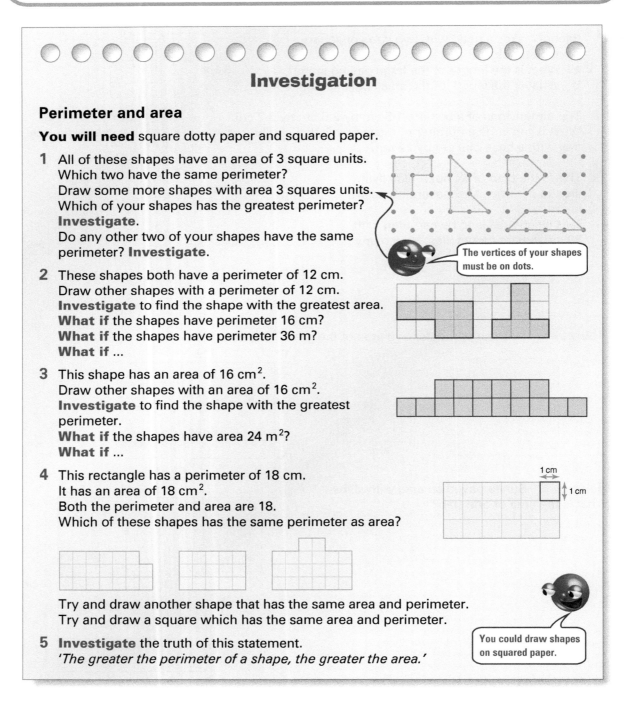

Investigation

Perimeter and area

You will need square dotty paper and squared paper.

1 All of these shapes have an area of 3 square units.
   Which two have the same perimeter?
   Draw some more shapes with area 3 squares units.
   Which of your shapes has the greatest perimeter?
   Investigate.
   Do any other two of your shapes have the same perimeter? **Investigate**.

   The vertices of your shapes must be on dots.

2 These shapes both have a perimeter of 12 cm.
   Draw other shapes with a perimeter of 12 cm.
   Investigate to find the shape with the greatest area.
   What if the shapes have perimeter 16 cm?
   What if the shapes have perimeter 36 m?
   What if ...

3 This shape has an area of 16 cm^2.
   Draw other shapes with an area of 16 cm^2.
   Investigate to find the shape with the greatest perimeter.
   What if the shapes have area 24 m^2?
   What if ...

4 This rectangle has a perimeter of 18 cm.
   It has an area of 18 cm^2.
   Both the perimeter and area are 18.
   Which of these shapes has the same perimeter as area?

   1 cm

   1 cm

   Try and draw another shape that has the same area and perimeter.
   Try and draw a square which has the same area and perimeter.

5 **Investigate** the truth of this statement.
   'The greater the perimeter of a shape, the greater the area.'

   You could draw shapes on squared paper.

Surface area

The **surface area** of a 3-D shape is the total area of all the faces.

Practical

You will need some unit cubes or centimetre cubes.

1   Janet made this shape from centicubes.
    Each face has an area of 1 cm².
    She counted 18 faces so the surface area of her shape is
    18 cm².
    Make some other shapes using cubes.
    Find the surface area of each.

2   Make as many different cuboids as you can each with 20 cubes.
    Do they all have the same surface area?

Discussion

Below is the net for the cuboid shown.

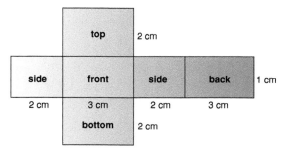

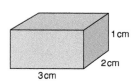

Opposite faces have
the same area.

Brandon wrote

top and   front
bottom   and back   sides
↓   ↓   ↓

Surface area of cuboid = 2(3 × 2) + 2(3 × 1) + 2(2 × 1)

Will this give the right answer for the surface area?
Discuss.

How could you use this to write a formula for the surface
area of this cuboid? **Discuss.**

What is the surface area of this cube? **Discuss.**
If each face is painted in the same pattern as the faces
shown, how much of the surface area is red? **Discuss.**

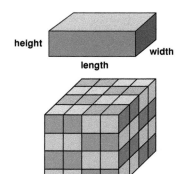

Surface area of cuboid = 2(length × width) + 2(length × height) + 2(height × width)

= 2 *lw* + 2 *lh* + 2 *hw*

= 2(*lw* + *lh* + *hw*)

Worked Example

Brad wanted to glue green cardboard onto the surface of this wooden cuboid.
What area of cardboard will he need?

Answer

Total surface area = 2 *lw* + 2 *lh* + 2 *hw*

= 2(12 × 10) + 2(10 × 5) + 2(12 × 5)

= 2 × 120 + 2 × 50 + 2 × 60

= 240 + 100 + 120

= **460 cm²**

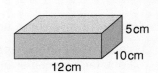

Worked Example

Thomas made a wooden T to put on his bedroom door.
He wanted to cover it with sticky red paper before he put it on the door.
What area of sticky paper will he need?

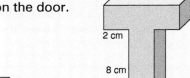

Answer

Divide the front face into rectangles.

Area of front face = area A + area B

= 6 × 2 + 8 × 2

= 12 + 16

= **28 cm²**

Area of back face = **28 cm²**

Area of top = 6 × 1

= **6 cm²**

Area of end = 2 × 1

= **2 cm²**

Area of other end = **2 cm²**

Area of side = 8 × 1

= **8 cm²**

Area of other side = **8 cm²**

Area of underside = 2 × 1

= **2 cm²**

Area of other underside = **2 cm²**

Area of bottom = 2 × 1

= **2 cm²**

Total Area = 28 + 28 + 6 + 2 + 2 + 8 + 8 + 2 + 2 + 2 = **88 cm²**

Exercise 10 **Only use a calculator if you need to.**

1   These shapes are made from unit cubes.
    Find the surface area of each shape.

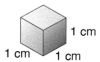
1 cm
1 cm   1 cm

a   b   c   d

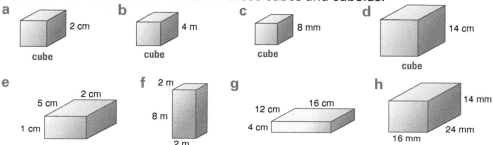

2   Calculate the surface area of each of these cubes and cuboids.

a   b   c   d

2 cm   4 m   8 mm   14 cm
cube   cube   cube
                                                               cube

e   f   2 m   g   h

        2 cm   16 cm   14 mm
5 cm   8 m   12 cm
1 cm   4 cm   16 mm   24 mm
             2 m

3   This small box is made from cardboard.
    a   If the box has a lid, how much cardboard is needed?
    b   How much cardboard is needed if the box has no lid?

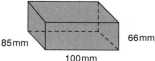
85 mm   66 mm
        100 mm

4   Sally places three cubes as shown to make a stand.
    The top cube is in the middle of the other two.
    She covers the stand with sticky paper.
    Sketch the shapes of the sticky paper she could
    use and say how many of each are needed.

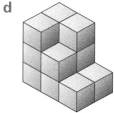

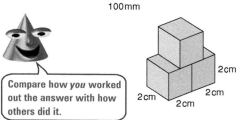
2 cm
2 cm   2 cm
2 cm

Compare how *you* worked
out the answer with how
others did it.

5   Calculate the surface area of these shapes.

a   b   c

        1 cm   5 mm   8 cm   9 cm
    4 cm   2 mm   3 cm
             3 cm   3 cm
                              6 mm
    2 cm   2 mm   10 cm
        1 cm   3 mm   3 cm

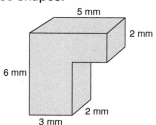   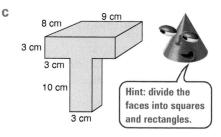

Hint: divide the
faces into squares
and rectangles.

6   a   Basil makes 12 coloured cubes.
        Each cube has 1 cm edges.
        Each is covered with sticky paper.
        How much paper is needed?
    b   Rosemary makes this shape from 12 cubes.
        She covers it with sticky paper.
        What area of paper did she use?
    c   What other shapes could Rosemary have made with 12 cubes?
        Which shape uses the least paper?

*7 The surface areas of the faces is given. Find the length of each edge.

a
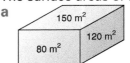

b
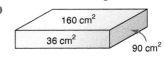

Review 1 Find the surface area of these shapes. Each is made from unit cubes of length 1 cm.

a  **b**   **c**

Review 2 Calculate the surface area of these cubes.

a 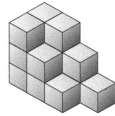 5 cm   **b** 11 mm

Review 3 Bon Bon the clown stands on this platform for his performance. He wants to paint it.
a What is the surface area of the platform?
b Bon Bon buys 1 ℓ of red paint. Will it cover 2400 cm²? Will he have enough paint?

80cm 20cm 200 cm

*Review 4 Mohammed built this platform to display the cups he had won.
Find the surface area of the platform.

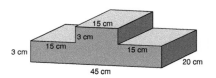

15 cm 3 cm 15 cm 15 cm 3 cm 20 cm 45 cm

Investigation

Surface area

You will need some unit cubes or centimetre cubes.

This shape is made with four centicubes.
The area of all of the surfaces (top, bottom, front, back and both sides) is 18 cm².

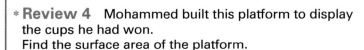

What is the surface area of these?

Can four centicubes be arranged in other ways?

Investigate to find the arrangement which has the smallest surface area.

Investigate to find the arrangement which has the greatest surface area.

What if we began with three cubes?

What if we began with five cubes?

Volume

Discussion

- For this cuboid: length = 5 cubes
                    width = 4 cubes
                    height = 3 cubes.

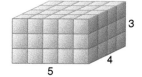

  What is the area of the base of this cuboid?
  Use this formula to find the volume.
    Volume = area of base × number of layers
  How many cubes are there altogether?

  What is the volume of a cube l units long, w units wide with h layers? **Discuss**.

- Archimedes noticed that when he got into his bath the water level went up. When he got out the water level went down. He shouted 'Eureka'.
  What had Archimedes discovered? **Discuss**.
  How could you use Archimedes' discovery to find the volume of some objects?
  Discuss.

Volume is a measure of the amount of space something takes up.
We measure volume in mm³, cm³, m³ or mℓ, ℓ, pints or gallons.

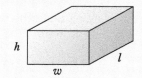

> For a **cuboid, volume = length × width × height**
> = $l \times w \times h$

Example   Volume = length × width × height
            = 8 m × 5 m × 3 m
            = 120 m³

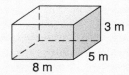

Exercise 11   **Only use a calculator if you need to.**

1 Find the volume of these. Estimate the volume of **e** to **h** first.

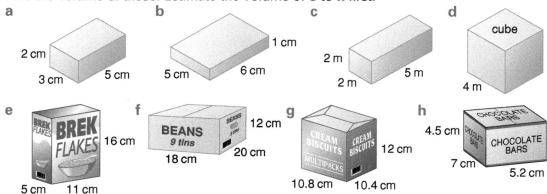

a 2 cm, 3 cm, 5 cm

b 1 cm, 5 cm, 6 cm

c 2 m, 2 m, 5 m

d cube 4 m

e BREK FLAKES 16 cm, 5 cm, 11 cm

f BEANS 9 tins 12 cm, 18 cm, 20 cm

g CREAM BISCUITS MULTIPACKS 12 cm, 10.8 cm, 10.4 cm

h CHOCOLATE BARS 4.5 cm, 7 cm, 5.2 cm

2 a What is the volume of the smaller fish tank?

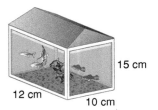

15 cm
12 cm
10 cm

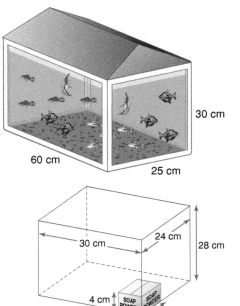

30 cm
60 cm
25 cm

b What is the volume of the larger fish tank?
c How many times bigger is the volume of the larger fish tank?

3 Hetty is packing packets of soap powder into a box as shown.
  a How many will fit along the 30 cm side?
  b How many packets of soap powder will fit in the box altogether?

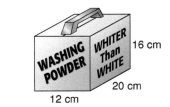

30 cm
24 cm
28 cm
4 cm
SOAP POWDER SOAP POWDER
5 cm
6 cm

4 A hotel buys washing powder in boxes which are 20 cm long, 12 cm wide and 16 cm high.
  a Calculate the volume of powder in the box.
  b The hotel puts the powder into smaller boxes which are cubes of side 4 cm.
     How many smaller boxes can be filled from one of the large boxes?

WASHING POWDER WHITER Than WHITE
16 cm
20 cm
12 cm

5 Pen was packing bars of chocolate in a box as shown.
  a How many chocolate bars will fit along the 20 cm side?
  b How many chocolate bars will fit on the bottom layer?
  c How many chocolate bars will fit in the box altogether?

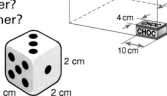

50 cm
20 cm
18 cm
4 cm
CHOC
CHOC
2 cm
10 cm

6 A box has dimensions 10 cm by 8 cm by 6 cm.
Dice like these are to be packed into this box.
What is the largest number of dice the box will hold?

2 cm
2 cm
2 cm

∗7 Tim buys a trailer load of soil. The trailer measures 170 cm by 150 cm by 40 cm.
  a What is the volume of the trailer in m^3.
  b The soil fills the trailer exactly. Tim puts the soil on a rectangular garden measuring 300 cm by 400 cm.
     How thick will the soil be if it is evenly spread on the garden?

∗8 Raji digs out an area of shingle to make a lawn.
The shingle measures 8 m by 6 m and has a depth of 0·08 m.
  a What is the volume of the shingle?
  b How many 2 m^3 trailers are needed to take the shingle away?
  c Raji gets 4·8 m^3 of soil delivered.
     He spreads it evenly over the area. How deep will the soil be?

∗9 'Jim's Rail' containers come in three lengths: 6 m, 12 m and 14 m.
Each is 2·4 m wide by 3·2 m tall.
What is the maximum number of boxes measuring 1·2 m by 1·2 m by 2·8 m that will fit in each container?

Review 1   Find the volume of these.

a

3 cm

b
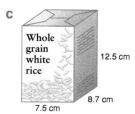
3 cm
6 cm
9 cm

c

Whole grain white rice
12.5 cm
8.7 cm
7.5 cm

Review 2   A 'square' baking dish is a cuboid with a square base, as shown in the diagram.
How much cake mixture could this 'square' dish hold if it was half filled?

6 cm
18 cm
18 cm

Review 3   Deborah built a sandpit for her sister.
She made it 350 cm long, 240 cm wide and 40 cm high. Deborah then filled it with sand to within 10 cm of the top.
a   What volume, in cm^3, of sand did Deborah put in this sandpit?
* b   What volume, in m^3, of sand is this?

Review 4   Rob buys chocolate pudding mix in large boxes 35 cm long, 21 cm wide and 14 cm high.
a   Calculate the volume of chocolate pudding mix in the box.
b   Rob puts the pudding mix into smaller boxes which are cubes of side 7 cm.
   How many smaller boxes can be filled from one large box?

Chocolate Pudding Mix
14 cm
21 cm
35 cm

○ ○ ○ ○ ○ ○ ○ ○ ○ ○ ○ ○ ○ ○ ○ ○ ○ ○ ○ ○

*Investigation

Making boxes

You will need 1 cm squared paper and thin card.

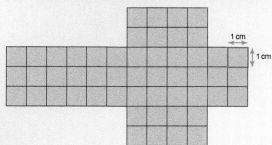
1 cm
1 cm

This net can be drawn on cardboard and folded to make a box of volume 24 cm^3.
The dimensions of this box are 3 cm × 4 cm × 2 cm.

What other dimensions could a box of volume 24 cm^3 have? **Investigate**.

Design nets for these other dimensions.
Which of these nets uses the smallest area of cardboard? **Investigate**.

What if the volume was 48 cm^3?
What if the volume was 30 cm^3?
What if the volume was 36 cm^3?
What if ...

Summary of key points

 You need to know these **metric conversions**.

length	mass	capacity (volume)	area
1 km = 1000 m	1 kg = 1000 g	1 ℓ = 1000 mℓ	1 hectare(ha) = 10 000 m^2
1 m = 100 cm	1 tonne = 1000 kg	100 cℓ = 1 ℓ	
1 m = 1000 mm		1 cℓ = 10 mℓ	
1 cm = 10 mm		1 ℓ = 1000 cm^3	
		1 mℓ = 1 cm^3	
		1000ℓ = 1 m^3	

Examples

$0 \cdot 62$ kg = $0 \cdot 62 \times 1000$ g   520 mℓ = 520 ÷ 1000 ℓ 42 cm = 42 ÷ 100 m

            = **620 g**   = **0·52** ℓ = **0·42 m**

52 800 m^2 = 52 800 ÷ 10 000 ha   3000 cm^3 = **3** ℓ or   4680 ℓ = 4680 ÷ 1000 m^3

            = **5·28 ha**   **3000 mℓ**   = **4·68 m^3**

 You need to know these **metric and imperial equivalents**.

length	mass	capacity
8 km ≈ 5 miles	1 kg ≈ 2·2 lb	4·5 ℓ ≈ 1 gallon
1 m ≈ 1 yard	30 g ≈ 1 oz	1 ℓ ≈ 1·75 pints
1 m ≈ 3 feet		
2·5 cm ≈ 1 inch		

 When **reading scales** you need to work out the value of each small division.

Examples   There are five small divisions between 0 and 1.

Each small division is $\frac{1}{5}$ or 0·2 kg.

The pointer is at 1·6 kg.

 When measuring we must choose the **degree of accuracy**, the **unit** and a suitable **measuring instrument**.

Example   When measuring the length of a pin we could measure it to the nearest mm using a ruler.

The **degree of accuracy** chosen depends on the situation.

Example   A jeweller might need to measure to the nearest mm.

A surveyor might need to measure to the nearest m.

 When we **estimate** a measurement it is a good idea to give **a range** for the estimate.

Example   200 g < mass of calculator < 500 g

Area is the amount of space covered by a shape.
Area is measured in km², m², cm², mm² or hectares.

Area of a triangle = $\frac{1}{2}$ area of rectangle

$\qquad\qquad\qquad = \frac{1}{2} \times$ base $\times$ height

$\qquad\qquad\qquad = \frac{1}{2}\,bh$

The height and base must be perpendicular.

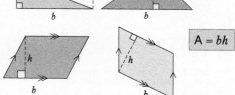

$A = \frac{1}{2}bh$

Area of a parallelogram = bh where b is the base of the parallelogram and h is the perpendicular height.

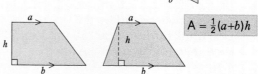

$A = bh$

Area of a trapezium = $\frac{1}{2}(a + b)h$ where a and b are the parallel sides of the trapezium and h is the perpendicular height.

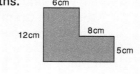

$A = \frac{1}{2}(a+b)h$

We can find the **perimeter** and **area** of a **shape made from rectangles**.

Example   To find the perimeter, we find the missing lengths.

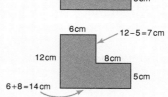

Perimeter = 6 + 7 + 8 + 5 + 14 + 12

$\qquad\qquad$ = 52 cm

To find the area we divide the shape into rectangles.

Total area = area of A + area of B

Area of A = 12 × 6$\qquad$Area of B = 8 × 5

$\qquad\quad$ = 72 cm²$\qquad\qquad\qquad$ = 40 cm²

Total area = 72 cm² + 40 cm²

$\qquad\qquad$ = 112 cm²

The **surface area of a cuboid**

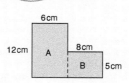

= 2(length × width) + 2(length × height) + 2(height × width)

= 2 lw + 2 lh + 2 hw

Example   Surface area = 2 lw + 2 lh + 2 hw

$\qquad\qquad\qquad$ = 2 × (6 × 5) + 2 × (6 × 3) + 2 × (3 × 5)

$\qquad\qquad\qquad$ = 60 + 36 + 30

$\qquad\qquad\qquad$ = 126 cm²

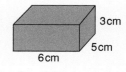

Volume is the amount of space taken up by a solid.
Volume is measured in mm³, cm³, m³ or ℓ, pints, gallons.

Volume of a cuboid = length × width × height

$\qquad\qquad\qquad\quad$ = lwh

Test yourself   **Only use a calculator if you need to.**

1 Find the missing numbers.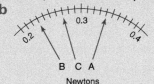
   a 32 mm = ___ cm   **b** 2800 m = ___ km   **c** 360 cm = ___ m
   d 680 mm = ___ m   **e** 760 mℓ = ___ ℓ **f** 47 g = ___ kg
   g 6852 mℓ = ___ ℓ **h** 4200 kg = ___ tonne   **i** 0·06 kg = ___ g
   j 5·86 cℓ = ___ ℓ **k** 42 kg = ___ tonne   **l** 70 mℓ = ___ cm^3
   m 4·6 ℓ = ___ cm^3 **n** 5420 ℓ = ___ m^3 **o** 45 000 m^2 = ___ ha

2 Grant is training for cycling.
   a He trains on a 2500 m track. Yesterday he travelled 40 km on the track.
         How many laps did he do?
   b One drink bottle holds 750 mℓ. In a 50 km race he needs to drink 300 cℓ.
         How many bottles will he need?
   c A road race is 56 km. About how many miles is this?
   d His bike weighs 10 kg. About how many lb is this?
   e This is a map for a road race.
         How long is the race?
         About how many kilometres is this?
   f The wheels on Grant's bike have a 16 inch diameter.
         About how many centimetres is this?
   g When training, Grant takes a 2 ℓ bottle of energy drink with him.
         Grant trains twelve times in a week.
         How many pints of energy drink does he need for a week?

10 miles
Park Lane   4 miles
8 miles
3 miles

3 Find the measurements given by pointers A, B and C. In **c** you will need to estimate.
   a   **b**   **c**

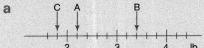
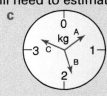

4 **a** What was the measurement you wrote down for question **3a** pointer A?
            About how many kilograms is this?

5 To what degree of accuracy do you think these are given?
      Choose from the box.
   a Mr Jones says his dog weighs 44·2 kg.
   b Bett says one of her fingernails is 16 mm long.
   c Reece says it takes him 10 minutes to walk to school.
   d Piers says he drinks 1500 mℓ of water each day.

to the nearest				
m	cm	mm	10 m	100 m
30 sec		min		sec
mℓ		ℓ		100 mℓ
g	kg		500 g	0.1 kg

6 Choose the best range.
   a A 0·5 m < length of a woman's arm < 1 m   **b** A 10 g < mass of pen < 1 kg
       B 1 m < length of a woman's arm < 1·5 m   B 10 g < mass of pen < 50 g
       C 1·5 m < length of a woman's arm < 2 m   C 10 g < mass of pen < 200 g

7 Suggest a metric unit, a measuring instrument, a degree of accuracy and a way to
      measure each of these.
   a the mass of an egg   **b** the length of your foot   **c** the capacity of an egg cup

8 Find the area of these triangular flags.

a 12 cm

6 cm

b 2.5 m

c 20 cm

8 cm

9 Estimate the height of your tallest school building.
What did you use as a benchmark? Give your answer as a range.

10 Find the area of these courtyards.

a

5 m 4.6 m

8 m

b

8.4 m

8.6 m

14.6 m

11 Patrick built a courtyard with a square garden in the middle.
  a What is the area of the courtyard?
  b Patrick wants to put a wooden border along the perimeter of the courtyard.
  How many pieces of wood, 125 cm long, will he need?

8 m

2 m
1 m
6 m

12 a What is the total surface area of this box?
  b A teacher wants to cover the sides and bottom with blue card and the rest with red card.
  How much of each colour card will she need?

16 cm

8 cm

side

6 cm

13 This trophy is to be silver plated.
Find the surface area of the trophy.

20 cm

4 cm

Most Improved
Tennis Player

3 cm

3 cm

3 cm

14 cm

14 A baking dish is shown.
What is the volume of the dish?

6 cm

10 cm 18 cm

15 a What is the volume of this swimming pool?
  b What volume of water is in the pool if it is filled to within 20 cm of the top?

25 m

8 m

1.4 m

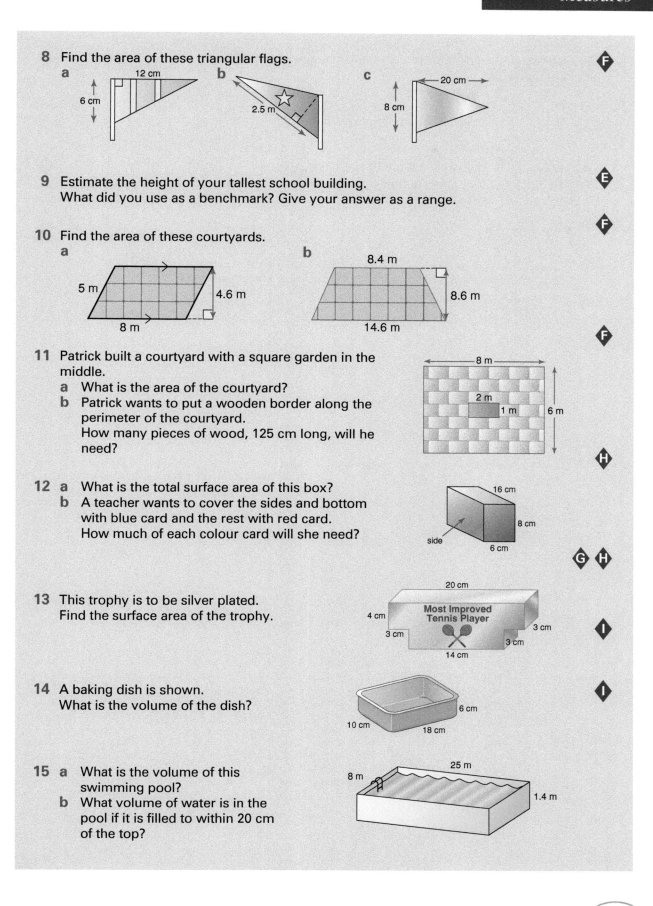

16 Twelve rectangles, all the same size, are arranged to make a square outline as shown.
Calculate the area of one rectangle.

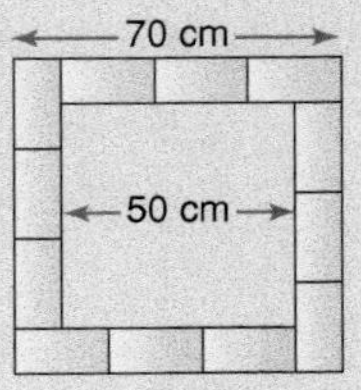

70 cm

50 cm

G

*17 A bandage pack measures 3·5 cm by 4·5 cm by 8·2 cm.
What is the largest number of these packs that can be put in a box 9 cm by 41 cm by 28 cm?

I

*18 Find the area of these shapes.

a

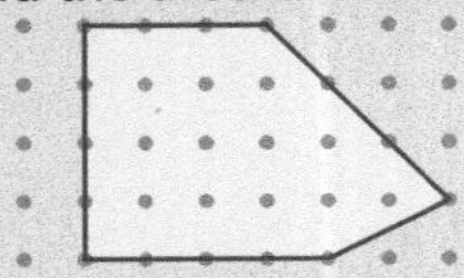

b

Hint: look back at questions 13 and 14 of exercise 8.

F

Handling Data Support

Collecting and displaying data

We collect data using a **collection sheet** or **questionnaire**.
The data can then be displayed on a graph.

Examples

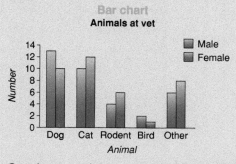

Bar chart

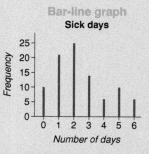

Bar-line graph

The length of the bar represents the frequency.

Graphs must have — what is being counted or measured on the horizontal axis
— values at equal intervals on the vertical axis
— the axes labelled
— a title.

Sometimes we draw bar charts with this on the vertical axis and the bars horizontal.

The owner of a supermarket wanted to know how many items each customer bought.
He did a survey.
This **frequency table** shows his results.
The number of items are grouped into equal class intervals.
Sometimes the last class interval is open.

Number of items	Tally	Frequency
1–5	IIII I	6
6–10	IIII IIII	9
11–15	IIII IIII	10
16–20	IIII II	7
21–25	IIII IIII	10
26+	IIII I	6

equal class intervals

open class interval

He drew this **bar chart for grouped data**.
The class interval is written under each bar.

In hospital, Jane's temperature was taken every 4 hours.

At 8 a.m. it was 37 °C, at noon it was 38 °C, at 4 p.m. it was 37·5 °C and at 8 p.m. it was 37·8 °C.

This **line graph** shows these temperatures.
It was drawn by plotting the temperatures at 8 a.m., noon, 4 p.m., 8 p.m. then joining the points with straight lines.

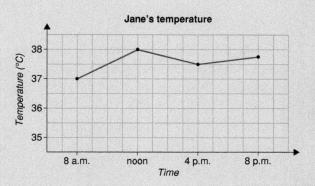

This **pie chart** shows what Sita spent her money on in July.
A pie chart is a circle graph.
It shows us the **proportion** Sita spent on each item.
The bigger the section of the circle the bigger the proportion spent on that item.
Sita spent the biggest proportion on food.
What else can you tell from this pie chart?

Spending for July

Rent
Food
Car
Going out
Clothes

Sita spent about $\frac{1}{3}$ on food.

Sita spent about 20% on clothes.

Practice Questions 2, 4, 6, 7, 9, 13, 15, 17, 18, 19

Databases

A **database** is a file of information.
It can be written on paper or stored on a computer.

Example Maths marks for Year 7 students.

Name	Age	Class	Test 1	Test 2	Test 3
D. Asch	11	7BN	8	9	7
B. Aitken	11	7TI	7	8	7
P. Archer	11	7AP	6	4	5

A **computer database** organises large amounts of information.

Examples of databases that your school might use are:
a database that stores details of each student
a database that stores details of each teacher
a database that stores details of the timetables of each student.

Practice Question 16

Sorting diagrams

A class was surveyed about board games they played one wet lunch time.
This **Venn diagram** shows that

4 played both Scrabble and Monopoly
6 played Scrabble but not Monopoly
9 played Monopoly but not Scrabble
2 played neither Scrabble nor Monopoly.

Scrabble   Monopoly

6 4 9
2

This **Carroll diagram** tells us about the weather in April.
It was cold and raining on 8 days, cold but not raining on 2 days, mild and raining on 5 days, mild and not raining on 15 days.

April weather

8	2	Cold
5	15	Mild

Raining   Not raining

Practice Questions 1, 5

Mode and range

The **range** is one way of measuring the spread of data.

Range = highest value – lowest value

The **mode** is the most frequently occurring data value.
Julia wrote down the masses of 20 small packets of crisps.

Sometimes there is more than one mode or no mode.

| 47 g | 50 g | 51 g | 50 g | 50 g | 48 g | 53 g | 54 g | 50 g | 51 g |
| 49 g | 51 g | 52 g | 50 g | 53 g | 55 g | 49 g | 50 g | 54 g | 52 g |

The range is **55 g – 47 g** = 8 g.
The mode is **50 g**.

Practice Question 11

Probability

Some events are certain, some **impossible**, some **likely** and some **unlikely** to happen.
It is certain that next year will have 52 weeks and impossible that it will have 13 months.
It is likely I will go to sleep tonight and unlikely that I will dream of skiing.

We also use the following words to say **how likely** an event is to happen.

> **good chance**
> **even chance or fifty-fifty chance**
> **poor chance**
> **no chance**

Examples   There is a poor chance I will fly to the moon before I die.
There is a good chance I will use the phone next week.
There is an even chance of getting a head when I toss a coin.

Sometimes we show probability on a scale.

← less likely			more likely →	
no chance	poor chance	even chance	good chance	certain

| I will fly to the moon | | I will get a head when I toss a coin | | I will use the phone next year |

Practice Questions 3, 8, 10, 12, 14, 20

Practice Questions

1   This diagram shows the people seen on one day by a dentist.
   a   How many were female adults?
   b   How many were adults?
   c   What else can you tell from this diagram?

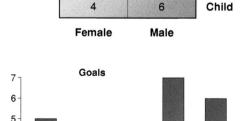

16	15	**Adult**
4	6	**Child**
Female	**Male**	

2   Beth's soccer team played against the 'Raiders' once each week.
   This bar chart shows the number of goals scored by her team.
   a   How many goals did Beth's team score in week 1?
   b   Which week did they score no goals?
   c   Which week did they score the most goals?
   d   How many more goals did they score in week 1 than in week 2?

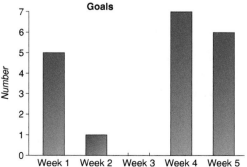

3 Write **impossible**, **unlikely**, **likely** or **certain** for each of these.
   a A rabbit will fly past your window today.
   b Tomorrow someone in the world will cry.
   c It will become dark tonight.
   d Someone in your class will be away one day next week.
   e Everyone in your class will be away one day next week.

4 For one week Sam wrote down the midday temperatures.
   He drew a graph.
   a What was the midday temperature on Saturday?
   b On which day was the midday temperature 8 °C?
   c Which was the coldest day?
   d Can you tell what the temperature was at midnight on
   Saturday? Explain your answer.

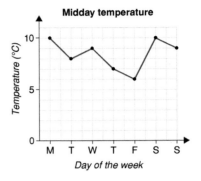
Midday temperature

5 This diagram shows the number of students who
   are interested in drama and music.
   How many are interested in
   a both drama and music
   b neither drama nor music
   c drama but not music?

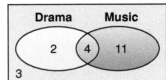
Drama   Music
2 4 11
3

6 These bar-line graphs show the number of hours of TV watched by four boys one week.

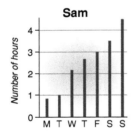
Sam

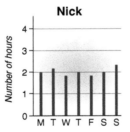
Nick

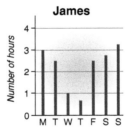
James

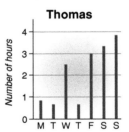
Thomas

Whose graph matches these comments?
   a I watched most TV at the beginning and end of the week.
   b I watched about the same amount of TV each day.
   c I watched quite a lot of TV on four days and not much on the other three days.
   d Each day I watched more TV than the day before.

7 This pie chart shows the type of pizza Year 9 students
   like best.
   a Which is the most popular pizza?
   b What goes in the gap?
   Vegetarian and ____ are liked best by about the
   same number of students.
   c What goes in the gap?
   Chicken and ____ together are liked by over half the
   students.

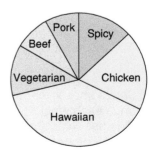
Pork Spicy
Beef
Vegetarian   Chicken
Hawaiian

8 Which of these have an even chance of happening?
   a The next baby born will be a boy.
   b The next time you walk up stairs you will trip.
   c The next time you roll a dice you will get an even number.
   d When this spinner is spun it will stop on blue.

T

9 7M had a cricket competition with another class.
This table gives the number of runs made by the students in 7M.
Use a copy of the bar chart.
Finish it.

No. of runs	Tally	Frequency
0–4	II	2
5–9	III	3
10–14	IIII I	6
15–19	IIII IIII	9
20–24	IIII IIII	10

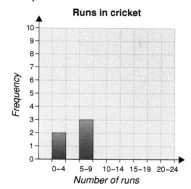

Runs in cricket

10 Draw a scale like this one.

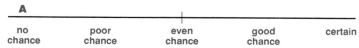

Write the letter which is beside each of these on the line to show how likely it is to happen.
A is done for you.
A A seal will drive a car down your street.
B You will watch TV tonight.
C It will snow next Christmas.
D You will grow taller.
E There will be no clouds in the sky tomorrow morning.
F You will visit a relative this weekend.
G You will see the city or town where your mother was born this year.

11 This table gives the time spent on homework by six girls on three nights.
  a What is the modal time spent on homework over all three nights?
  b What is the modal time spent on Wednesday?
  c What is the modal time spent on Thursday?
  d What is the range of the times spent on Wednesday?

Time spent on homework		
Tuesday	Wednesday	Thursday
1 hour	50 min	20 min
50 min	45 min	15 min
20 min	50 min	10 min
45 min	50 min	15 min
30 min	30 min	30 min
55 min	45 min	25 min

12 Madhu puts five purple beads and one pink bead in bag.
Madhu takes a bead without looking.
What colour bead is she more likely to get?
Explain why.

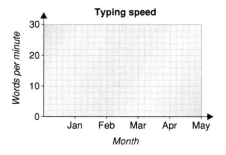

T

13 Tom was learning to type.
The table shows how many words per minute he could type at the end of each month.
Use a copy of this grid.
Draw a line graph for the data in the table.

Month	Jan	Feb	Mar	Apr	May
Words per minute	8	12	18	24	28

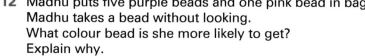

Typing speed

14 All of these spinners are spun at the same time.
   a Which one has the greatest chance of stopping on red?
   b Which has the greatest chance of stopping on green?

    Spinner A   Spinner B   Spinner C   Spinner D

15 There are 360 people in a sports club.
Each person plays just one sport.
This pie chart shows the sports played.
   a What fraction play basketball?
   b What percentage play basketball?
   c What percentage play netball?
   d How many play hockey?
   e How many play squash?
   f How many play netball?

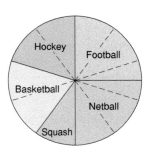

16 Amanda gathered data from her friends about the board games they played one holiday.
This is what she put into a database.

Name	Monopoly	Pictionary	Draughts	Trivial Pursuit	Chess
Marie	N	N	Y	Y	Y
Beth	Y	N	N	Y	N
Riffet	N	N	Y	Y	Y
Holly	Y	Y	Y	Y	N
Chelsea	N	Y	Y	N	N
Sarah	Y	N	Y	Y	N
Heather	N	Y	N	N	Y
Pam	Y	N	N	N	N
Amanda	Y	N	N	Y	N

Amanda printed these lists of names:
   a those who played Trivial Pursuit
   b those who did not play Monopoly
   c those who played both Pictionary and Monopoly
   d those who played either Chess or Draughts
   e those who played Monopoly but not Trivial Pursuit.
Write down the lists that were printed.

17 Omar's class got these marks for a project.
   19   16   45   43   40   39   36   30   28   42   35   40
   32   38   41   48   27   18   29   38   42   26   41   35
   a Use a copy of this table. Fill it in.

Mark	11–20	21–30	31–40	41–50
Tally				
Frequency				

   b Draw a bar chart for this data.

18 Julia wrote down the number of letters in the surnames of the teachers in her school.

5 6 8 9 7 7 4 5 6 8 5 7 5
6 8 4 7 6 6 8 6 9 8 5 6 5

a Copy and fill in this frequency table.
b Draw a bar-line graph for this data.

Number of letters	4	5	6	7	8	9
Tally						
Frequency						

19 Hal did a survey to find out the most popular activity of Year 7 pupils.
He asked this question.
Which of these activities do you enjoy doing most?

   playing sport   watching TV or videos   doing homework
   talking with friends   doing an artistic activity (dancing, drama, art, craft)

a Write down two things Hal might find out from his survey.
b Which of these collection sheets do you think is best? Say why?

Name	Class	Activity

Sheet 1

Activity	Tally	Number
Sport		
TV / Videos		
Homework		
Friends		
Artistic		

Sheet 2

20 Write some sentences. Use the following in your sentences.

poor chance	good chance	even chance
very unlikely	no chance	probable
unlikely	likely	very likely
possible	certain	impossible

15 Planning and Collecting Data

You need to know

✓ collecting and displaying data   page 395

✓ databases   page 396

Key vocabulary

class interval,   **continuous data,**   **database,**   **discrete data,**   **data,**
grouped data,   **data collection sheet,**   **experiment,**   **frequency,**
frequency chart,   **interval,**   **primary source,**   **questionnaire,**   **sample,**
secondary source,   **survey,**   **table,**   **tally**

 Viewing figures

Television has become a large part of many people's lives.
There are lots of questions we might like answered about
TV.

How much TV do children under 5 watch each day?
What percentage of households watch TV while eating
dinner?
What programmes do 12 year olds like best?

These questions could be answered by doing a survey.
Think of a survey you might like to do to find out
something about TV.
What might you find out in your survey?

Surveys — planning and collecting data

Before you carry out a **survey** you need to **plan it**.

Step 1 **Decide on the purpose of your survey.**
What is the problem or the question you would like answered?

Example How will the volume of traffic on motorways change over the next 20 years compared with traffic on other roads?

Are there any related questions?

Example Is this different for different parts of the country?
Is this different for lorries and cars?

Step 2 **Decide what data needs to be collected.**
It is useful to think of possible results you might get from your survey.
This helps you decide what data needs to be collected.

Example Reuben wanted to know at what times of the day pupils were most likely to use the internet during the school week.

> If the data is measurement data you need to decide the accuracy you want.

He thought of these related questions.
Are there differences for different aged pupils?
What is the best time to use the internet so it is not slowed down by overuse?
What is the internet used for most?

He thought of some of the results he might get.

Younger pupils are more likely to use the internet between 4 p.m and 7 p.m.
The hour when the internet is used most by pupils is 5 p.m to 6 p.m.
The best time to use the internet is after 9 p.m.
The internet is used most for e-mail.

From this he decided to collect the following data from each pupil.

Age	Time of day internet used this week
< 10 ☐	Put a tick beside any time you have used the internet in the last week.
10 – 12 ☐	7 a.m.– 8 a.m. ☐ 8 a.m.– 9 a.m.☐ 3 p.m.– 4 p.m.☐ 4 p.m.– 5 p.m.☐
13 – 15 ☐	5 p.m.– 6 p.m. ☐ 6 p.m.– 7 p.m.☐ 7 p.m.– 8 p.m.☐ 7 p.m.– 8 p.m.☐
16 – 18 ☐	9 p.m.– 10 p.m.☐ after 10 p.m. ☐ other ☐
> 18 ☐	What did you use the internet for?

Step 3 **Decide how to collect the data and how much to collect.**
You could collect data from these sources.

a A **survey of people**. Sometimes it is not practical to ask every possible person to answer the survey. Instead we choose a **sample** of people.
The sample size should be as large as is sensible.

Example If we want to know what music 12 year olds like we can't ask **every** 12 year old. We ask a sample.
The sample size will depend on cost, time, who the survey is for, etc.

> If the sample size is too small it is not representative of the large group.

b An **experiment** where you observe, count or measure something, or use technology such as a data logger.

Examples If we want to know if boys or girls have longer arms, we have to measure.
If we want to know the cooling rate of different metals we could use a data logger attached to a computer.

c **Other sources** such as reference books, newspapers, internet websites, historical records, databases, CD-ROMs, ...

Example If we want to know the birth rate over the last 10 years we have to use historical records.

a and b are called **primary sources** because you collect the data yourself.
c are called **secondary sources** because someone else has collected the data.

Discussion

For each of the questions given below suggest:

some possible results
other related questions you could explore
what data you would need to collect
how you would collect the data
how much data to collect (a sensible sample size)
how you could use ICT to help.
Discuss

- Does the school need more rubbish bins? If so where?

- How do pupils travel to school? Why do they travel this way?

- Are there different numbers of invertebrate communities in different parts of a stream?

- Do different types of magazine have different word or sentence lengths?

- Do shoppers use different modes of transport to different local shopping centres?

- How will the number of cars in more developed countries change over the next 50 years compared with the number of cars in less developed countries?

- Are there different light intensities in different parts of your school?

- At what time during a sports match are goals/points most likely to be scored?

- What improvements could be made at the local gymnasium?

- Which is your best catching hand?

- Is the population ageing?

Once you have planned your survey, design a **data collection** sheet or **questionnaire**.
The data collection sheet could be a **tally chart** or **frequency table**.

Examples

Tally chart   *Survey on rubbish*

Pieces of rubbish		
Place	Tally	Frequency
Outside Block A	卌 卌 卌 II	17
Outside Block B	卌 II	7
Outside Library	II	2
In Playing Fields	卌 卌 卌 卌 卌 卌 I	31

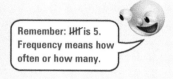

Remember: 卌 is 5.
Frequency means how
often or how many.

Frequency chart

Survey on ways of travelling to school

Time(min)	Car	Bus	Train	Walk	Other
1–5	3	2	0	11	0
6–10	18	7	3	25	1
11–15	12	16	14	19	1
16–20	18	24	26	10	0
20+	14	21	14	5	0

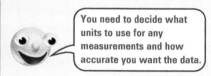

You need to decide what
units to use for any
measurements and how
accurate you want the data.

Discussion

Discuss how to design a collection sheet to collect the data needed for the questions
given in the discussion on page 404.
As part of your discussion, design a collection sheet for each.
Remember to decide what units to use for measurements and how accurate you want
the measurements to be.

Exercise 1

1 For each of these questions, what would be the best way to collect the data. Choose from
   the box.

   > A questionnaire or data collection sheet   B experiment   C other source

   a   What percentage of motor vehicle accidents in Britain occur after 6 p.m.?
   b   How often do people go to the cinema? At what times?
   c   Do most people in your class have a reaction time of less than $\frac{1}{2}$ second?
   d   Which items does the canteen run out of most often at lunch time?
   e   What sport do people in your class watch most often?
   f   Is the height of most people the same as three times the distance round their head?
   g   How has the birth rate increased or decreased over the last 20 years?

2 Choose two of the questions **A** to **F** on the next page. For each
   a   suggest other related questions you could explore
   b   write down some possible results
   c   write down what data you would need to collect and, if relevant, the accuracy needed
   d   design a collection sheet or questionnaire to collect the data
   e   suggest a suitable sample size if this is relevant to the question
   f   say whether your data source is a primary or secondary source.

A Do right-handed people in your class catch better than left-handed people?
B Which month of the year has the highest rainfall?
C Can taller people hold their breath longer than shorter people?
D How long do pupils spend moving between classes each day?
E How much sport do teenagers and adults watch?
F What factors affect the growth of water-cress? Light, water, warmth, ...

Review   Danya wanted to known how many times each week pupils used the school library and when they used it.
a Write down any related questions Danya could explore.
b Write down some possible results Danya could get from this survey.
c What data does she need to collect?
d Design a collection sheet or questionnaire to collect the data.
e Explain how she could collect the data. Mention sample size in your explanation.

Discrete and continuous data

Discrete data can only have certain values.
It is often found by counting
If we collect data about the number of people at bus stops we might get   24, 12, 16, 3, 18,   but not $26\frac{1}{2}$, $19\frac{1}{4}$, $3\frac{1}{3}$, ...   because it is only possible to have whole numbers of people.

Examples   the number of goals scored by a soccer team this season
the number of trees in the gardens in your street
shoe sizes of the rugby team

Continuous data is data which is measured and can have any values within a certain range. If we collected data on the shoulder height of the big cats at London Zoo, we might get any height between 0·8 m and 1·2 m. A height might be 0·94 m or 1·07 m.

Examples   the masses of people in a weight lifting class
finger nail length of netball players

Exercise 2

1 Which of the following are discrete data and which are continuous data?
a the number of teeth people have
b the height of new-born lambs
c the number of windows in houses
d the number of pupils in classes
e the lengths of classrooms
f the area of playgrounds at schools
g the marks of pupils in a science test
h the masses of eggs
i the shoe sizes of pupils
j the number of pupils who wear glasses
k the cost of fruit
l the number of rooms in houses
m the amount of cola in bottles
n the lengths of cats' tails

Review   Is the following data discrete or continuous?
a mass of letters in a postbag
b number of passengers on buses
c age of students when they leave school
d daily rainfall
e amount of money made at cake stalls.

Grouping continuous data

Remember   **Discrete data** is sometimes grouped.

Example   This frequency table shows the amount spent at the canteen one lunchtime.

Amount spent	Tally	Frequency
£0.01–£2	IHH	5
£2.01–£4	IHH II	7
£4.01–£6	IHH I	6
£6.01–£8	III	3
£8.01–£10	IIII	4
over £10	I	1

When we collect **continuous data** we usually group it into equal class intervals, (equal-sized groups) on a **frequency table**.

Example   Libby drew this frequency table to show the times taken to run the cross-country course.

Time (t min)	Tally	Frequency
$10 < t \leqslant 12$	I	1
$12 < t \leqslant 14$	II	2
$14 < t \leqslant 16$	IHH IHH II	12
$16 < t \leqslant 18$	IHH IHH IHH I	16
$18 < t \leqslant 20$	IIII	4

equal class intervals

We usually have between 5 and 10 class intervals.

$10 < t \leqslant 12$ means 'times greater than 10 minutes but less than or equal to 12 minutes.'
If we want to know how many pupils had a time less than or equal to 16 minutes we must add the frequencies for the first three class intervals.

$1 + 2 + 12 = 15$

Fifteen had a time less than or equal to 16 minutes.

Discussion

- Notice the difference between how the class intervals for discrete data and the class intervals for continuous data are written.
  Why is this? **Discuss.**

- In the example above about Libby's cross country, in which class interval would a time of 18 minutes be put? **Discuss.**
  What if the intervals had been
  $10 \leqslant t < 12$, $12 \leqslant t < 14$, $14 \leqslant t < 16$, $16 \leqslant t < 18$, $18 \leqslant t < 20$?

- Would it have been more useful if Libby had grouped the data in intervals of 4 minutes? **Discuss.**
  What about intervals of 30 seconds? **Discuss.**

Exercise 3

1   Derek wrote down the masses, to the nearest gram, of some apples.

| 155 | 172 | 186 | 162 | 168 | 196 | 153 | 160 | 201 |
| 185 | 175 | 168 | 180 | 194 | 186 | 163 | 170 | 210 |

   a   Use a copy of this table.
   Fill it in.

   b   How many apples had a mass in the class interval $160 < m \leqslant 170$?

   c   How many apples had a mass less than or equal to 160 grams?

   d   How many apples had a mass of more than 180 g?

Mass of apple (g)	Tally	Frequency
$150 < m \leqslant 160$		
$160 < m \leqslant 170$		
$170 < m \leqslant 180$		
$180 < m \leqslant 190$		
$190 < m \leqslant 200$		
$200 < m \leqslant 210$		

e Regroup this data using the class intervals $150 < m \leqslant 170$, $170 < m \leqslant 190$, $190 < m \leqslant 210$.
f Regroup the data using the class intervals $150 < m \leqslant 155$, $155 < m \leqslant 160$, $160 < m \leqslant 165$, $165 < m \leqslant 170$, ... $205 < m \leqslant 210$.
g Which class intervals do you think give the most useful information about the data?

2 The lengths, in metres, of the shot put throws of 20 people at a 'have a go' competition are listed.

| 8·2 | 16·0 | 5·3 | 6·8 | 9·5 | 15·7 | 12·0 | 17·4 | 16·3 | 15·3 |
| 3·9 | 18·6 | 7·9 | 8·5 | 10·5 | 8·0 | 11·6 | 12·1 | 13·2 | 14·8 |

a Use a copy of this frequency chart and fill it in.
b How many throws were greater than or equal to 4 metres but less than 8 metres?
c How many throws were less than 12 m?
d Explain why the class intervals given on the table are more useful than the intervals $0 \leqslant x < 1$, $1 \leqslant x < 2$, ...

Length of throw (x m)	Tally	Frequency
$0 \leqslant x < 4$		
$4 \leqslant x < 8$		
$8 \leqslant x < 12$		
$12 \leqslant x < 16$		
$16 \leqslant x < 20$		

Notice where the $<$ and $\leqslant$ are.

3 This data is the time, in minutes, it took for 30 students to jog or run around the park.

| 22 | 16 | 21 | 24 | 23 | 15 | 17 | 19 | 32 | 28 | 22 | 27 | 34 | 18 | 22 |
| 33 | 25 | 31 | 17 | 27 | 19 | 25 | 34 | 21 | 15 | 29 | 16 | 30 | 24 | 18 |

Draw a frequency chart for this data. Use the class intervals 14–, 18–, 22–, ...

14– means the same as $14 \leqslant t < 18$
18– means the same as $18 \leqslant t < 22$.

4 This data gives the handspan, in mm, of some 13-year-old girls.

171	182	202	164	198	211	164	173	177	173	192	205	228
223	189	186	191	194	197	205	192	207	169	173	184	209
176	181	207	164	196	203	221	198	190	169	184	207	197

Draw a frequency chart for this data. Choose suitable class intervals.

Review The amount of ice cream, in millilitres, in 20 containers is given below.

| 1985, | 1860, | 2004, | 2103, | 2004, | 1996, | 1987, | 2000, | 2150, | 2190 |
| 1994, | 1832, | 2100, | 2096, | 1900, | 2056, | 1934, | 1864, | 1999, | 2001 |

a Use a copy of this frequency table. Fill it in.
b How many containers had more than 1900 mℓ but less than or equal to 2000 mℓ?
c How many containers had more than 2000 mℓ?
d How many containers had less than or equal to 2100 mℓ?
e Explain why the class intervals given on the table are more useful than the intervals $1000 < x \leqslant 1500$, $1500 < x \leqslant 2000$, $2000 < x \leqslant 2500$.

Amount of ice-cream (x mℓ)	Tally	Frequency
$1800 < x \leqslant 1900$		
$1900 < x \leqslant 2000$		
$2000 < x \leqslant 2100$		
$2100 < x \leqslant 2200$		

Practical

Plan a survey.

You could choose one of the surveys already mentioned in this chapter **or** you could use one of the suggestions below **or** you could make up your own.

Check your choice with your teacher.

Follow the steps given on page 403.

Design a collection sheet or questionnaire. Remember you may need to group the data. Decide if it is continuous or discrete data. Collect the data.

Suggestions

What extra items should the canteen sell?

What is the life expectancy of people in different countries? Has this changed in the last 20 years?

Do right-handed people spell better than left-handed people?

Why do we put insulation in our ceilings? You could model this by putting warm items in an insulated and an uninsulated box.

Could the local bus service be improved?

What factors affect the growth of weeds in a garden?

How do the communities in two different habitats differ?

Summary of key points

Use these steps to **plan a survey**.

Step 1 Decide on the purpose of the survey.

Think of related questions you might want to explore.

Step 2 Decide what data needs to be collected.

Think of possible results you might get from your survey.

Step 3 Decide how to collect the data and how much to collect.

You could collect the data by

a doing a survey using a questionnaire or collection sheet

b doing an experiment where you count or measure something

c gathering data from other sources such as books, newspapers, the internet, historical records or a database.

The **sample size** should be as large as is sensible. If it is too small it is not representative. How big it is depends on time, money and who the survey is for.

> You need to decide how accurate the data needs to be.

Once a survey has been planned we often have to design a **collection sheet** or **questionnaire**.

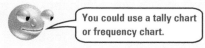

> You could use a tally chart or frequency chart.

Discrete data can only have certain values.

Example The number of crisps in crisp packets.

Continuous data can have any values within a certain range.

Example The diameter of crisps at their widest part.

D When we collect continuous data we usually group it into **equal class intervals** on a **frequency chart**.

Example This frequency chart shows the speed of cars passing a school.

Speed (s) mph	Tally	Frequency
$10 < s \leqslant 15$	IIII IIII III	13
$15 < s \leqslant 20$	IIII IIII IIII II	17
$20 < s \leqslant 25$	IIII II	7
$25 < s \leqslant 30$	IIII	4
$30 < s \leqslant 35$	II	2

Test yourself

1 For each of these questions, how would you collect the data? Choose from the box.

> A questionnaire or data collection sheet
> B experiment
> C data someone else has already collected

  a What percentage of accidents are caused by drunk drivers?
  b Do boys or girls have longer big toes?
  c Which fast food is the most popular with pupils at your school?

2 Mendel wanted to know how often people eat takeaways and on which days.
  a Write down two things Mendel might find out from his survey.
  b What data does Mendel need to collect?
  c Design a data collection sheet or questionnaire for this data.
  d How could he collect the data?
  e What does Mendel need to consider when choosing a sample size?

3 Which of these are discrete data and which are continuous data?
  a the masses of dogs
  b the number of fillings pupils have in their teeth
  c the shirt size of men
  d the length of material used for dresses

4 Rodney wrote down the time taken by 20 pupils to carry out a task.
These are the times in seconds.

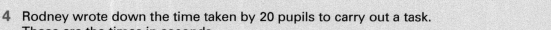

| 33 | 39 | 38 | 34 | 31 | 28 | 35 | 34 | 27 | 34 |
| 32 | 21 | 32 | 26 | 30 | 35 | 28 | 36 | 25 | 23 |

  a Use a copy of this frequency table. Fill it in.
  b How many pupils took more than 25 seconds but less than or equal to 30 seconds?
  c How many pupils took less than or equal to 35 seconds?

Time (t seconds)	Tally	Frequency
$20 < t \leqslant 25$		
$25 < t \leqslant 30$		
$30 < t \leqslant 35$		
$35 < t \leqslant 40$		

  d How many pupils took more than 30 seconds?
  e Melissa wrote down the same data. She chose these class intervals. $20 < t \leqslant 22$, $22 < t \leqslant 24$, $24 < t \leqslant 26$, ..., $38 < t \leqslant 40$.
    Are the class intervals in the table or Melissa's intervals more useful? Explain.

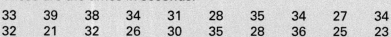

5 Choose a survey topic.
Plan your survey using the steps on page 403.
Design a data collection sheet or questionnaire to collect the data.
You could choose a topic already mentioned in this chapter or choose one of your own.

16 Mode, Range, Median, Mean. Displaying Data

You need to know

✓ mode and range   page 396
✓ collecting and displaying data   page 395

page 396
page 395

Key vocabulary

average,   bar chart,   bar-line graph,   frequency,   frequency diagram, interval,   label,   mean,   median,   mode,   modal class/group, pie chart,   range,   represent,   statistic,   title

Averaging it out

- We often hear the word 'average' used.

  The average house price is ...
  The average wage is ...
  The average annual rainfall in York in January is ...
  The average student at our school ...

  Think of other examples where average is used.
  Make a collection of labels from packets where the word average is used.
  Make a collection of articles from newspapers or magazines where the word average is used.

- The 'average' is a way of summarising the data into a single number.

  The age of the pupils in 7P is about 11.
  The height of the pupils in 7P is about 1·5 m.
  The number of children in the families of 7P is about 2·4.

  Estimate some 'averages' for your class.

Mode and range

Remember
The **mode** is the most commonly occurring data value.
The **range** is the difference between the highest and lowest data values.

> Sometimes there is more than one mode.

If a frequency table has grouped data we find the **modal class** or **modal group**.
This frequency table shows the times people spent looking around an
art exhibition.

Time (minutes)	$5 \leqslant t < 10$	$10 \leqslant t < 15$	$15 \leqslant t < 20$	$20 \leqslant t < 25$
Frequency	8	16	5	2

The modal class is $10 \leqslant t < 15$ because this class interval has the
highest frequency.

For continuous data the **range** is the
highest rounded value–lowest rounded value.

Example   The longest time, rounded to the nearest minute, someone spent at the art
exhibition was 24 minutes. The shortest time, rounded to the nearest minute,
was 3 minutes.
The range is 24 – 3 = 21 minutes.

Discussion

● The art exhibition organisers in the example above wanted to know which section of
the exhibition was most popular so they could make postcards of it.
Think of other instances when it would be useful to know the mode. **Discuss**.
● When might it be useful to know the range of a set of data? **Discuss**.

Exercise 1

1 These tables give the marks Sally's class got in two tests.
Write down the modal group for the each of them.

a   Maths

Score	Tally	Frequency
1–10	III	3
11–20	ЖΙ	6
21–30	ЖΙ II	7
31–40	ЖΙ ЖΙ II	12
41–50	III	3

b   Science

Percentage	Tally	Frequency
41–50	ЖΙ I	6
51–60	ЖΙ ЖΙ	10
61–70	I	1
71–80	ЖΙ ЖΙ	10
>80	III	3

> Sometimes there are two modes.

2 a Write down five pieces of data which have mode 6.
b Write down seven pieces of data which have no mode and range 10.
c Write down six pieces of data which have modes 8 and 10 and range 8.

3 This table shows the temperature of the 100 hottest days one year.
   a What is the modal class for temperature?
   b If the lowest temperature was 22·6 °C and the highest was 34·6 °C, what is the range?

Temperature (°C)	Frequency
22 ≤ T < 24	5
24 ≤ T < 26	27
26 ≤ T < 28	26
28 ≤ T < 30	31
30 ≤ T < 32	8
32 ≤ T < 34	2
34 ≤ T < 36	1

*4 This table gives the fastest and slowest times for a fun run.

This year (Hours : minutes)	**Last year** (Hours : minutes)
Fastest 1 : 20	Fastest 1 : 19
Slowest 2 : 10	Slowest 2 : 04

   a What was the range for this year?
   b What was the range for last year?
   c Why might it be useful for the organisers to know these ranges?

Review 1 Fifty students each tossed a dice 100 times.
This table shows the number of sixes tossed. What is the modal class?

Number of sixes	0–4	5–9	10–14	15–19	20–24	25–29
Frequency	3	8	12	18	7	2

Review 2 This table gives the number of minutes late 'Trans Tours' buses arrived.
   a What is the modal class?
*b If the shortest time was 1 minute 30 seconds and the longest time was 8 minutes 20 seconds, what is the range?

Minutes late	0–	2–	4–	6–	8–
Frequency	12	16	10	16	4

Practical

1 Find out the ranges for the daily maximum and minimum temperatures last month for your local area.

2 Find out the times of the top 100 males or females in the London marathon.
You could try using the internet.
Draw a frequency table for the data.
Write down the modal class for marathon times of the males and of the females.

Mean

Olivia	Joseph	Emily	Thomas	Rebecca
2	5	8	3	2

These pictures show the number of Easter eggs given to five children.
By sharing the Easter eggs equally between the children we find the **mean** number of eggs given to each child.

Altogether the children were given 20 Easter eggs.
If these were shared equally between the children each would have 4.

20 ÷ 5 = 4

The mean number of Easter eggs given to each child is 4.

The mean is often called 'the average'.

$$\text{mean} = \frac{\text{sum of data values}}{\text{number of values}}$$

Worked Example
This data gives the points scored by the horses in a gymkhana.

24·9   28·9   23·2   27·4   29·6   29·7   20·6   25·3

Find the mean number of points scored.

Use a calculator.

Answer

Mean = $\frac{24·9 + 28·9 + 23·2 + 27·4 + 29·6 + 29·7 + 20·6 + 25·3}{8}$

= **26·2**

Another way of calculating the mean is using an **assumed mean**. To do this

1 assume the **mean is a particular value**
2 subtract the mean from each data value
3 find the mean of the differences you found in **2**
4 add your answer to **3** to the assumed mean

Make a good guess at the mean to get the assumed mean.

To find the mean of the data above

1 assume the mean is 27
2 ⁻2·1,   1·9,   ⁻3·8,   0·4,   2·6,   2·7,   ⁻6·4,   ⁻1·7
   ↑   ↑   ↑   ↑   ↑   ↑   ↑   ↑
24·9 − 27 28·9 − 27 23·2 − 27 27·4 − 27 29·6 − 27 29·7 − 27 20·6 − 27 25·3 − 27

3 The mean of the differences = $\frac{⁻2·1 + 1·9 + ⁻3·8 + 0·4 + 2·6 + 2·7 + ⁻6·4 + ⁻1·7}{8}$

= $\frac{⁻6·4}{8}$

= ⁻0·8

Link to adding and subtracting integers page 35.

4 Actual mean = 27 + ⁻0·8

= **26·2**

Discussion

● Three meals have a mean price of £5.
  What might the price of these meals be? **Discuss**.

● In five tests, Tina's mean mark was 66.
  What might Tina's marks be? **Discuss**.

● Ben's marks for his first four maths tests were   72,   64,   63,   69.
  What was Ben's mean mark?
  What mark does Ben need to get in his next test to raise his mean to 70? **Discuss**.

Exercise 2

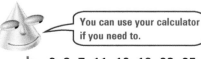
You can use your calculator if you need to.

1 Calculate the mean for each set of data.
   a 2, 2, 3, 4, 4, 4, 5, 5, 7, 7, 8, 9
   b 2, 3, 7, 11, 12, 19, 23, 25, 29, 31, 36
   c 4, 7, 7, 7, 7, 8, 8, 9, 10, 10, 11, 13, 14, 14, 15
   d 22, 19, 14, 16, 27, 11, 32, 41, 17, 43
   e 5·2, 4·6, 2·7, 7·1, 11·4, 5·3, 6·2, 2·3
   f 8·7, 1·4, 3·7, 2·6, 8·9, 5·9, 1·2, 0·9, 4·3, 6·2
   g 24·6, 30, 25·3, 82·4, 63·9, 51·1, 80, 32·7, 48·75

2 Calculate each of the means for question **1** using the *assumed mean* method.

3 Shannon wrote down how much she spent each day of her holidays.

	Mon	**Tues**	**Wed**	**Thur**	**Fri**	**Sat**	**Sun**
Week 1	£2·50	£3·60	£0·65	£2·55	£3·05	£2·60	£3·25
Week 2	£8·20	£2·25	0	£3·60	£0·80	£4·25	£3·30

Find
a her mean daily spending in week 1
b her mean daily spending in week 2
c her mean daily spending for the two weeks.
Could you find this by finding the mean of the answers to **a** and **b**?

4 Paula was studying beetles in science.
She wrote down the lengths, in cm, of four beetles.
     2·5,   2·1,   2·7,   2·2
a Find the mean length of the beetles.
Paula measured another six beetles she had collected from another place.
     2·0,   1·9,   2·1,   2·1,   1·8,   2·6
b Find the mean length of these beetles.
c Find the mean length of all 10 beetles.
*d Could you find the answer to **c** by finding the mean of the answers to **a** and **b**?

5 Jan wrote down her results for some of her subjects.

ENGLISH	68	75	71	74		
MATHS	81	62	79	82	86	
SCIENCE	70	75	58	68	72	47
FRENCH	50	46	58	55	56	

a What is her mean for Science?
b In which subject does Jan have the highest mean?
c Jan has to drop one of these subjects next year. Which one do you think she should drop? Why?

6 Matthew weighed himself five times on a set of very accurate scales. The readings were
61·380 kg,   61·284 kg,   60·982 kg,   61·034 kg,   61·000 kg
Find the mean of the five readings.

7 The mean number of hours of training done by a group of ten swimmers last week was 12.
a How many hours training did the ten swimmers do in total last week?
b Another two swimmers were asked how much training they did last week. One said 14 hours and the other 10 hours.
Calculate the new mean number of hours training for all 12 swimmers.

8 What three numbers could go in these boxes so that the mean of the three numbers is 7?

9 Jack has four number cards.
Their mean is 5.

Jack picks another card.
The mean of his five cards is 6.
What number is on the new card?

10 Asad got these marks in his first four mental tests.
8   7   9   8
The mean is 8.
After his fifth test, the mean of his marks was still 8.
What mark did he get in his fifth test?

11 Tamara is playing a game.
To win she has to get a mean score of exactly 80 in three rounds.
In the first two rounds she scored 84 and 75.
What must she score in her third round to win?

12 Broomsdale School bought 28 computers.
The costs are given by this table.
a Find the total cost for the 28 computers.
b Find the mean cost of each computer.

Item	Cost
28 computers @ £1425 each	___
Installation of 28 computers	£236
Total freight	£ 64
Insurance for 28 computers	£194
Total cost	___

*13 Olivia and Nishi played three rounds of a computer game.
The mean of their scores was the same.
The range of Olivia's scores was twice the
range of Nishi's scores.
What are Nishi's two missing scores?

	Game 1	Game 2	Game 3
Nishi		100	
Olivia	40	100	160

*14 Sonny got a mean of 40 in three rounds of 9-hole mini golf.
His range was 10. He scored 38 in one of his rounds. What did he score in the other two rounds?

Review 1
a Who has the highest mean mark? By how much?
b After another test Callum's mean was 67.
What did he score in this test?

Maths test marks

Callum	68	73	55	74	65
Dylan	93	39	65	73	70

Review 2
Joshua wrote down how late the bus was each morning.

	Monday	Tuesday	Wednesday	Thursday	Friday
Minutes late	1	8	5	2	

The mean for Monday to Thursday is 4 minutes.
After Friday the mean for the 5 days is 5 minutes.
How many minutes late was the bus on Friday?

*Review 3
Sam and Adam played three rounds of golf. The mean of their scores was the same. The range of Sam's scores was half the range of Adam's.
Sam got 72, 78 and 78. Adam got 76 in one game.
What did he get in the other two?

Data is sometimes given in a **frequency chart**.
We can find the mean using the chart.

Example This frequency chart shows how many sick days
Pamela's friends had last term.

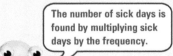

The number of sick days is found by multiplying sick days by the frequency.

$$\text{mean} = \frac{\text{total number of sick days}}{\text{total number of students}}$$

$$= \frac{4+8+6+4+5}{4+4+2+1+1} \quad \longleftarrow \text{ sum of number of sick days}$$
$$\longleftarrow \text{ sum of number of frequencies}$$

$$= \frac{27}{12}$$

$$= 2 \cdot 25$$

Sick days	Frequency	Number of sick days
1	4	4
2	4	8
3	2	6
4	1	4
5	1	5

The mean number of sick days is **2·25**.

Exercise 3   **Only use a calculator if you need to.**

1 The number of goals scored by a hockey team are given in
this table.
Find
  a the mode
  b the range
  c the mean.

Number of goals	0	1	2	3
Frequency	1	3	3	5
Total goals				

2 This frequency table gives the class sizes at Aleisha's school.
Find
  a the modal class size
  b the range in class size
  c the mean class size.

Class size	28	29	30	31	32
Number of classes	1	0	2	5	2
Number of pupils	28	0			

3 Joel sold some raffle tickets to his neighbours.
He drew this table to show how many each
bought.
  a What was the modal number of tickets
     bought?
  b What was the range?
  c What was the mean number of tickets
     bought?

Number of tickets per person	Frequency	Number sold
0	3	0
1	11	11
2	10	
3	8	
4	5	
5	3	

Review   This table shows the marks scored in a parents'
quick quiz at a school mathematics evening.
a What was the range of marks?
b What was the modal mark?
c What was the mean mark?

Number of marks	Number of parents	Total marks
0	2	
1	7	
2	5	
3	2	
4	3	
5	1	

Practical

You will need a spreadsheet package or graphical calculator.

* 1 Use the spreadsheet or calculator to find the answers to Exercise 3 questions
    1c, 2c, 3c, and Review **c.**

*2 Kylie tossed a dice with eight triangular faces.
   It had the numbers 1 to 8 on it.
   This table gives her results.

Score	1	2	3	4	5	6	7	8
Number of throws	64	72	58	69	57	61	70	65

   Use a spreadsheet to find the mean.

Median

Mr Bradley's class had a hopping competition.
The distances they hopped, to the nearest 10 m, were

80 m	120 m	60 m	200 m	150 m	170 m	190 m	205 m	90 m
60 m	140 m	170 m	190 m	160 m	100 m	110 m	205 m	

He put the distances in order from shortest to longest.

60 m	60 m	80 m	90 m	100 m	110 m	120 m	140 m	(150 m)
160 m	170 m	170 m	190 m	190 m	200 m	205 m	205 m	

Mr Bradley said 'There are 17 results. The middle of these is the 9th result, which is 150 m. Half of the class hopped 150 m or more'.
Mr Bradley had found the **median**.

The median is the middle value when a set of data is arranged in order of size.

When there is an even number of values the median is the mean of the two middle values.

Worked Example
Find the median of 21, 10, 72, 15, 43, 20, 56, 83.

Answer
Put the values in order.

10, 15, 20, **21, 43**, 56, 72, 83

There are two middle values, 21 and 43.
The median is the mean of these two values.

mean $= \frac{21 + 43}{2}$

$= \frac{64}{2}$

$= 32$

The median is **32**.

Exercise 4

1 Find the median of
   a   1, 5, 8, 12, 16, 24, 29   **b**   3, 9, 5, 12, 16, 12, 7   **c**   13, 24, 16, 33, 17, 31, 28
   d   80, 20, 30, 50, 40, 10, 32   **e**   4, 9, 5, 9, 7, 12   **f**   42, 38, 16, 57, 32, 41, 28, 52

2

Wei-Hsin	Rajiv	Victoria	Rebecca	Caleb	Charlotte	Jeremy	Anna	Sandra
1·47 m	1·57 m	1·61 m	1·39 m	1·65 m	1·62 m	1·76 m	1·53 m	1·49 m

What is the median height of these students?

3 This bar chart shows the number who were late to school each day one week.
What is the median number of pupils late per day?

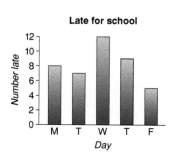

Late for school

4 Fleur's class had a charity competition to see who could walk 100 m backwards the fastest.
This list gives their times (in seconds).

25	29	26	28	32	37	29	36	31	46	33	39
46	28	34	30	29	31	37	42	56	48	29	36

a Find the median time.
b What goes in the gap?
About half of Fleur's class walked 100 m backwards in less than _____ seconds.

5 James wrote down the prices of nine different CD players sold at 'Rock CD'.

£69   £87   £54   £99   £68   £72   £83   £91   £64

a What is the median price?
b 'Rock CD' had a sale and took £5 off the price of all CD players.
What is the new median price?
* **c** If the same amount, x, is subtracted from all the data values in a set of data, what happens to the median?

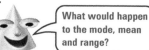

What would happen to the mode, mean and range?

Review   Tina counted the people at the gym on Wednesday and Friday lunchtimes.

Wednesdays	26	28	21	18	22	16	18	21	12	12	18	24	36	12	17	23
Fridays	35	37	14	30	28	24	27	19	21	8	21	19	34	16	21	

a Find the median for   **i** Wednesdays   **ii** Fridays.
b Find the median for both days together.

Practical

Collect some number data on a topic that interests you.
Find the median of your data.

Ideas
Times of cars in Monte Carlo Rally.
Times taken to run the New York or London marathon.
Cost of a packet of crisps at various shops.
Ages of horses at a race meeting.
Marks of Year 8 pupils in a test.
Runs made by cricketers in a test series.
Midday temperatures or hours of sunshine in your area.
Prices of cars of a particular year, make and model.

You could use the internet or a CD-ROM or database.

Mean, median, mode and range

Discussion

- When might it be useful to know the median? **Discuss**.
- When might it be useful to know the mean? **Discuss**.
- If a set of data has a large range, would it be better to use the mean or median to describe the data?
- If a set of data has a lot of values clustered around one value and then a few very large data values, is it better to use the mean or median to describe the data? **Discuss**.
- Stan and Bob have both just joined Russley Golf Club.
  The club has a place for just one of them in its 9-hole competition team.
  Stan and Bob have a play off.
  This table shows the results.

Hole	1	2	3	4	5	6	7	8	9
Stan	3	4	4	3	4	3	4	4	21
Bob	4	5	5	4	5	4	5	5	4

Work out the mean, median, and mode for each player.

Is it better to use the mean, median or mode to decide who is the better player?
Discuss.

Investigation

* **Basketball scores**

You will need a calculator.
There are four basketball teams in a competition.
The heights of the team members are given in this table.

Team	Heights (cm)							
Green Valley	149	152	160	158	160	162	158	156
Red Ferns	155	156	154	158	152	157	160	157
St Peters	143	147	150	152	155	149	150	154
Albion	150	146	159	152	144	158	157	158

Find the modal height, range, mean and median for each team.

This table gives the scores for the last eight games of each team.

Team	Scores							
Green Valley	35	17	33	35	36	16	27	41
Red Ferns	16	24	22	31	29	15	31	32
St Peters	18	21	31	22	12	39	20	19
Albion	14	16	18	31	31	32	30	22

Find the modal score, range, mean and median for each team.

Does there seem to be any relationship between the heights of the players in a team and the scores? **Investigate**.

Bar charts and line graphs

We often draw **graphs** to show the results of our data collection.

A picture can be worth a thousand words.

Bar charts are usually used to show **categorical data**. This is non-numerical data.

Callum collected data on the number of junior, intermediate and senior tennis players in a club over the last 5 years.
He drew this **compound bar chart** to show the data.

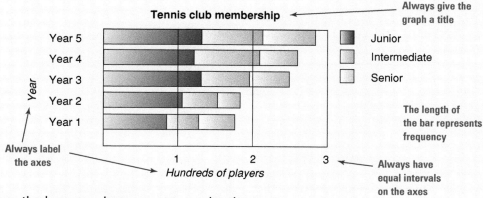

Tennis club membership ← Always give the graph a title

Junior
Intermediate
Senior

The length of the bar represents frequency

Always label the axes

Hundreds of players

Always have equal intervals on the axes

Sometimes the bars are drawn next to each other.

Example

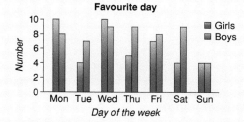

Favourite day

Girls
Boys

Day of the week

A **line graph** can be used to show trends over time.

Worked Example
This table shows the percentage of children immunised against measles, mumps and rubella (MMR) and whooping cough by their second birthday.

Year of 2nd birthday	'89	'90	'91	'92	'93	'94	'95	'96	'97	'98	'99
Percentage MMR	80	84	87	90	92	91	91	92	92	91	88
Percentage whooping cough	75	79	84	88	92	93	93	94	94	94	94

Draw a line graph to show this data.

Answer
We join the points with straight lines.
This makes it easier to see trends.
Sometimes the points in between the plotted ones have meaning and sometimes they don't.
On this graph they don't.

Percentage immunised by 2nd birthday

MMR
Whooping cough

Percentage

Use a key if there is more than one set of data on the same grid.

Why not?

This tells us the scale doesn't start at zero

Year

Exercise 5

T

1   Asha asked twenty Year 7 and twenty Year 11 pupils how many children were in their family.
This table shows her results.

Number of children	1	2	3	4	5	6
Year 7	3	5	8	1	3	0
Year 11	4	6	5	2	2	1

Use a copy of the bar chart below. Finish it.
Shade the Year 7 and Year 11 bars differently and fill in the key.

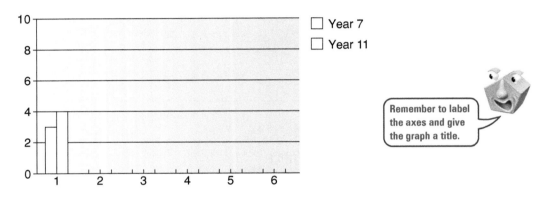

Remember to label the axes and give the graph a title.

T

2   Use a larger copy of the grid.
William and Joshua went to a typing class.
At the end of each fortnight they tested their speed.
This table shows the number of words typed per minute.

Week	2	4	6	8	10	12
William	12	14	18	24	30	42
Joshua	16	16	20	24	32	48

Remember to have a key.

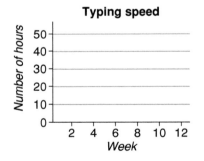

a   Draw a line graph with both sets of data on the same grid.
b   Which boy do you think made the better progress?
c   From the graph, would you be able to estimate William's typing speed after nine weeks?

T

3   This table shows the number per thousand of population of male and female criminal offenders.

Year	1900	1925	1950	1975	2000
Male	2	1·3	5·5	20·8	19·6
Female	0·5	0·1	0·8	4	4·1

a Use a copy of this compound bar chart. Finish it.

b Use a copy of this line graph. Finish it.

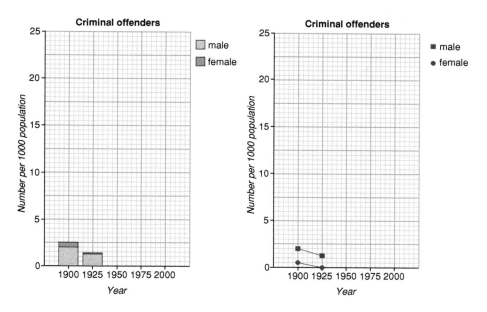

c Do the intermediate points of the line graph have meaning? Explain.

d Which graph, the compound bar chart or the line graph, do you think shows the trends better? Give a reason.

e What do the graphs tell you about criminal offenders?

*4 This table gives the air emissions, in thousand tonnes, for carbon monoxide, sulphur dioxide and black smoke.

Year	1973	1983	1993	1995	1998
Carbon Monoxide	9000	8000	6500	5700	5000
Sulphur Dioxide	5900	3900	3100	2400	1700
Black Smoke	900	500	500	300	300

To find what the horizontal axis must go up to, find the highest *total* of emissions.

Draw a compound bar chart with horizontal bars for this data.

*5 Jack did an experiment. He put hot water in two test tubes. He surrounded one of the test tubes (tt_1) with other test tubes full of hot water. The other (tt) he left by itself. He took the temperature of both test tubes at the end of each minute.
This table gives his results.

Time (minutes)	0	1	2	3	4	5	6	7
Temperature tt (°C)	90	70	48	34	28	24	20	18
Temperature tt_1 (°C)	90	78	68	58	50	44	40	38

This experiment simulates animals huddling together.

a Draw a line graph with both sets of data on the same grid.

b Do the intermediate points have meaning? Explain.

c What conclusions might Jack make from looking at the graph?

d Do you think this experiment simulates the huddling behaviour of some animals well?

Review This table gives the attendance in millions, at three of London's most popular tourist attractions.

Year	1991	1993	1995	1997	1999
Madame Tussauds	2·1	2·5	2·7	2·8	2·8
Tower of London	2·6	2·3	2·5	2·6	2·6
London Zoo	1·7	0·9	1·0	1·1	1·1

a Use a copy of this compound bar chart. Finish it for the data.
b Draw a line graph with all three sets of data on the same grid.
c Do the intermediate points on the line graph have meaning?

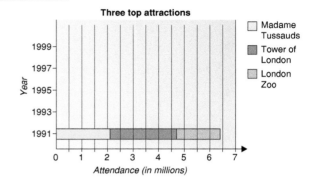

Practical

Collect some data that can be displayed on a compound bar graph or a line graph. You could use the government statistics website or you could do an experiment or collect data using a collection sheet or questionnaire.

Ideas
Number of students away from school each day in each year 7 class.
Time taken by two students to do homework on 5–10 nights.
You could do the cooling experiment in Exercise **5** question **5**.

Frequency diagrams

Charlotte wanted to know if her thumb was longer than most other girls her age.
She collected this data from some year 7 girls.

The data she collected is continuous data.

> Remember continuous data can take any value within a given range.

She graphed it on this **frequency diagram**.

Class interval (l is in mm)	Frequency
$30 \leqslant l < 35$	2
$35 \leqslant l < 40$	2
$40 \leqslant l < 45$	4
$45 \leqslant l < 50$	9
$50 \leqslant l < 55$	10
$55 \leqslant l < 60$	2

Thumb lengths

> This is very important to remember.

For **continuous data**, we **label the divisions between the bars**.

Exercise 6

T

1 This gives the size of the last 24 strong earthquakes in the South Pacific. Use a copy of the frequency diagram and complete it.

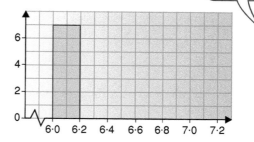

Remember to give the graph a title and label the axes.

Magnitude	Tally	Frequency
$6.0 \leqslant m < 6.2$	HH II	7
$6.2 \leqslant m < 6.4$	III	3
$6.4 \leqslant m < 6.6$	HH II	7
$6.6 \leqslant m < 6.8$	III	3
$6.8 \leqslant m < 7.0$	III	3
$7.0 \leqslant m < 7.2$	I	1

T

2 This frequency chart gives the time taken for students to evacuate the school during a fire drill.

Time taken (min)	0–	1–	2–	3–	4–	5–	6–	7–8
Number of students	39	54	123	347	29	42	21	16

Use a copy of this grid.
  a Draw a frequency diagram for this data.
  b How many students evacuated the school in under 3 minutes?

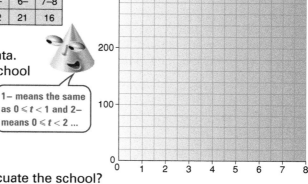

1– means the same as $0 \leqslant t < 1$ and 2– means $0 \leqslant t < 2$ ...

  c How many students took at least 5 minutes to evacuate?
  d Once all the students were out of the school, a count was made. How many students were counted?
  e In which minute did most pupils evacuate the school?
 *__f__ Comment on what this graph tells you about the fire drill at this school.

3 This table gives the time, in minutes, that 50 pupils took to run the cross-country.
  a Draw a frequency diagram for this data.
  b How many pupils took less than 55 minutes?
  c How many pupils took at least 40 minutes?
  d Can you tell how many pupils took more than 55 minutes? Explain.

Class interval (t is time, in min)	Tally	Frequency
$30 \leqslant t < 35$	HH	5
$35 \leqslant t < 40$	HH III	8
$40 \leqslant t < 45$	HH HH I	11
$45 \leqslant t < 50$	HH HH II	12
$50 \leqslant t < 55$	HH IIII	9
$55 \leqslant t < 60$	HH	5

*__4__

Number of females of different ages (in thousands)																		
Age	0–	5–	10–	15–	20–	25–	30–	35–	40–	45–	50–	55–	60–	65–	70–	75–	80–	85–90
Wales	91	88	84	105	113	109	89	94	97	80	77	78	83	84	68	60	42	29
Northern Ireland	67	65	62	67	64	60	53	48	48	42	39	38	37	36	29	25	16	10

  a Draw a frequency diagram for the age of females in   **i** Wales   **ii** Northern Ireland.
  b Regroup the data into class intervals 0–, 10–, 20–, ...
  c Draw two more frequency diagrams with these new class intervals.
  d Which set of class intervals do you think is most useful?
  e What do your graphs tell you about the ages of females in Wales and Northern Ireland?

*__5__ Draw a frequency diagram for each of the sets of data given in **Chapter 15 Exercise 3**.

425

T

Review 1 This table gives the injury time played in football games last season for Burnside Club.

Injury Time (t) (minutes)	Frequency
$0 \leqslant t < 2$	4
$2 \leqslant t < 4$	5
$4 \leqslant t < 6$	8
$6 \leqslant t < 8$	4
$8 \leqslant t < 10$	2

a Use a copy of this grid. Draw a frequency diagram for the data.

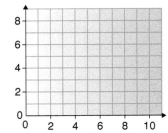

Remember to give the graph a title and label the axes.

b Rory drew the graph for this data and said, 'My graph shows injury time is usually 5 minutes'. Is he correct?

Review 2 This table shows the heights of the members of a ballet company.

Height	140–	150–	160–	170–	180–
Frequency	3	9	12	8	5

a Draw a frequency diagram for this data.
b How many members of the ballet company were shorter than 160 cm?
c How many were at least 150 cm?
d Can you tell how many were taller than 160 cm? Explain.

 Practical

Collect some continuous data about a topic or question that interests you.

Ideas
At what time in a first division match is a goal likely to be scored?
Collect data from an experiment in one of your other subjects.
Amount of time the pupils in your class spend on the computer each week.
Times of the tracks on your five favourite CDs.

Draw a frequency diagram for your data.

Drawing pie charts

A **pie chart** is a graph shown by a circle. It shows how something is shared or divided.
The bigger the section of the circle the bigger the proportion it represents.

Kilojoules measure energy.

kJ in Big Mac Combo

Coke 12%

Med Fries 31%

Big Mac 57%

Example This pie chart shows the proportion of kilojoules in a Big Mac Combo.

Each section is called a **sector**.
The whole pie chart has an angle of 360° at the centre.
The angle at the centre of each sector is a fraction of 360°.
Pie charts are mostly suitable for categorical (non-numerical) data.

sector
sector
360°
sector
sector

Example   A class of 30 students produced their own play. Four produced the play, six were back stage, 15 acted in it and five made the costumes.
A pie chart is a good way to show this data.
To draw a pie chart follow these steps.

Step 1   Find the angle at the centre of each sector.

4 out of 30 or $\frac{4}{30}$ produced the play

$\frac{4}{30}$ of 360° = $\frac{4}{30}$ × 360°

$= 48°$

> $\frac{4}{30}$ is the proportion that produced the play.

48° is the angle at the centre of the 'producers' sector.

We can work out the other angles the same way.

backstage $\frac{6}{30}$ × 360° = 72°

actors $\frac{15}{30}$ × 360° = 180°

costumes $\frac{5}{30}$ × 360° = 60°

Check the angles add to 360°.

Step 2   Draw a pie chart with these angles at the centres of the sectors.

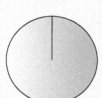  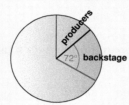 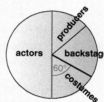

> It is best to draw the smaller sectors first.

Draw a circle. Draw a radius on your circle. Start with a vertical or horizontal line.

Use a protractor to measure an angle of 48° from the radius. Draw a sector and label it.

Use a protractor to measure an angle of 72° from the bold line. Draw a sector and label it.

Use a protractor to measure an angle of 60° from the bold line. Draw a sector and label it. Label the last sector.

Exercise 7

1   Which of these sets of data would be most suitable to display on a pie chart?
   a   the temperature of salt water each minute as it is heated
   b   the proportion of money a family spends on housing, food, clothing, heating and other
   c   the distances jumped in the final of the long jump competition
   d   the number of students leaving school at the end of each year
   e   the proportion of students at a school born in England, Scotland, Wales, Ireland, overseas.

2   This table shows how Jake spent the last 24 hours.

Activity	Sleeping	Playing sport	Watching TV	Eating	Friends
Hours	9	5	2	1	7

He wanted to draw a pie chart.
Make a copy of this and fill in the missing numbers to calculate the angles of the sectors.

sleeping $\frac{9}{24}$ × 360° = 135°   playing sport $\frac{5}{24}$ × 360° = ___ °

watching TV $\frac{2}{\_}$ × 360° = ___ °   eating $\frac{\_}{\_}$ × 360° = ___ °

friends $\frac{\_}{\_}$ × ___ ° = ___ °

Draw a circle with radius 5 cm.
Show Jake's information on a pie chart.

3 A farmer had 360 acres.
120 acres were in crops and the other 240 acres had sheep on them.
He wanted to draw a pie chart to show this.
  a What angle does the sector for crops need to be?
  b What angle does the sector for sheep need to be?

T

4 A shop that sold mobile phones noted the colours of the phones sold one month.
The owner of the shop wanted to draw a pie chart to show the results.
  a Use a copy of the table below.
    Calculate and then fill in the pie chart angles.

Mobile phone colour	Number of mobile phones	Pie chart angle
black	75	
yellow	5	
grey	25	
blue	15	
Total	**120**	**360°**

  b Draw a circle of radius 5 cm.
    Represent the information in the table on a pie chart.

∗5 **a** Work out the angles for each sector.
    Do the angles add to 360°? Why not?
  b Draw a pie chart for this data.

Number of pupils by school type

Type of school	Thousands
State nursery	137
State primary	5345
State secondary	3886
Non-maintained schools	618
Special schools	114
All schools	**10 100**

Review 1 This table shows the grades
120 students got in a test.

Grade	A	B	C	D
Number of students	24	58	33	5

1 Make a copy of this and fill in the missing numbers to calculate the angles of the sectors.

grade A $\frac{24}{120} \times 360° =$ ____ °   grade B $\frac{58}{120} \times 360° =$ ____ °

grade C $\frac{}{120} \times$ ____° = ____ °   grade D   — × ____° = ____ °

Draw a circle with radius 4 cm.
Show the information in the table on a pie chart.

Review 2 Reena recorded the ages of 36 Beatle fans.

Age	<20	20–	30–	40–	⩾50
Frequency	3	5	9	14	5

Draw a circle of radius 5 cm.
Draw a pie chart to show the results in the table.
Label it clearly.

Practical

You will need a spreadsheet program or statistical package.

1 This table shows the percentage of pupils at a school who have 0, 1, 2, 3 and 4 brothers.

Number of brothers	0	1	2	3	4
Percentage	20	40	27	10	3

Put this data into a spreadsheet or statistical package.

	A	B	C	D	E	F
1	Number of brothers	0	1	2	3	4
2	Percentage	20	40	27	10	3

Use the graph function of your spreadsheet to draw a pie chart.

2 Choose some other data that could be displayed on a pie chart.
Enter the data into the spreadsheet or statistical package. Use the graph function to draw a pie chart.
You could enter the data from one of the questions in **Exercise 7**.

Summary of key points

A The **range** is   **highest data value – lowest data value**.

B The **mode** is the most commonly occurring data value.

If a frequency chart has grouped data we find the **modal class**.

Example

Age	0–	5–	10–	15–	20–
Frequency	2	3	19	26	21

The **modal class** is 15– because this class interval has the highest frequency.

C   **Mean = $\dfrac{\text{sum of data values}}{\text{number of values}}$**

Example   25·6, 28·7, 36·4, 24·9, 19·3   mean $= \dfrac{25\cdot6 + 28\cdot7 + 36\cdot4 + 24\cdot9 + 19\cdot3}{5}$

$$= \textbf{26·98}$$

We can find the mean using an **assumed mean**.

1 Assume the mean is a particular value.

2 Subtract the assumed mean from each data value.

3 Find the mean of the differences you found in 2.

4 Add your answer to 3 to the assumed mean.

For the last example   1 assume the mean is 25

                2 0·6, 3·7, 11·4, ⁻0·1, ⁻5·7

                3 mean $= \dfrac{0\cdot6 + 3\cdot7 + 11\cdot4 + ⁻0\cdot1 + ⁻5\cdot7}{5} = 1\cdot98$

                4 25 + 1·98 = 26·98

Sometimes the data is given in a frequency chart.

Example

Number of scars	0	1	2	3	4
Frequency	5	8	4	2	1
Total number of scars	0	8	8	6	4

Multiply the number of scars by the frequency to get the totals

$$\text{mean} = \frac{0+8+8+6+4}{5+8+4+2+1} \begin{matrix} \leftarrow \text{sum of total numbers of scars} \\ \leftarrow \text{sum of frequencies} \end{matrix}$$
$$= 1\cdot3$$

 The **median** is the middle value when a set of data is arranged in order of size. When there is an even number of values, the median is the mean of the two middle values.

Example 5, 8, 3, 7, 9, 2, 1, 10

In order of size these are 1, 2, 3, **5, 7**, 8, 9, 10

$$\text{median} = \frac{5+7}{2}$$
$$= \mathbf{6}$$

There are two middle data values.

E We often display data on a **graph**.

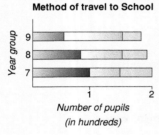

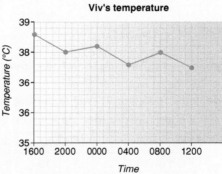

F For **continuous** data we draw a **frequency diagram**.

We label the divisions between the bars.

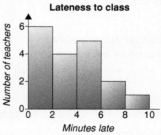

 A **pie chart** is a graph represented by a circle.

The bigger the sector of the circle the bigger the proportion it represents.

To draw a pie chart
  find what fraction of the whole each sector should represent
  multiply this fraction by 360° to find the angle at the centre of the sector.

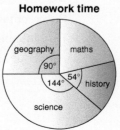

Test yourself

1 This table shows the marks that class 9L got in a test.
What is the modal group?

Mark	Tally	Frequency
31–40	II	2
41–50	IIII	4
51–60	HHT II	7
61–70	HHT HHT III	13

2 The heights, in metres, of 20 students are given.

| 1·58 | 1·72 | 1·59 | 1·64 | 1·67 | 1·83 | 1·73 | 1·64 | 1·59 | 1·72 |
| 1·66 | 1·67 | 1·69 | 1·72 | 1·58 | 1·60 | 1·71 | 1·70 | 1·68 | 1·68 |

a What is the mean height of these students?
b What is the range of the heights?
c What is the median height?

3 Paul got these marks in three tests: 72, 80 and 82.
He had one more test to sit.
He wanted a mean of 80 for the four tests.
What mark must he get in the fourth test?

4 Kate got a mean of 10 and a range of 4 on her first six maths mentals tests.
What marks did she get on her last two tests?

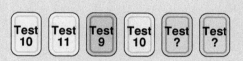

Test 10   Test 11   Test 9   Test 10   Test ?   Test ?

5 Jake and Jenni had a competition to see who could not blink for the longest number of seconds. They decided to find the mean of three tries.

| **Jenni's** results were | 8 sec | 12 sec | 16 sec |
| **Jake's** results were | ◯ | 12 sec | ◯ |

They both had the same mean. Jake's range was half of Jenni's.
What are Jakes's two missing scores?

6 Sinead ordered eight large pizzas at £4·80 each and three large fries at £1·40 each.
The delivery cost was £5.
If four people shared the cost, what is the mean price each paid?

7 Find the mean of these.
a number of pets   b number of goals   c marks in a test.

Number of pets	Number of families
0	12
1	11
2	4
3	2
4	0
5	1

Number of goals	Number of pupils
0	5
1	8
2	4
3	1
4	0
5	1
6	2

Mark in test	Number of pupils
10	5
20	3
30	10
40	8
50	4

8 This bar chart shows the number who attended five trainings on mouth-to-mouth resuscitation from January to August.
What is the median number who attended the training?

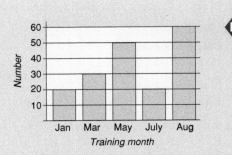

Training month

T **9** This table gives Abbie's temperature every four hours when she was in hospital.

Time	4 p.m.	8 p.m.	Midnight	4 a.m.	8 a.m.	12 p.m.
Temperature (°C)	38·5	38	38·3	37·6	38	37·5

Use a copy of this grid.
a Draw a line graph for this data.
b Do the intermediate points have meaning? Explain.

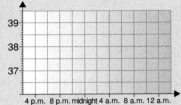

10 This table gives the number of times each number on a dice was tossed by three pupils doing an experiment.

Number	1	2	3	4	5	6
Jessica	50	46	58	42	51	53
Nandoor	41	57	43	58	51	50
Rosie	58	48	49	51	48	46

Draw a compound bar graph to show this data.

T **11** Use a copy of this.
The table shows the grouped frequency distribution of handspans of the pupils in Simon's class.

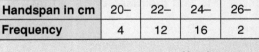

Handspan in cm	20–	22–	24–	26–
Frequency	4	12	16	2

a Use your grid to draw a frequency diagram.
b How many pupils had a handspan of 24 cm or more?

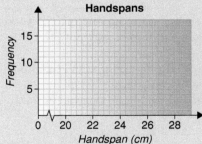

12 The table shows how much water is used in a household during one day.

a

Use	Washing	Showers	Dishes	Garden	Toilet	Total
Amount	240 ℓ	80 ℓ	10 ℓ	290 ℓ	100 ℓ	720 ℓ
Pie chart angle						360°

Work out the pie chart angle for each sector.
b Draw a circle of radius 5 cm.
Show the information in the table on a pie chart.

17 Interpreting Graphs. Comparing Data

You need to know

✓ displaying data   page 395

✓ mode and range   page 396

Key vocabulary

bar chart, bar-line graph, interpret, mean, median, mode, pie chart, population pyramid, range, survey

A date with data

What are some of the things you can tell from these bar charts?
Find some other published data.

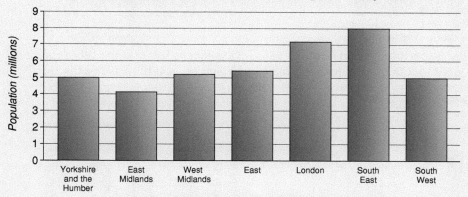

Resident populations of various areas of England in the year 2000

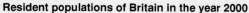

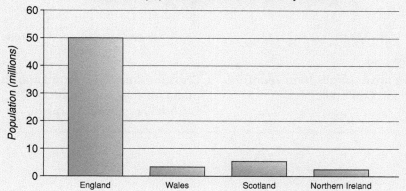

Resident populations of Britain in the year 2000

Interpreting graphs

Robyn found these pie charts in a book.
They show the proportion of foot, car, taxi and
cycle traffic for two cities of similar population.

She used the graphs to make some comparisons.

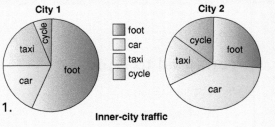

Inner-city traffic

There was about twice as much foot traffic in city 1.
There was about half as much car traffic in city 1.
There was about the same amount of taxi traffic in both cities.
There was about twice as much cycle traffic in city 2.

Discussion

Discuss the question given under each graph.

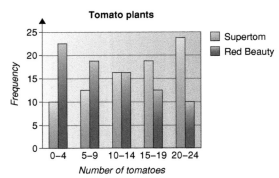

Would you buy Supertom or Red Beauty
plants if you wanted lots of tomatoes? Explain.

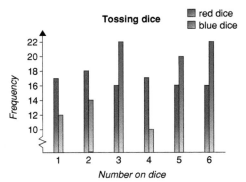

One dice is fair and the other unfair.
Which do you think is the fair dice? Why?

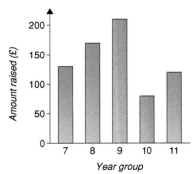

Does year 7 have fewer pupils than year 9?
A. Yes   **B**. No   **C**. Can't tell.
Explain.

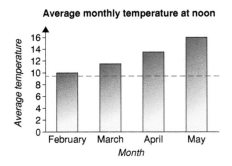

Christine said 'The dotted line shows the
mean for the four months.'
Is Christine correct? Explain.

Exercise 1

1   Rickey wanted to know if girls or boys were better
    at blind tasting. He surveyed 40 girls and 40 boys.
    He gave each pupil five different foods to taste
    while blind-folded.
    This bar chart shows his results.

    Compare the results for the boys and girls.

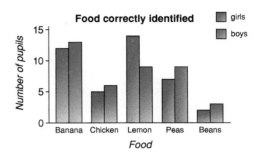

2   This graph shows the amount of pocket money
    Julia's friends get.
    a   What is the modal amount of pocket money?
    b   Julia gets £1·50 pocket money.
        She asked her mother to increase it to £3.
        She told her mother that most of her friends
        got from £2 to £4.
        Is Julia's mother being mislead?
        Give a reason.

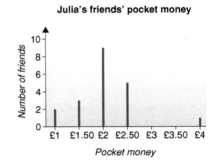

3   a   There were 1200 people at the concert. Copy
        and fill in this table to show the number in
        each age group.

Age	12–15	16–20	21–25	26–30	30+
Number					

    b   Why do you think there were fewer 30+
        people at the concert than other age groups?

4   This pie chart shows the number of men and
    women at a conference.
    a   About what percentage were women?
    b   About what percentage were men?
    c   There were 200 people at the conference.
        Use your percentage from **b** to work out about
        how many men were at the conference.
    d   The topic for the conference was one of these.
        Which do you think is more likely? Give a reason.
            Designing Racing Cars.
            Back to work after raising children.

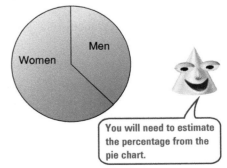

5

Sentence length

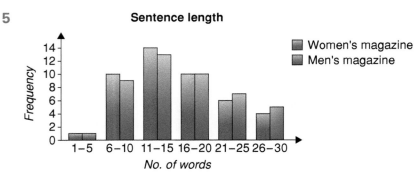

This graph shows the number of words in the first 45 sentences of two magazines.
a Geoff says
  '25 out of 45 sentences in the womens' magazine have fewer than 16 words. This shows that over half of the sentences in the magazine have fewer than 16 words'. Explain why he might be wrong.
b Compare the number of words in the first 45 sentences of the two magazines.

6 These pie charts show the items sold at two cafés. Café Roma sold 5000 items altogether and Café Sun sold 2000.
  a Roughly what percentage of items sold at Café Roma were muffins?
  b About how many muffins did Café Roma sell?
  c Jon thought the charts showed that Café Sun sold more sandwiches than Café Roma. Explain why the charts don't show this.

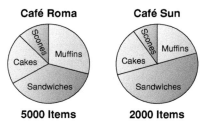

7 This compound bar chart shows the ages of workers at an electronics company.
  a About what percentage of female workers are
    i 40–49   **ii** 30–39   **iii** 50+
  b About what percentage of male workers are
    i 50+   **ii** 20–29   **iii** 40–49?
  c There are about 200 male workers aged 40–49. Estimate the number aged 50+.
  d There is about the same total number of male workers as female workers. Which of these is true?
    A Generally, the male workers are younger than the female workers.
    B Generally, the female workers are younger than the male workers.
    Explain how you used the chart to decide.

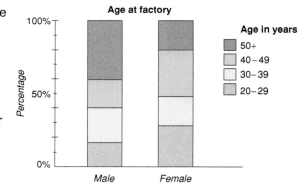

8 Camilla likes the weather to be still and hot when she goes on holiday.
These line graphs give the average wind speed and average temperature for the place where Camilla would like to holiday.
Use the graphs to choose which month you think Camilla should go on holiday. Explain your reasons.

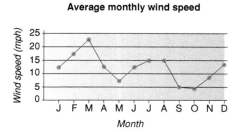

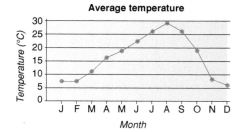

9

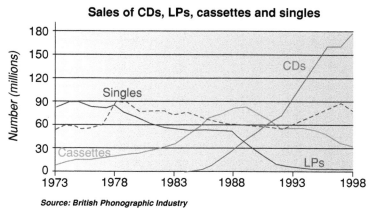

Sales of CDs, LPs, cassettes and singles

Source: British Phonographic Industry

a   How have the sales of CDs, LPs, cassettes and singles changed over the years?
b   Using the data on the graph, predict what the sales will look like this year. Predict what they will look like in 10 years time.

10 In a survey people were asked what they did now to keep healthy and what they think they might do in the future.

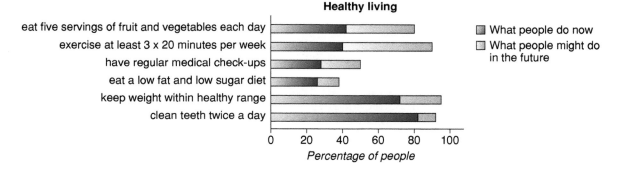

You have been asked to make an advertisement which promotes healthy lifestyles.
a   Choose two issues to be in your advertisement based on the information in the chart.
b   Explain how you chose each issue using **only** the information in the chart.

* **11** These pie charts show the fuel used in electricity generation. Describe the differences between the two years.

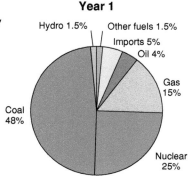

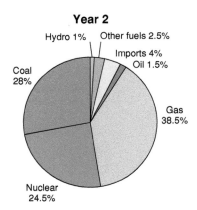

*12

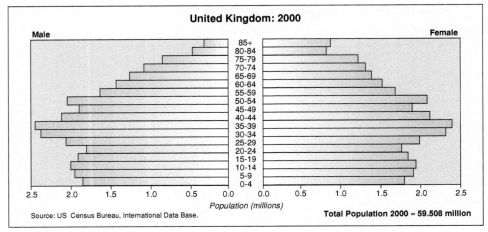

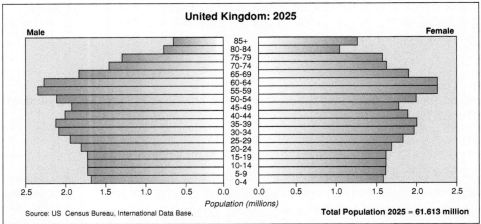

Link to percentages on page 138.

These diagrams show a **population pyramid** for the United Kingdom for 2000 and a predicted population pyramid for 2025.

a   About what percentage of the population was 70 or older in 2000?
b   About what percentage of the population is predicted to be 70 or older by 2025?

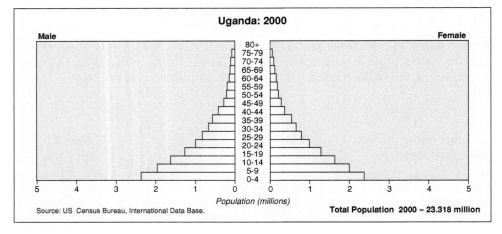

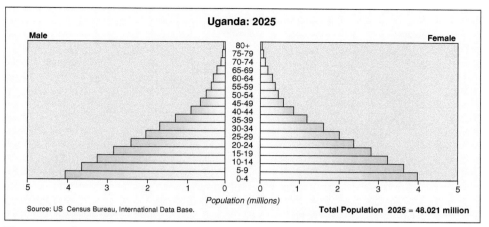

Uganda: 2025

Male | Female

80+
75-79
70-74
65-69
60-64
55-59
50-54
45-49
40-44
35-39
30-34
25-29
20-24
15-19
10-14
5-9
0-4

Population (millions)

Source: US Census Bureau, International Data Base.

Total Population 2025 = 48.021 million

These are the population pyramids for Uganda for 2000 and 2025.

c   About what percentage of the population of Uganda was under 30 in 2000?
d   How is the population of Uganda predicted to change by the year 2025?
e   Write a short report outlining the differences between the population pyramids of the United Kingdom and Uganda. Are there any differences for males and females?

Review 1

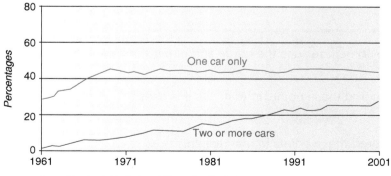

Bookworm

The Bookshop

200 Cards

600 Cards

These pie charts show what sorts of cards were stocked in two bookshops.

Bookworm had 200 cards and The Bookshop had 600.

a   Roughly what percentage of cards at Bookworm were birthday cards?
b   Use your percentage from **a** to work out roughly how many birthday cards there were at Bookworm.
c   Ryan thought that the charts showed that there were more birthday cards at Bookworm than at The Bookshop.
   Explain why the charts do not show this.

Review 2

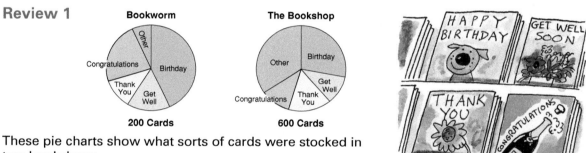

Households with regular use of a car

One car only

Two or more cars

Source: Department of the Enviroment, Transport and the Regions

a   How has the number of cars per household changed over the years?
b   What might affect whether or not these trends continue?

Review 3

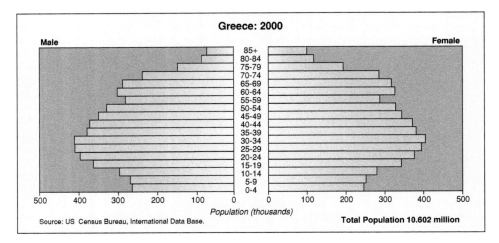

This shows a population pyramid for Greece in the year 2000.

a Which age group has the greatest number of people (males and females)?
b About what percentage of the population of Greece was under 20 in the year 2000?
c What differences are there in the population of males and females?

 ### Practical

Find some published graphs.
Good places to look are newspapers, books on national statistics or the internet.

Choose a graph to interpret. Try to choose a graph that you are familiar with like a bar graph, pie chart or line graph.
Write a short report which includes

> the graph
> what the graph is about
> what you can tell from the graph

Comparing data

To compare data we sometimes use the range and either the mean, the median or the mode.

Worked Example
Brad's rugby team scored these points in matches this season.

   21   24   19   22   32   22   28

Zak's rugby team scored these points in matches this season.

   0   19   49   19   60   7   14

a Find the mean and range of both sets of data.
b Which team should be picked to play a visiting team? Explain your choice.

Answer

a Mean for Brad's team $= \frac{21 + 24 + 19 + 22 + 32 + 22 + 28}{7}$

$= \mathbf{24}$

Range for Brad's team $= 32 - 19$

$= \mathbf{13}$

Mean for Zak's team $= \frac{0 + 19 + 49 + 19 + 60 + 7 + 14}{7}$

$= \mathbf{24}$

Range for Zak's team $= 60 - 0$

$= \mathbf{60}$

b Both teams have the same mean.
Zak's team has a much greater range.
This means their play is less consistent.
Sometimes they score very high points and sometimes very low points.
Brad's team has a smaller range. This tells us their play is more consistent.
We could choose Zak's team because there is more chance of a high score than with Brad's team. We could choose Brad's team because they are more consistent.

Discussion

The mean is the most useful average for describing data that is fairly evenly spread.

The median is the most useful average for describing data that has some extreme values.

The mode is most useful for describing non-numeric data if you want to know the most common or most popular. It is also useful for describing data where one value occurs much more often than others.

Discuss these statements.

Exercise 2

1 This table gives the mean and range of the scores of two girls in a gymnastics competition.
Only one girl can be chosen to go to a national competition.
Use the mean and range to decide which girl you would choose.
It doesn't matter which one you choose but you must explain why you have chosen her.

	Mean	Range
Jackie	7·9	0·8
Elaine	8·1	4·3

2 Ross could catch either the 83 or the 64 bus home.
He caught the 83 one week and the 64 the next.
This table shows the time for the journeys.

	M	T	W	T	F
83 Bus (time in min)	25	28	26	25	26
64 Bus (time in min)	20	34	19	40	17

   a Find the mean time for the 64 bus. Do you think the mean represents the data well? Give a reason.

   b Find the median of the 83 bus times.

   c Find the median of the 64 bus times.

   d Find the range of each set of data.

   e Decide which bus it would be more sensible to catch. Explain why.

3 The mean and range of the typing speeds of ten students at the beginning and end of a typing course are given.
Write two sentences comparing the results.

	Mean	Range
Beginning of course	44	20
End of course	56	35

4 The mean and range of the number of letters in 100-word samples from two newspapers are given. Write two sentences comparing the results.

Newspaper type	Mean	Range
Tabloid	4·2	8
Sunday	4·0	15

5 Ryan did an experiment to see which was his better catching hand. He threw five cubes in the air 100 times and wrote down how many of these he caught each time.
Which do you think is Ryan's better catching hand?

	Mode	Range
Left hand	3	3
Right hand	2	5

6 This data gives the number of hours two brands of light bulbs lasted.

Brand A   204   202   186   12   152   86   173   204   197   183
              62   179   194   217   203   196   184   51   172   168

Brand B   187   172   164   172   187   163   164   173   174   178
             182   180   183   173   185   164   172   183   185   186

a Find the mean, median and range for each brand.
Write a brief report comparing the brands using the range and one of the mean, median or mode.

b Which brand would you buy? Explain your answer using the range and one of the mean, median or mode. It doesn't matter which one you choose as long as you give the reasons why you chose it. Explain why you chose to use the mean, median or mode for your comparison.

*7 These tables give the average temperatures and average rainfalls in London and in Bangkok (the capital of Thailand).
Calculate the mean, median, and range of all sets of data.

a Compare and contrast the weather in these two places.
b Does the mean describe the data for Bangkok well? Explain.
c Based on the temperature only, where would you rather live? Give a reason using the range and one of the mean, median or mode. Explain why you chose the mean, median or mode.

London	Average temperature											
	Jan	Feb	Mar	Apr	May	Jun	Jul	Aug	Sep	Oct	Nov	Dec
°C	4.9	4.6	7.1	9.0	12.6	15.6	18.4	17.8	15.2	12.0	7.7	6.1

	Average rainfall												
	Jan	Feb	Mar	Apr	May	Jun	Jul	Aug	Sep	Oct	Nov	Dec	Year
mm	61.5	36.2	49.8	42.5	45.0	45.8	45.7	44.2	42.7	72.6	45.1	59.3	590.4

Bangkok	Average temperature											
	Jan	Feb	Mar	Apr	May	Jun	Jul	Aug	Sep	Oct	Nov	Dec
°C	25.9	27.6	29.2	30.1	29.6	29.0	28.5	28.4	28.1	27.7	26.8	25.5

	Average rainfall												
	Jan	Feb	Mar	Apr	May	Jun	Jul	Aug	Sep	Oct	Nov	Dec	Year
mm	10.6	28.2	30.7	71.8	189.4	151.7	158.2	187.0	319.9	230.8	57.3	9.4	1445

Review 1   The median and range of the time taken (in minutes) to travel to school by bus and by car over a two week period are given.

	Median	Range
By car	15·6	5·3
By bus	13·8	9·7

Write two sentences comparing the results.

Review 2 Hamish sells two breeds of pet mice.
This data gives the mass in grams of the two breeds of mice.

Breed A (total 24)

104	102	116	83	42	124	125	43	61	84	92	84
51	41	119	120	53	62	84	104	112	79	51	42

Breed B (total 23)

87	76	84	87	88	91	73	62	58	73	94	82
89	96	82	87	91	88	83	62	94	93	86	

a   Find the mean, median and range for breed A.
b   Find the mean, median and range for breed B.
c   Write a brief report comparing the masses of the breeds.
*d   Hamish wants to sell just one breed of mice. He has found that smaller mice are more popular. Which breed do you think he should choose to sell? Explain your answer using the range and one of the mean, median or mode. Explain why you chose the mean, median or mode.

Practical

Collect two sets of data you can compare.
Find the mean, median, mode and range of your data.
Use these to compare the data.

Suggestions
goals scored in matches last season by two teams or by first and second division teams
numbers of hours of TV watched by two different classes or year groups or ages each day for a month
number of fast food meals eaten each week by two different age groups
weather patterns in two different cities or countries.

Surveys

Investigation

Census

What is a census?
Who organises a census?
When was the last full census taken?
Are there different sorts of census?
How is the data gathered?
What data is gathered?
What is the information used for?
Who is able to use the information?
Investigate these and other questions to do with a census.

Remember
Before you carry out a survey you need to think about these.

- What do you want to find out?
- Are there any related questions?
- What might you find out?
- What data do you need to collect?
- How will you collect the data and who from?
- How many pieces of data do you need? (sample size)
- When and where will you collect the data?
- How will you display the data?
- How will you interpret the data? Is finding the mean, median, mode or range appropriate?

 Practical

1 a Think of a question you want answered or a statement you want to test.
  b Plan a survey. Think about all of the points given above.
     Try to choose something that interests you.

Suggestions

Questions
- What colour would students in this class like the walls painted? Why?
- Where are more rubbish bins needed around the school?
- What are the television viewing habits of Year 7 pupils?
- Which sport do people like watching most on TV? Why?
- What is the most common month for people in this class to have a birthday?
- How old are the grandparents of students in this class?
- How do pupils at this school travel to school? Why?
- Which foods should the canteen stop or start selling?
- Which road rules do people obey least?
- Which things do people do to look after their health (or environment) and what might they do in the future?
- How much pocket money do Year 7 pupils get?
- What are the food likes and dislikes of Year 7 pupils?
- What are the music likes and dislikes of Year 7 pupils?
- What types of video are watched by Year 7 pupils and how often do they watch them?
- How many calculations can Year 7 pupils do in 10 minutes?
- What type of product is advertised most on TV?
- How will the population of different countries change over the next 50 years?
- Do magazines or newspapers use words or sentences of different length?
- How do communities of animals/plants in two habitats differ?
- How do methods of transport to and from two different schools or shopping centres or ... differ?

You could use some published data or the internet.

Statements to test
- More 12-year olds like cats than dogs.
- More boys are left-handed than girls.
- Adults' pulse rates are slower than 5 to 16 year olds' pulse rates.

c   Carry out the survey.
d   Display the results on a table and graph.
e   Analyse the data using the mean, median or mode
    and the range if appropriate.

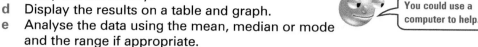

You could use a
computer to help.

f   Write some conclusions. This means, write some sentences about what
    you found out.
    Make sure your conclusions relate to the questions you asked or the
    statements you tested.
g   Write about any difficulties you had and how you solved these difficulties.
    Justify your choices for **d** and **e**.

2 Find some published data which is given in table form.
   Here is an example.

Adults living with their parents: by gender and age

England   Percentages

	1977–78	1991	1998–99
Males			
20–24	52	50	56
25–29	19	19	24
30–34	9	9	11
Females			
20–24	31	32	38
25–29	9	9	11
30–34	3	5	4

Source: National Dwelling and Household Survey and Survey of English
Housing, Department of the Environment, Transport and the Regions;
Labour Force Survey, Office for National Statistics

The internet is a good place to find up to date data.
Try the website   http://www.statistics.gov.uk/.
Use your data to draw some graphs.
Write a short report which includes your graphs and what you found out.

Summary of key points

 Interpreting graphs

We often use a graph to help us **interpret data**.

Example   This bar chart tells us that Year 10
            students watch more television than
            Year 7 students. The amount of TV
            watched varies each week for both Year groups.

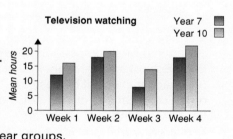

 To **compare data** we can use the **range** and one of the **mean, median** or **mode**.

Example   Melissa and Yolande both wanted to be chosen to represent the school
            in the National cross-country competition.

            The mean and range of their last ten races is

	Mean	Range
Melissa	1 hour 46·62	25 minutes
Yolande	1 hour 47·25	4 minutes

            We could choose Melissa because her mean time is better or Yolande
            because her times are more consistent.

Test yourself

1 This bar chart shows the activities done on Saturday and Sunday at a sports centre.

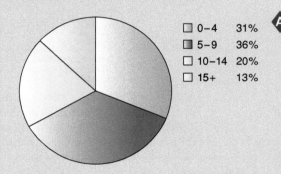

   a On one day the weather was warm and on the other it was cool.
      Which day do you think was warm? Explain.
   b Which of these is the most likely reason nobody did aerobics on Sunday?

   A Everyone slept in.
   B The centre didn't have aerobics classes on Sunday.
   C An earthquake destroyed the aerobics room.

   c The sports centre is thinking of closing the exercise room and squash courts on either Saturday or Sunday. Which day do you think they should close them? Why?

2 Kirsty's drama school is putting on a play for the students' families.
   This pie chart shows the ages of the students' brothers and sisters.
   There are 200 brothers and sisters.

☐	0–4	31%
◼	5–9	36%
☐	10–14	20%
☐	15+	13%

   a Copy and fill in this table to show the number in each group.

Age	0–4	5–9	10–14	15+
Number				

   b Do you think it would be better to start the concert at 6 p.m. or 8 p.m.? Explain your answer.

3 Marjorie was trying to organise a summer sports competition between Eden School and Farnsdown School. She said 'cricket will be easy to organise because both schools have the same number of cricketers' Explain why Marjorie is wrong.

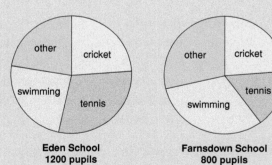

Eden School
1200 pupils

Farnsdown School
800 pupils

4 Pete and Kishan both want to represent the school in the public speaking competition.
   The mean and range of the scores they have achieved in the last ten speeches is given. Both Peter's and Kishan's scores were fairly evenly spread.
   Who would you choose to represent the school?
   Use the mean and range to justify your choice.

	Mean	Range
Pete	73	15
Kishan	73	35

5 Lee Valley School have two good problem solving teams.
Here are the last 15 results for both teams.

Toby's team	4	4	4	6	7	8	8	9	4	7	3	5	6	9	8
Gretchen's team	6	6	6	7	6	7	7	6	6	8	6	6	5	5	5

Find the mean, median, mode and range for each team.
Which team would you choose to enter in the competition? Explain why using the range and one of the mean, median or mode.

*6 This graph shows the amount spent by a school on computing equipment, paper and video/camera equipment.

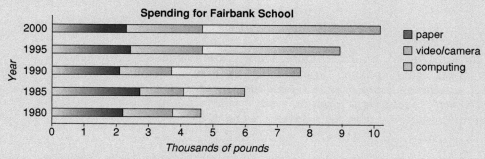

a How has the spending changed over the last 20 years?
b Using the data in the graph, predict what the results might look like this year and in ten years time.

18 Probability

You need to know

✓ probability page 397

Key vocabulary ·····························

certain, chance, dice, doubt, equally likely, even chance, fair, fifty-fifty chance, good chance, impossible, likelihood, likely, no chance, outcome, poor chance, possible, probability, probability scale, probable, random, risk, spin, spinner, theoretical probability, uncertain, unfair, unlikely

---

 Can I play?

Some games are said to be pure chance.
Think of some.
Some games are said to be pure skill.
Think of some.
Some games are unfair.
Is this game fair or unfair?

MOVE FOUR

START	1	2	3	4	5	6	7	8	9	10	11	12
26	25	24	23	22	21	20	19	18	17	16	15	14	13
27	28	29	30	31	32	33	34	35	36	37	38	STOP

Find a partner, a dice and 2 counters.
Decide who is player A and who is player B.
Take turns to move your counters.

Rules for moving Player A throws the dice and moves the number shown.
Player B moves 4 squares every time.

> A game is fair if all players have an equal chance of winning.

Make up a game that is unfair. Use dice, counters, spinners, cards or boards.

How likely. Probability scale

Discussion

- Think of an event that   is certain to happen
  is impossible
  has an even chance of happening
  is likely to happen
  is unlikely to happen.

- **Discuss** the chances of getting injured while doing each of these. Which has the highest injury risk?

  horse-riding   playing cricket/rugby/netball/badminton   skating
  climbing   skiing   sailing

- **Discuss** the chances of having a road accident. What factors affect this?

Some events are **more likely** to happen than others

Example   These shapes are put in a bag.
One is taken without looking.
It is more likely to be a triangle than a circle because there are
more triangles in the bag.

Some events have an **equal chance** of happening.

Example   These shapes are put in a bag.
One is taken without looking.
There is an equal chance of it being a triangle
or a circle.

There are the same number of circles as triangles.

Probability is a way of measuring the chance of a particular outcome.
We can show probabilities on a **probability scale**.
Impossible is at one end and **certain** is at the other.

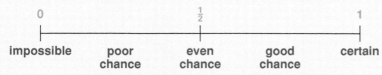

The probability of an event that is **certain** to happen is **1**.

The probability of an event that is **impossible** is **0**.

The probability of all other events is between 0 and 1. The more likely an event is to happen, the closer the probability is to 1.

Example

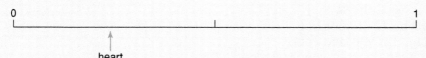

There are 13 cards in each suit.

Jake has a pack of 52 playing cards. He takes one without looking. The probability it is a heart is $\frac{1}{4}$.

Exercise 1

1

The numbers 1 to 10 are written on cards. The cards are shuffled then put face down in a line.

a The first card is turned over. It is a 6. Is the next card to be turned over likely to be higher or lower than this? Why?

b The next card turned over is a 3. Is the next card likely to be higher or lower than 3? Why?

2 Gwen mixes these cards up then puts them face down.

a She turns one over.

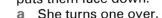

Is the next card she turns over more likely to be or ? Explain why.

b Gwen wanted to put just and in a pile so that it was more likely she would choose a than a .

What cards could she put in the pile?

3 Children at a Christmas party are given a red or a white balloon. They are not allowed to choose which colour.
Celia has 10 red and 12 white balloons to give out.
Karen has 12 red and 24 white balloons to give out.

a Katya's little brother Beau really wants a red balloon. From which person is he more likely to get a red balloon?

b Samantha then also gives out balloons. Samantha has 6 red balloons and 5 white balloons. Who is Beau most likely to get a red balloon from now?

T

4 Use a copy of this probability scale.

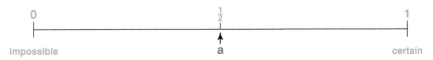

Show where each of these events might be on the scale.
The first one is done.

a You will get a tail when you toss a coin.

b Your teacher will live for 200 years.

c Next year will be 1954.

d There will be 12 months in the year next year

e The red marble will land in box A.

f I will get a red counter when I take one of these counters without looking.

g I will get a 6 when I toss a dice.

5 Yoghurt is sold in packs of 12.
Mel is going to take one without looking.
The scale shows the probability of getting each flavour.

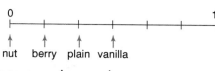

Use the probabilities to work out how many of these flavours are in a pack.
a vanilla **b** plain **c** nut **d** berry.

6 Martha uses a regular hexagon as a spinner for a game.
On a probability scale like this one below, draw an arrow to show how likely Martha is to get these when she spins the spinner.
a 2 **b** a number less than zero **c** a number greater than 1.

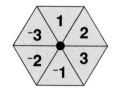

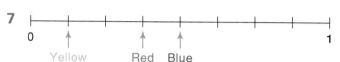

7

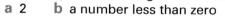

Yellow   Red   Blue

In a game Todd spins an arrow. The arrow stops on one of sixteen equal sectors of a circle. Each sector of the circle is coloured. The probability scale shows how likely it is for the arrow to stop on any one colour.
How many sectors are
a coloured red **b** coloured blue **c** coloured yellow?

Review 1   Sam has these cards.

He mixes the cards up and turns them face down.

He turns one over. It has a ③ on it. He turns another card over.
Is it more likely he will get a number smaller than 3 or a number bigger than 3? Explain.

Review 2   Tabitha picked a pair of socks at random from her sock draw. From which of these drawers would she be more likely to pick a pair of white socks? Explain.

6 pairs white
9 pairs other
colours
Drawer 1

4 pairs white
5 pairs other
colours
Drawer 2

Review 3   Use a copy of this probability scale.

impossible   certain

Put an arrow on the scale to show the probability of these.
a the spinner shown stopping on red
b the spinner shown stopping on green
c the next baby born being a boy
d a black bead being taken, without looking, from this bag
e a red bead being taken, without looking, from this bag.

Ups and downs – a game for 2 players

You will need 10 cards numbered 1 to 10.
You could use playing cards and use the ace as 1.

To play
● Shuffle the cards.
● Place them face down in a line.
● Turn over the first card.
● Player 1 predicts whether the next card will be higher or lower than the first card.
● The card is turned over. Player 1 gets 1 point if the prediction was correct.
● Take turns to predict if the next card will be higher or lower until all the cards are turned over.
● Play the game four times, taking turns to start.
● The player with the most points wins.

Link to place value on page 000.

Outcomes

If we toss a coin, we can get a head or a tail.
These are called **outcomes**.

Example The possible outcomes of throwing a dice are
          1, 2, 3, 4, 5 or 6.
Each of these outcomes has the same chance of happening.
We call these **equally likely outcomes**.

Example If we toss a coin and then roll a dice, the possible outcomes are
          H1 (heads and 1), H2, H3, H4, H5, H6, T1, T2, T3, T4, T5, T6.

Exercise 2

1 Bob spun this spinner.
   a Copy and finish this list of possible outcomes.
      10, 15, __, __, __, __
   b Do each of these outcomes have the same chance of happening? Explain.

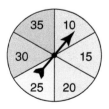

2 Alex was invited to four friends' houses one Saturday.
   She couldn't decide which to go to.
   She put the names in a hat.
      Sarah   Amelia   Charlotte   Kate
   She pulled one out without looking.
   a Copy and finish this list of possible outcomes.
      Sarah __
   b Do each of the friends have the same chance of being chosen?

3 A multi-pack of crisps has 4 salt and vinegar, 3 ready salted,
2 barbecue and 3 sour cream and chives.
Kanta takes one without looking.
   a Write down the list of possible outcomes.
   b Are these equally likely outcomes? Explain.

4 In a contest there were three people, Jan, Laura and Caryl.
There was a prize for first and second.
Copy and finish this table for who could come first and who could come second.

1st	Jan	Jan				
2nd	Laura	Caryl				

5 What are the possible outcomes when
   a you choose flavours for two ice creams from vanilla, chocolate, raspberry
   b you choose two pieces of fruit from a bowl of bananas, apples and oranges
   c a woman has twins (the order they are born doesn't matter)
   d you choose two cards from a pile of Jacks, Queens and Kings?

6 Copy and finish this list for the outcomes if a coin is tossed twice. HH, HT, __, __.

*7 Dave spun these two spinners.
He added the numbers together.
He drew this table to show the possible answers he could get.

+	1	2
3	4	__
4	__	__

   a Make a copy of the table.
     Fill in the missing answers.
   b Dave spun the two spinners
     He multiplied together the numbers he spun.
     Make a copy of this table.
     Fill it in to show all the possible answers.

×	1	2
3	__	__
4	__	__

Review 1   Melanie buys a pack of Christmas bells. There are 5 red, 2 green, 4 blue and
3 yellow ones. She takes one without looking.
a Write down the list of possible outcomes.
b Do each of the bells have an equally likely chance of being chosen?

* Review 2   Emily spun these two spinners and added the two numbers together.

+	2	4	6
2	4	6	8
4			
6			

Copy this table. Fill it in to show all the possible answers.

Calculating probability

We can **calculate the probability** of events which have equally likely outcomes.

Example   Harry buys three tickets in a raffle. 100 tickets are sold altogether.
Each ticket has an equally likely chance of being chosen.
Harry has 3 out of 100 chances of winning first prize.
The probability is $\frac{3}{100}$ or 3% or 0·03.

Probabilities can be written as fractions, decimals or percentages.

For **equally likely outcomes**,

Probability of an event = $\dfrac{\textbf{number of favourable outcomes}}{\textbf{number of possible outcomes}}$

For the example above
probability of Harry winning = $\frac{3}{100}$ ← number of ways a winning ticket could be chosen.
                                    ← total number of tickets that could be chosen.

Worked Example
The letters of the word CHANCE are put in a bag.
Mandy took one without looking.
What is the probability it is the letter   a E   b C?

Answer
a   There is one way the letter E can be chosen.
The number of favourable outcomes is 1.
The total number of possible outcomes is 6.
The probability of getting E is $\frac{1}{6}$.
b   There are two ways the letter C can be chosen.
The number of favourable outcomes is 2.
The total number of possible outcomes is 6.
The probability of getting a C is $\frac{2}{6}$ or $\frac{1}{3}$.

Mandy didn't look when she chose a letter. We could have said 'Mandy chose a letter at random'.
Choosing at **random** means every item has the same chance of being chosen.

Discussion

● Is this true? **Discuss**.
It is certain that an event will either happen or will *not* happen.
The probability of an event happening and the probability of it not happening must add to 1.
P (not happening) = 1 – P (happening).

● The probability of Brett winning his tennis match is 0·6. What is the probability he will *not* win? **Discuss**.

Exercise 3   **Only use a calculator if you need to.**

1   Sami has 15 socks in his drawer, 4 of which are blue. He pulls out a sock at random.
    What is the probability that the sock he has pulled out is blue?

2   Sarah puts 4 purple and 6 blue balls in a bag.
    She takes one without looking.
    a   What is the probability she takes a purple one?
    b   Marie thinks there is a 60% chance of getting a blue ball.
        Is she right? Explain.

3   A letter of the word IMAGINATION is picked at random.
    What is the probability of getting an
    a M   b N   c A   d I?

4   a   The probability that Janet wins her tennis match against Julia is 0·75.
        What is the probability Janet will **not** win against Julia?
    b   The probability that Maria will pass her next maths test is 0·68.
        What is the probability that Maria will **not** pass her next maths test?

5   A fair dice is rolled.
    What is the probability of rolling
    a   a six   b   an odd number
    c   a number greater than 4   d   a factor of 8
    e   a prime number   f   zero
    g   7   h   a number between 0 and 7
    i   not a 4   j   a 4 or a 6
    k   neither a 2 nor an odd number?
    Mark these probabilities on a scale.

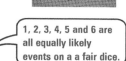

1, 2, 3, 4, 5 and 6 are all equally likely events on a a fair dice.

6   There are 15 bikes in a bike stand. Seven are red, 6 are green and 2 are blue.
    Calculate the probability that the next bike to be taken from this stand will be
    a   red
    b   green
    c   neither red nor green
    d   red or blue.

7   Sophie has a pack of 52 playing cards. Calculate the probability that if she takes one card
    without looking it will be
    a   a black suit
    b   a diamond
    c   a King
    d   not a King
    e   a picture card (King, Queen, Jack)
    f   either a diamond or a heart
    g   either an Ace or a Queen
    h   an odd numbered red card.

8 Ben and Sarah are playing a word game.
Letters are chosen at random from a bag.
There are 4 letters left in the bag.

A M E T

Ben needs the T to make a word and Sarah needs the
A or the E.
What is the probability that the next letter chosen will be
a the letter Ben needs
b one of the letters Sarah needs?
One of the letters is chosen at random by Ben's sister.
She tells them it is **not** a T.
What is the probability that the letter Ben's sister chose is
c the one Ben needs
d one Sarah needs?

9 Each person at a concert was given a coloured ticket.
The number of tickets of each colour given out were

1000 tickets were given out in total.

green	120	orange	100
blue	250	purple	280
yellow	180	red	70

What is the probability that a person picked at random has
a a green ticket   **b** a blue or purple ticket
c neither an orange nor a yellow ticket   **d** not an orange ticket?

10 Five cards numbered 1 to 5 are shuffled.
One is chosen at random.
a What is the probability it will be a prime number?
b Repeat **a** for six cards numbered 1 to 6.
c Repeat **a** for seven cards numbered 1 to 7.
d Draw a bar-line graph showing the probability of getting a prime number on one card
chosen at random from 5 up to 15 cards numbered 1, 2, 3, 4, 5, ...

Hint: you will need to change the fractions into decimals to 2 d.p.

*11 A school is opening a new gymnasium.
Someone is to be chosen to speak at the opening.
The names of all the pupils, all the staff and all the parents are put into a box.
One name is taken out at random.
Relah says 'there are only three choices, a pupil, a staff member or a parent.
The probability of it being a pupil is $\frac{1}{3}$'.
Relah is wrong. Explain why.

*12 Benjamin has a rectangular-shaped dice. Would the probability
of getting a 6 be the same as on a cube-shaped dice? Explain.

Link to shape, space and measure.

*13 A coin is tossed twice.
What is the probability of getting
a two heads   **b** a head and a tail?

*14 This spinner is spun twice. The numbers are added.
a Copy and complete the table to show all the
possible outcomes.
b Find the probability the total will be
   i 6   **ii** 14   **iii** 7   **iv** at least 8
   v more than 7   **vi** less than 7
   vii an even number   **viii** a prime number   **ix** not more than 6   **x** not less than 9.

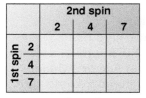

		2nd spin	
	2	**4**	**7**
1st spin **2**			
4			
7			

*15 Two dice are tossed and the numbers obtained are added.
    a Copy and complete the table to show all the outcomes.
    b Find the probability of getting a total of
        i 2   **ii** 5   **iii** 9   **iv** 12   **v** 1   **vi** less than 7
        vii at least 7   **viii** more than 7   **ix** exactly 8
        x at least 8   **xi** less than 8   **xii** more than 8.
    c What is the most likely total?
    d What is the least likely total?
    e What is the probability of getting the same number on both dice?
    f What is the probability the number on one dice will be double the number on the other?

DICE 2

+	1	2	3	4	5	6
1						
2						
3						
4						
5						
6						

DICE 1

*16 There are only toffees and mints in a bag.
The probability of getting a toffee is $\frac{1}{2}$.
    a Tamsen takes a sweet at random from the bag and eats it. It is a toffee.
       What it the smallest number of mints that could be in the bag?
    b Tamsen takes another sweet at random and eats it. It is also a toffee. Now, what is the smallest number of mints that could be in the bag?

Review 1 What is the probability that your maths teacher's birthday is on the 30th of February?

Review 2 Free gifts were given to customers at a store.
Customers were equally likely to get one of four things.

Jenni wants the perfume. Cass wants the notebook or the rose.
What is the probability that
a Jenni will get the perfume
b Cass will get the notebook or the rose?

Review 3 Twenty cards are numbered 1 to 20.
One card is chosen at random.
What is the probability it will be
a a prime number   **b** a non-prime number
c a number from 1 to 9   **d** an odd number
e not a number less than 11   **f** neither a 4 nor an odd number?

Review 4 Bowl A has 10 white chocolates and 15 dark chocolates.
Bowl B has 12 white chocolates and 22 dark chocolates.
Thomas is going to take a chocolate at random.
He only likes white chocolates. Which bowl should he take from?
Justify your choice.

A   B

Review 5 Families with two children are studied. One of these families is chosen for further study.
Find the probability that in this chosen family
a the children will both be boys
b one child will be a boy and one a girl
*c at least one of the children will be a girl
*d not more than one of the children will be a boy.

Probability from experiments

Discussion

Roseanne made a biased dice.
How might she have done this. **Discuss.**

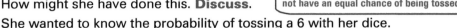

A biased dice is not fair. Each number does not have an equal chance of being tossed.

She wanted to know the probability of tossing a 6 with her dice.

Is it possible for her to calculate the probability of tossing a 6 using this formula?

$$\text{probability} = \frac{\text{number of favourable outcomes}}{\text{number of possible outcomes}} \cdot \textbf{Discuss.}$$

Often we cannot calculate the probability of an event happening. But we can sometimes get an **estimate of the probability** by doing an **experiment**.

Example If we want to know the probability of someone catching a ball we need to do an experiment to find an estimate.

The more times the experiment is repeated the better the estimate will be.

Example Owen tossed a biased dice 100 times. He got a four 47 times.
From this he estimated the probability of getting a 4 as

$$\text{probability of getting a 4} = \frac{\text{number of times 4 was tossed}}{\text{total number of tosses}}$$

$$= \tfrac{47}{100}$$

We call this the experimental probability.

He said 'The probability of getting a 4 on my dice is just under $\frac{1}{2}$'.
If Owen wanted a more accurate estimate he would need to toss the dice more times.

Practical

1 **You will need** a bag and a mixture of coloured counters.
   a Ask someone to put ten of the counters in the bag without you seeing.
   b Without looking, take a counter out of the bag, write down its colour, then put it back.
   c Repeat **b** fifty times. Use a tally chart like this to record your results.

Colour	Tally	Frequency

Which colour counter do you think there is most of in the bag? Explain.
Which colour do you think there is fewest of in the bag? Explain.
Estimate the probability of each colour counter.
Use the formula

$$\text{probability of \_\_\_\_\_} = \frac{\text{number of times colour pulled from bag}}{50}$$

Now empty the bag.
Are **your estimated** probabilities about right?

2 **You will need** two fair dice.
Toss the two dice 50 times.
Add the numbers and record the total on a frequency table.
Use your results to estimate the probability of getting each total.

Total	2	3	4	5	6	7	8	9	10	11	12
Tally											
Frequency											

Toss the two dice another 50 times.
Combine the results together.
Estimate the probability of getting each total using the combined results.
Would you expect these probabilities to be more or less accurate than the first ones?

3 Make a biased dice. Use a net for a closed box.
Before you fold the net, tape a small piece of Blu-tac to the inside of one face.
Toss your dice 100 times and record the results in a tally chart.
Estimate the probability of getting each number.

* 4 **You will need** four different cards.
Shuffle the cards. Turn them face down.
Take a card without looking. Before you look at it,
write down which card you think it is.
Keep the card in your hand and repeat with
the other cards.
Design an experiment to give you an estimate of the
probability of guessing each card correctly.

> Remember, the more times you repeat the experiment the more accurate the estimates of the probability.

Exercise 4

1 Hayley has woken at 7 a.m. on 33 out of the last 40 days.
Estimate the probability that she will wake at 7 a.m. tomorrow.

2 100 out of the last 500 cars that passed the school gate were white. Estimate the probability that the next car to go past will be white.

3 It has been cloudy on 150 out of the last 200 days.
Estimate the probability it will be cloudy tomorrow.

4 Ben has won 17 out of the last 20 tennis games he has played with Rob. Estimate the probability he will win the next one.

5 29 of the last 50 people to buy a burger have bought a cheeseburger. Estimate the probability that the next person to buy a burger will buy a cheeseburger.

6 a Faith asked 50 people entering a supermarket what colour eyes they had.
Use the table to estimate the probability that the next person to enter the supermarket will have
  i blue eyes   ii brown eyes   iii hazel eyes.

Blue	19
Brown	16
Hazel	6
Other	9

b Faith then asked another 50 people entering the same supermarket.
The results are given in this table.
Use all 100 results to estimate the probability of the next person entering the supermarket having   i blue eyes   ii brown eyes   iii hazel eyes.

Blue	21
Brown	17
Hazel	5
Other	7

c Would the probabilities you found in a or b be more accurate? Why?

7 This table shows the results of a survey of the arm lengths of 100 fifteen-year-old boys.
Another fifteen-year-old boy's arm is measured.
Use the information in the table to estimate the probability that his arm
a has a length of 62 cm or more but less than 64 cm
b has a length of 58 cm or more but less than 60 cm.

Length of arm	Frequency
58–	5
60–	26
62–	54
64–	15
Total	100

Review 1 It has rained on Mr Wilson's birthday 17 out of 35 times.
Estimate the probability it will rain on his next birthday.

Review 2 These were the flavours of the last 50 ice creams sold at Delicious Delights.
i Estimate the probability that the next ice cream sold will be
  a chocolate
  b vanilla.
ii If the table had given the results for the last 200 ice creams sold, would the estimated probabilities be more accurate? Explain.

Chocolate	27
Berry	8
Vanilla	1
Banana Choc Chip	14

Comparing calculated probability with experimental probability

Practical

1 You will need a coin.
  a What is the probability of tossing a head?
  b Toss the coin 10 times. How many heads did you get?
  c Toss the coin 100 times. How many heads did you get?
  d Compare the calculated probability from **a** with the experimental probabilities you got in **b** and **c**.

> We call this the calculated probability of getting a head.

2 You will need a dice.
  Toss a dice 60 times.
  Record the results in a tally chart.
  Draw a bar chart to show your results.
  Are the results what you would expect?
  If you repeated this experiment would you expect to get the same results?
  What results would you expect if you tossed the dice 600 times?

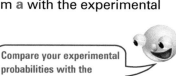

> Compare your experimental probabilities with the calculated probabilities.

  *Repeat but toss two dice and add the numbers. You could use the results you got in question **2** of the Practical on page 459.

T **3** **You will need** a computer software package that simulates tossing a coin or a dice.

a Simulate tossing a coin 50 times.
How many heads do you get? How many heads would you expect to get based on a calculated probability?
Repeat the simulation.
Did you get the same result?
Simulate tossing a coin 1000 times, then 2000 times, then 3000.
What do you notice?

> You could use the random (RND) function in Excel.

b Simulate rolling a dice 60 times. How many 3s would you expect to get based on a calculated probability?
Compare your results with this.
Compare your results with the results of others.
Did you all get the same results? Explain why or why not.

Add your results for rolling a dice 60 times to nine other people's results.
How many heads were tossed altogether out of 600 tosses?
What do you notice?
Add the results of the whole class together.
What proportion of heads were tossed?
What do you notice?

*4 Tasha and Simone are playing a game with a 10-sided fair dice.
It has the numbers 0–9 on it.
Tasha needs a 4 to win.
What is the theoretical probability she will win?
Devise an experiment to estimate the probability she will win.
Compare the theoretical and experimental probability.

*5 In the National Lottery there are 49 balls numbered 1 to 49.
What is the probability that the first ball is a 26   b even
c greater than 20   d a multiple of 3   e not a prime number?
Design an experiment to estimate these probabilities.
Compare the theoretical and experimental probabilities.

> The historical results of the National Lottery are on the internet.

Exercise 5

1 Brian spins a fair coin four times. He gets a head three times.
Brian says that if he spins it another four times, he'll get three heads again.
Explain why Brian is wrong.

2 Olivia, Ben and Joshua each rolled a different dice 360 times. Only one of the dice was fair.
Whose was it? Explain your answer.

Number	Olivia	Ben	Joshua
1	27	58	141
2	69	62	52
3	78	63	56
4	43	57	53
5	76	56	53
6	67	64	5

3 Mr Peters gave his class this homework.
  Roll a dice 100 times. Record your results on this chart.

Number on dice	1	2	3	4	5	6
Number of times rolled						

Two students handed in exactly the same results.
Mr Peters asked if one had copied the other's results without rolling the dice.
Explain why he might have thought one had copied.

4 There are 100 sweets in a box. Eric takes a sweet without looking inside the box. He writes down what sort it is and then puts it back.
He does this 100 times.
This chart shows Eric's results.

toffee	20
mint	38
jelly	14
choco	25
caramel	3

  a Eric thought there must be exactly 20 toffees in the box.
     Explain why he is wrong.
  b What is the smallest number of caramels that could be in the box?
  c Is it possible there is any other sort of sweet in the box? Explain.
  d Eric starts again and does the same thing another 100 times.
     Will his chart have exactly the same numbers on it?
  e Eric's friend takes a sweet. What sort is he most likely to get?

Review Helal, Michelle and Millie each rolled a fair dice 300 times.
Here are the results they wrote down.
Only one of them had written down the results correctly.
Which one? Explain.

Number	Helal	Michelle	Millie
1	52	50	25
2	47	50	32
3	58	50	28
4	46	50	89
5	46	50	63
6	51	50	63

Boxes – a game for a group

Draw a box like this.

The leader calls out six digits from 0 to 9 inclusive, each one different.
As each digit is called write it in one section of your box.
The winner is the person who has made the largest possible number.
The winner becomes the leader for the next round.

Remember your place value, page 1.

Summary of key points

 A Some events are **more likely** to happen than others.

Example You are more likely to get a heart than a club if you take one of these cards without looking.

 B **Probability** is a way of measuring the chance or likelihood of a particular outcome. We can show probabilities on a probability scale.

```
0                    ½                    1
├──────────────────────┼──────────────────────┤
impossible           even                 certain
                     chance
```

All probabilities lie from 0 to 1.

C This spinner is spun.

It could stop on red, green or yellow.

These are called the possible **outcomes**.

If the spinner is spun twice, the possible outcomes are:

    green green, green red, green yellow, red red, red green,

    red yellow, yellow yellow, yellow red, yellow green.

Equally likely outcomes have an equal chance of happening.

Example The spinner above is divided into three equal sections of red, green and yellow.

Stopping on red, stopping on green and stopping on yellow are equally likely outcomes.

D For equally likely outcomes

$$\text{probability of an event} = \frac{\text{number of successful outcomes}}{\text{total number of possible outcomes}}$$

Example The names of 5 boys and 3 girls were put in a box.

Tyrone took one at random.

Probability of a girl = $\frac{3}{8}$ ← 3 possible ways of getting a girl
                           ← total numbers of possible outcomes

Sometimes we cannot calculate the probability, but we can **estimate** it by **doing an experiment**.

Example If we wanted to know the probability that a car at a particular intersection will turn right we would need to do an experiment.

We could find out how many, out of say 1000 cars, turned right at the intersection.

From this we could estimate the probability that the next car will turn right from

$$\text{probability} = \frac{\text{number of cars that turned right}}{1000}$$

Example Robert found by experiment that he kicked a rugby ball over the post 61 times out of 100.

He estimated the probability of his kicking the ball over the post as $\frac{61}{100}$ or about 0·6.

F When we repeat an experiment the results are often slightly different.

Example Josh tossed a coin 20 times and got 11 heads.

He tossed the same coin another 20 times and got 8 heads.

The more trials Josh does each time, the more similar the results would be to the theoretical probability of $\frac{1}{2}$.

Test yourself

1 Gareth takes a marble at random from each of these bags. From which bag is he most likely to get a yellow marble?

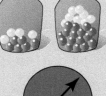

2 Draw a probability scale.
   Put an arrow on the scale to show the probability of these.
   a getting an odd number when you throw a dice
   b the spinner shown stopping on red
   c the spinner shown stopping on blue
   d getting a 3 when you throw a dice

3 Ellie had these ten number cards.

Her friend took one card without looking.
What is the probability the card will have
   a an odd number
   b an even number
   c a number less than 10
   d the number 3
   e a number less than 16
   f a number greater than 6
   g **not** the number 4
   *h neither an odd number nor the number 2?

4 Joseph did an experiment to estimate the probability of a drawing pin landing on its head when dropped.
He dropped it 50 times. It landed on its head 5 times.
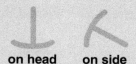

on head **on side**

  a Estimate the probability that the next time it is dropped it will land on its head.

  b He dropped it another 50 times. It landed on its head 3 times.
  Use all 100 results to estimate the probability of it landing on its head.

  c Which probability is more accurate. Why?

5 Bill put 8 purple counters and 8 black counters in a bag.

He took a counter without looking and then **put it back**.
He did this 16 times.
Bill took out purple counters 10 times and black counters 6 times.
Bill said 'I must have made a mistake and put 10 purple and 6 black counters in the bag.'
Explain why Bill is wrong.

6 Raffle tickets are numbered 1 to 300.
Lena buys 5 tickets and gets the numbers 8, 9, 10, 11, and 12.
Frederick buys 5 tickets and gets the numbers 18, 29, 182, 207, 234.
One ticket is chosen at random.
Is Lena or Frederick more likely to have the winning number? Explain your answer.

7 There are 6 counters in a box.
The probability of taking a green counter out of the box is 0·5.
A green counter is taken out of the box and put to one side.
Jason now takes a counter from the box at random.
What is the probability it is green?

8 Amy and Tim both spun their spinners.
Copy this table.

  a Finish it to show all the possible outcomes.

Tim's spinner   Amy's spinner

Tim's spinner	red	red	red	red	green			
Amy's spinner	1	2	3	4	1			

  b Are the outcomes red 1 and red 4 equally likely?

  c What is the probability of getting
   i red and 2
   iii green and a prime number
   v neither green nor a number bigger than 3
   ii red and an odd number
   iv green and a number bigger than 2
   vi not red?

Test yourself answers

Chapter 1 page 32

1   a 8 hundreds or 800   b 8 units or 8   c 8 tenths or 0·8   d 8 hundredths or 0·08
    e 8 thousandths or 0·008
2   87 050·4   3   a 3 500 004   b 2 249 993
4   a 5·375   b 8·985   c 15·24   d 15·589   e 4·242   f 17·278
5   0·004   6   a >   b <   c <
7   a 4·0741, 4·2741, 4·72, 4·7241   b 0·0078, 0·06, 0·6892, 0·798, 0·8692
8   a Hackl   b Zoeggeler   c Holm
9   a 4·65   b 0·075   c ⁻0·85
10 a 48 000   b 9·6   c 0·54   d 0·083   e 57   f 89   g 520   h 0·006 37   i 0·052   j 15 000   k 30
    l 720   m 0·008   n 0·0056
11 £3·65   12 a x is greater than 1·0 kg and less than or equal to 1·5 kg.   b Blue
13 a i 58 623 000   ii 4 932 000   iii 7 004 000   iv 886 000
    b i 58 600 000   ii 4 900 000   iii 7 000 000   iv 900 000
    c i 59 000 000   ii 5 000 000   iii 7 000 000   iv 1 000 000
14 a 7500   b 8499
15 a 54·39   b 72·35   c 4·03   d 16·3   e 4·6   f 186·1   g 5·10   h 16   i 105·00   j 10·00
16 About 4·8 days

Chapter 2 page 59

1   a ⁻3   b ⁻3 and ⁻1, or 2 and ⁻6   c ⁻3 and ⁻1, or 2 and ⁻6   d ⁻6
    e ⁻1 and 3, ⁻3 and 1, 1 and 5, ⁻2 and 2, ⁻6 and ⁻2   f Same as e   g ⁻6 + ⁻3 + ⁻2 = ⁻11   h 5 − ⁻6 = 11
2   a ⁻4   b ⁻11   c ⁻8   d ⁻1
3   a 4   b ⁻2 × 7 = ⁻14, ⁻14 ÷ ⁻2 = 7, ⁻14 ÷ 7 = ⁻2
4   a 600, 640   b 108, 117, 216, 387, 600, 702   c 108, 216, 600, 640   d 108, 216, 600, 702   e 216, 600, 640
    f 108, 117, 216, 387, 702
5   a 432, 3240   b 1035, 3150, 3240   c 432, 3150, 3240
6   a 79 and 3, 71 and 11, 59 and 23, 53 and 29
    b 2 and 5 are the only two prime numbers less than 100 that have a difference of 3.
7   A is 2, B is 58, C is 2, D is 29
8   a $84 = 2^2 \times 3 \times 7$ b $126 = 2 \times 3^2 \times 7$ c $364 = 2^2 \times 7 \times 13$
9   a 12   b 8
10 a   300   96   b 12   c 2400

11 a 8   b 5   c 9   d 4   e 3   f 27
12 a 16   b 10   c 49   d 7   e 125   f 1000   g 64   h 2   i 3   j 0·25   k 0·6
13 a 46·24   b 841   c 2·9   d 6·3   14   a 4·6   b 5·7   c 10·4   15   a 3600   b 40 000   c 60   d 400
16 a 2007   b 1936

Chapter 3 page 85

1   a 21   b 70   c 20   d 7   e 50   f 217   g 32   h 4816   2   a 37   b 44   c 248
3

1·4	2·1	1·6
1·9	1·7	1·5
1·8	1·3	2·0

4   a 60   b 130   c 148   d 410   e 450   f 39   g 32   h 800   i 3600   j 902   k 420   l 130   m 1800   n 925
5   a 2·6   b 21·5   c 39·2   d 2·4   e 0·7   f 2·4   g 5·4   h 0·2   i 0·6   j 0·7   k 56·1   l 129   m 265   n 61·1
6   a 9·5   b 388   c 485   7   a 120   b 264   c 282   d 117
8   a 4·9   b 0·24   c 0·05   d 0·12   e 0·003   f 0·06   9   a 130   b 116   c 12·1   d 50
10 145 years   11   a £3·71   b £0·90
12 a 11   b 14   c 3   d 18   e 11   f 60   g 3   h 4   i 50   j 24   k 4   13   a Exact number
14 Possible answers are: a 200 × 20 = 4000   b 800 ÷ 40 = 20   c 30 × 7 = 210   d 400 ÷ 5 = 80   e $\frac{4 \times 4}{4} = 4$
    f $\frac{4+7}{5-3} = \frac{11}{2} = 5·5$ g 5 ÷ (9 − 3) = 5 ÷ 6 ≈ 6 ÷ 6 = 1
15 A possible answer is 20 × 20 = 400 g

Chapter 4 page 112

1   a 2312   b 85 082   c 16·26   d 24·64   e 42·509   f 9412   g 820   h 1·36   i 88·82   j 295·06   k 796·81
2   a 65·49   b 20·8   3   £15·10
4   Possible estimates are given first.   a 35, 32·2   b 100, 91·5   c 20, 15·12   d 35, 33·81   e 100, 82·1
    f 120, 109·44   g 800, 1141·6   h 500, 313·8   i 1·6, 1·708

5 a 1904 b 4368 c 15 525 d 23 052 e 50 641 f 44 184 g 17 366 h 33 558 i 19 363
j 49 494 k 19 346
6 a 93·6 b 175·94 c 309·96 d 36·48   7 a £9·07 b 93p or £0·93   8 a 53 b 1
9 a 13·8 b 14·25   10 4472   11 a 0·8 b 2·3 c 14·6 d 21·3 e 23·56
12 a £20·48 b £4·52   13 £5·45   14 a 630 b 7100   15 a £112·50 b £106 c World d Us
16 66·35 m
17 There are many possible answers. One is
a Key 914 ⊖ 56 ⊜ b Key 939 ⊖ 93 ⊜ or Key 733 ⊕ 113 ⊜
18 18   19 a 5550 b 10·13 c 2·2 d 4·8
20 a Incorrect. All of the amounts are less than £15 and $6 \times £15 = £90$ so the total should be well below £90.
b Incorrect. 52 is about 50 and 204 is about 200. $200 \div 50 = 4$.
21 a $35\ 112 \div 42 = 836$ or $35\ 112 \div 836 = 42$ b $192 \times 1\cdot9 = 364\cdot8$ c $28\cdot08 - 19\cdot72 = 8\cdot36$ or $28\cdot08 - 8\cdot36 = 19\cdot72$
22 Possible answers are: a $(600 \times 27) + (3 \times 27) = 16\ 281$
b $(40 \times 59) + (2 \times 59) = 2478$ or $(42 \times 60) - 42 = 2478$ c $(632 \div 2) \div 2 = 158$.
23 a $423 - 156 + 56 = 323$ b $589 + 86 - 185 = 490$ 24 a 7 days 18 hours b 6 minutes 13 seconds

Chapter 5 page 134

1 a About $\frac{1}{4}$ b About $\frac{1}{6}$ 2 a $\frac{3}{20}$ b $\frac{1}{2}$ 3 $\frac{4}{5}$ 4 a $\frac{171}{200}$ b $\frac{9}{250}$
5 a 0·52 b 0·4 c 0·86 d 4·5 e 3·4   6 0·3   7 a 0·3 b 0·27   8 a > b < c > d >
9 Jenni   10 a $\frac{3}{4}, \frac{2}{3}, \frac{3}{8}, \frac{1}{4}, \frac{1}{6}$ b $1\frac{7}{8}, 1\frac{4}{5}, 1\frac{3}{4}, 1\frac{5}{8}$ 11 a $\frac{6}{7}$ b $\frac{3}{8}$ c $1\frac{2}{5}$ d $\frac{1}{4}$ e $\frac{1}{2}$ f $\frac{3}{4}$ g $\frac{10}{13}$
12 a $\frac{1}{2}$ b $\frac{3}{8}$ c $1\frac{1}{2}$ d $\frac{1}{8}$
13 a £42 b 9 m c 20 ℓ d 15 km e £16 f 35 cm g 9 m h 140 g i £4·50 j 24 m k 5 m
14 a 50 b 6 c £20 d 60
15 a 8 b 4 c 12 d 21 e 30 f $\frac{2}{5}$ g $\frac{10}{3}$ or $3\frac{1}{3}$ h $\frac{20}{3}$ or $6\frac{2}{3}$ i $10\frac{1}{2}$ j 20 k 63
16 a $9\frac{3}{4}$ m b $13\frac{1}{3}$ km c $10\frac{1}{2}\ \ell$ d $28\frac{1}{3}$ g   17 $99\frac{1}{5}$ m   18 a $6 \div \frac{1}{8} = 48$ b $5 \div \frac{1}{3} = 15$
19 84   20 6

Chapter 6 page 150

1 a $\frac{13}{20}$ b $\frac{2}{3}$ c $1\frac{7}{25}$ d $\frac{17}{200}$
2

Decimal	Percentage	Fraction
0·85	85%	$\frac{17}{20}$
0·35	35%	$\frac{7}{20}$
0·65	65%	$\frac{13}{20}$
0·08	8%	$\frac{2}{25}$

Decimal	Percentage	Fraction
1·4	140%	$1\frac{2}{5}$
0·3̇	$33\frac{1}{3}$%	$\frac{1}{3}$
1·48	148%	$1\frac{12}{25}$
2·625	262·5%	$2\frac{5}{8}$

3 a 63% b 84% c 5% d 11% e 149% f 3%   4 a About 50% b About 90% c About 25%
5 $\frac{1}{4}, \frac{7}{25}, 30\%, \frac{3}{8}, 0·42$ 6 Ruth Street, 64%
7 a 135 b 78 c 24 d 135 e 62 f 392 g 8·8 h 30·6   8 38·5 kg
9 a £6·65 b 8·16 ℓ c 34·2 m d £1387·76 e £0·52   10 C   11 a 136 cm b 169·2 cm
12 Mary   13 £4648   14 a B + M by £1 b £54·40

Chapter 7 page 167

1 a £10·40 b £15·60 c £20·80   2 a £20·25 b £14·25   3 a 4 : 7 b 7 : 4   4 2 : 1
5 a 12 b 3 c 20   6 a 4 : 5 : 6 b 2 : 8 : 5 c 5 : 6 d 18 : 5   7 a $\frac{13}{20}$ b 35% c 7 : 13
8 15 ℓ 9 a 12 b 8   10 3 kg   11 550 g   12 Natasha got 300 g and her sister got 400 g
13 35   14 a £170, £340 and £340 b 675 g, 540 g and 405 g   15 0·25 cubic metres

Chapter 8 page 209

1 a C b A c B
2 a $3y$ b $4b$ c c^2 d $4(x+3)$ e $6(y-4)$ f $2p^2$ g $3b^2$ h $(5y)^2$ or $25y^2$ i $\frac{s+t}{r}$ j $20m$ k $56r$
3 a $4x$ b $11a$ c $5b$ d $15m$ e $7n$ f $11p$ g $5q$ h $14p+4$ i $5y+5$ j $5x+4$ k $6m+3$ l $a+2b$
m $2x$ n $5p+q$ o n
4 $12x+4$ 5 a 5 b $4n$ c $4x$ d 30 e $\frac{3a}{2}$ f $2b$ 6 a $5x+15$ b $7n-28$ c $16a+32$ d $8y-12x$
7 a D b G c H d F e C f A g B h E
8 a 6 b 5 c 12 d 16 e 6 f 32 g 27 h 2 i 2 j 12 k 30
9 a $y + x = 6, 6 - y = x, 6 - x = y$ b x and y can take any values that add to 6
10 a £340 b £600 c £437·50   11 a 90 Joules b 25 000 Joules
12 a i Terry has $n+2$ counters. ii Joshua has $4n$ counters.
b i Andy has $m-2$ counters. ii Joshua has $4(m-2)$ or $4m-8$ counters.
c $\frac{p}{4}+2$
13 a $\frac{m}{6} = 4$ b $3m = 450$ c $2m + 4 = 20$ d $8(m+1) = 88$ 14 $p + q = 136$
15 a 12 b 13 c 10 d 36 e 5 f 8 g 6·4 h 2 i 6 j 1
16 a $4n + 9 = 37, n = 7$ b $5n - 6 = 51, n = £11·40$ c $x = 20°$ d $4n - 1 = 127, n = 32$

17 a $n + n + 4 = 78$; $n = 37$.

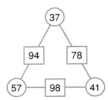

b x and y could be any two numbers which add to **73**.

18 a 4 **b** 24 **c** 2 **d** 5 **e** 9 **f** 3 **g** 8

Chapter 9 page 242

1 a 1, ⁻2, ⁻5, ⁻8, ⁻11, ⁻14 **b** 5, 6, 8, 11, 15, 20   **2 a** 140, 130, 120, 110, 100, ... **b** 90 km
3 a i 81, 100, 121 **ii** 32, 16, 8 **iii** 15, 21, 28   **b i** B **ii** D **iii** B
4 a 1000, 200 and 8 **b** 5, 8 and 17 **c** 3, 5 and 21
5 There are many possible answers. One is **a first term** − 1, **term-to-term rule** add 4
   b first term 40, term-to-term rule add 20 **c first term** − 19, **term-to-term rule** add 50
6 a 6, 7, 8, 9, 10 **b** ⁻1, 1, 3, 5, 7 **c** 18, 16, 14, 12, 10 **d** 0·4, 0·8, 1·2, 1·6, 2
7 a 20 **b** 27 **c** ⁻10 **d** 6
8

Term number	1	2	3	4	5	6	...	n
T(n)	6	12	18	24	30	36	...	$6n$

9 a There are 14 blocks in bridge 4.   **b**

Bridge number	1	2	3	4	5
Number of blocks	5	8	11	14	17

   c There are 5 blocks in bridge 1. Each bridge has 3 more blocks than the previous bridge.
   d There are 3 times n blue blocks in bridge n. There are 2 red blocks in every bridge.
      There are 3 times the bridge number plus two blocks in the nth bridge   **e** $3n + 2$
   f Yes, because if we take off 2 for the 2 blue blocks, this leaves 36 and 36 is divisible by 3.
10 a $2n − 1$ **b** $18n + 82$ **11 a** 19, 28, ⁻2, 11·5 **b** 10, 26, ⁻4, 3·2
12

x	1	2	3	4	5
y	9	11	13	15	17

13

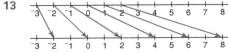

14 a $x \rightarrow 3x + 4$ **b** $y = 4(x − 3)$ **c** $y = \frac{x}{4} + 3$ **d** $x \rightarrow 5(x + 2)$
15 a

$x \rightarrow$ | multiply by 2 | $\rightarrow$ | subtract 4 | $\rightarrow y$ **b** $x \rightarrow$ | subtract 4 | $\rightarrow$ | multiply by 2 | $\rightarrow y$

   c No. When the operations are reversed the function is different. $x \rightarrow 2x − 4$ is different from $x \rightarrow 2(x − 4)$.
16 a $x \rightarrow x + 4$ or $y = x + 4$ **b** $x \rightarrow \frac{x}{3}$ or $y = \frac{x}{3}$ **c** $x \rightarrow 2x + 4$ or $y = 2x + 4$ or $x \rightarrow 2(x + 2)$ or $y = 2(x + 2)$
   d $x \rightarrow \frac{x}{2} + 3$ or $y = \frac{x}{2} + 3$
17 a 6, 10 **b** 30, 24   **18 a** ⁻10 **b** 1·5   **19** | multiply by 2 | $\rightarrow$ | subtract 7 |
20 a $x \rightarrow \frac{x+2}{3}$ **b** $x \rightarrow 4(x + 7)$ **c** $x \rightarrow \frac{x}{3} − 4$ **d** $x \rightarrow \frac{x}{5} + 3$ **e** $x \rightarrow 3x + 5$ **f** $x \rightarrow 5x − 2$ **g** $x \rightarrow \sqrt{x − 4}$

Chapter 10 page 268

1

x	⁻2	0	2
y	⁻3	3	9

x	⁻1	2	4
y	⁻6	0	4

x	⁻1	1	3
y	5	1	⁻3

2 a (⁻1, 5), (1, 1), (3, ⁻3)   **b**
   c No
   d There are many possible
      answers.
      All points which lie on
      the line will satisfy the
      equation $y = 3 − 2x$.

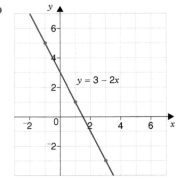

3 **a** Any number greater than 2 is the correct answer.   **b** $y = x + 3$ **c** Only C
4 **a** and **c**
5 **a**

Weeks	2	6	10
Charge (£)	140	220	300

b

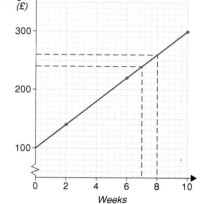

Computer Charges

c ⌇ means the numbers on the axis do not start at zero.
d Go up from 9 weeks to the graph. From there go across to the vertical axis.
  The charge for 9 weeks is £280.
e Go across from £240 to the graph. From there go down to the horizontal
  axis.
  She hired the computer for 7 weeks.

6 **a** May   **b** March, April, June

Chapter 11 page 304

1 **a** ∠Q **b** ∠M
2 **a** ∠PQR or ∠RQP or PQ̂R or RQ̂P   **b** ∠LMN or ∠NML or LM̂N or NM̂L
3 **a** AD and BE, AB and DE   **b** AC and CB, AB and BE, AD and DE, BE and ED, AB and AD
4 A rectangle   **5** **a** $a = 51°$ **b** $m = 151°$ **c** $p = 55°$
6 **a** $a = 88°$, $b = 88°$ **b** $c = 87°$, $d = 79°$, $e = 110°$ **c** $f = 128°$, $g = 128°$
7 **a** 75°   **b** $y = 70°$, $z = 55°$ **c** $m = 38°$, $p = 31°$ **d** $q = 73°$, $r = 153°$ **e** $s = 65°$ **f** $e = 73°$, $f = 57°$, $g = 57°$, $h = 50°$
  g $x = 63°$, $y = 30°$ **h** $s = 111°$ $k = 48°$, $l = 34·5°$

Chapter 12 page 327

1 Possible answers are: **a** △PQR or △PRQ   **b** △ABC or △ACB   **c** △LMN or △LNM
2 **a** p **b** c **c** m **3** **a** True   **b** True   **c** False
4 **a** Rectangle, rhombus   **b** Arrowhead   **c** Square, rhombus
5 **a** $p = 130°$ **b** $q = 140°$, $r = 40°$ **c** $s = 126°$, $t = 54°$ **6** $x = 130°$, $y = 65°$, $z = 50°$
7 No, because if it has two lines of symmetry, all three angles must be equal which means it would be an equilateral
  triangle with 3 lines of symmetry.
8 A right angle, 60°, 30°

9 **a** A rhombus   **b** 8
10 **a** and **c** will tessellate   **11** **a** 68° **b** 52°   **12** 3   **13**

14 Possible answers are given. The nets are not full size.
  a

  b

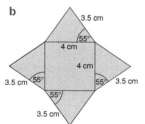

  c

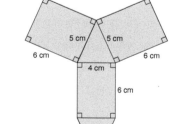

Test yourself answers

Chapter 13 page 351

1   b 4
    c There are many answers. Two are (0, 3) and (3, 1) and (⁻2, 0) and (1, ⁻2). Yes, it is possible to make a square because only two points are given. We can plot the other two points at (⁻1, 1) and (2, ⁻1) to make a square.
    d Second

2   a   b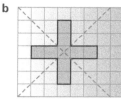

3 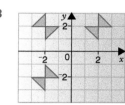   c Rotation 180° about (0, 0)
    d Yes

4   a 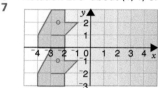   b 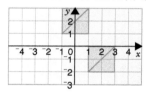   c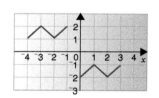

5   B and D
6   Rotation 270° about (1, 1) or a rotation 90° clockwise about (1, 1)
7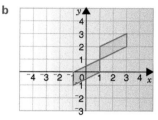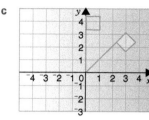

8   a 5 units left and 3 units up
    b 4 units right and 6 units down
9   a A′(2, 0)   B′(4, ⁻2)   C′(2, ⁻4)   D′(0, ⁻2)   A″(3, 5),   B″(5, 3),   C″(3, 1),   D″(1, 3)
    b 5 units right and 3 units up
10   a i A ii D iii E   b i C ii A iii D iv B
11   a 2 lines of symmetry, rotation symmetry of order 2   b 3 lines of symmetry, rotation symmetry of order 3
    c 4 lines of symmetry, rotation symmetry of order 4
12   or   13   a A′(10, 6)   b A′(7, 4)

Chapter 14 page 392

1   a 3·2 b 2·8 c 3·6 d 0·68 e 0·76 f 0·047 g 6·852 h 4·2 i 60 j 0·0586 k 0·042
    l 70 m 4600 n 5·42 o 4·5
2   a 16 b 4 c 35 d 22 e 50 miles or 80 km f 40 cm g 42
3   a A is at 2·2 lb, B is at 3·4 lb C is at 1·8 lb   b A is at 0·34 newtons, B is at 0·23 newtons, C is at 0·27 newtons
    c A is at about 0·5 kg, B is at about 1·8 kg, C is at about 3·2 kg
4   a About 1 kg
5   a To the nearest 0·1 kg   b To the nearest mm   c To the nearest min or to the nearest 30 sec
    d To the nearest 100 ml
6   a A b B
7   Possible answers are: a Use kitchen scales to measure to the nearest 10 g
    b Use a ruler to measure to the nearest cm by making a mark at the ends of your foot and measuring the distance between the marks.
    c Use a measuring jug to measure to the nearest 10 ml by filling the egg cup with water and then pouring it into a measuring jug.
8   a 36 cm² b 1·25 m² c 80 cm²   10 a 36·8 m² b 98·9 m²   11 a 46 m² b 23
12   a 544 cm² b Blue 352 cm², red 192 cm²   13   406 cm²   14   1080 cm³
15   a 280 m³ b 240 m³   16   200 cm²   17   80   18 a 18·5 units² b 8 units²

Chapter 15 page 410

1   a C b B c A
2   a What factors affect how often people eat takeaways? Hours of work? Live close to takeaway shop?
    b Most people have takeaways more than once a month. Friday night is the most popular day for takeaways.

c Data needed for each person: how often have takeaways and which day of the week?
d About how many times each month do you have takeaways? Which night/day do you most often have them?
e Cost, time and who the survey is for
3 **a** Continuous **b** Discrete **c** Discrete **d** Continuous
4 **a**

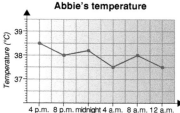

Time (seconds)	Tally	Frequency
$20 < t \leqslant 25$	III	3
$25 < t \leqslant 30$	JHt	5
$30 < t \leqslant 35$	JHt IIII	9
$35 < t \leqslant 40$	III	3

b 5 **c** 17 **d** 12 **e** The class intervals on the table are more useful because Melissa's class intervals would distribute the data too widely.

Chapter 16 page 431

1 61–70 **2 a** 1·67 m **b** 0·25 m **c** 1·675 m **3** 86 **4** 8 and 12 **5** 10 sec and 14 sec
6 £11·90 **7 a** 1 **b** 1·7 (1 dp) **c** 31 **8** 30
9 a

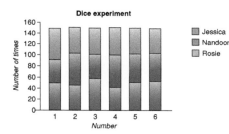

Abbie's temperature

b The plotted points are 4 hours apart and so intermediate points don't have any accurate meaning. It is possible to get a very rough estimate of a temperature at times between those plotted.

10

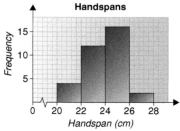

Dice experiment

11 a

Handspans

b 18

12 a

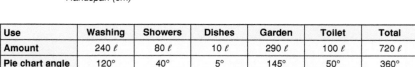

Use	Washing	Showers	Dishes	Garden	Toilet	Total
Amount	240 ℓ	80 ℓ	10 ℓ	290 ℓ	100 ℓ	720 ℓ
Pie chart angle	120°	40°	5°	145°	50°	360°

b

Chapter 17 page 446

1 **a** Saturday because more people went swimming and fewer had a spa.
This is what you would expect on a warm day.
b B
c Sunday because fewer people use them on a Sunday.
2 **a**

Age	0–4	5–9	10–14	15+
Number	62	72	40	26

b 6 pm because most of the brothers and sisters are under 10 and 8 pm would be too late for young children.

3   The charts show the proportion who play each summer sport and not the actual numbers. There are 1200 pupils at Eden School and only 800 at Farnsdown. Eden School will have more cricketers.

4   One possible answer is: I would choose Pete because he has the same mean as Kishan but his range is much smaller which means he is more likely to get a mark fairly close to 73. Kishan has a larger range so he might get a mark much higher than 73 but he also might get a mark much lower than 73.

5   **Toby's Team** mean = 6·1, median = 6, mode = 4, range = 6
    Gretchen's Team mean = 6·1, median = 6, mode = 6, range = 3
    Gretchen's Team because both teams have the same mean and median but Gretchen's team has a smaller range, indicating that team is more consistent. Gretchen's team also has a higher mode.

6   a   The amount spent on paper has remained about the same. The amount spent on video and camera equipment has increased in the last 10 years. The amount spent on computing equipment has increased a great deal over the last 20 years. It has increased the most.

Chapter 18 page 464

1   Bag 1

2

3   a $\frac{6}{10}$ or $\frac{3}{5}$ b $\frac{4}{10}$ or $\frac{2}{5}$ c $\frac{7}{10}$ d 0   e 1   f $\frac{6}{10}$ or $\frac{3}{5}$ g $\frac{9}{10}$ h $\frac{3}{10}$

4   a $\frac{5}{50}$ or $\frac{1}{10}$ b $\frac{8}{100}$ or $\frac{2}{25}$ c The probability estimated in **b** will be more accurate because the number of trials (times he dropped the drawing pin) is greater.

5   Although each colour has an equal chance of being selected it does not mean they will be selected equally. Bill might have taken out some counters more than once and others not at all.

6   They have the same chance of winning because they both have 5 tickets.

7   $\frac{2}{5}$

8   a

Tim's spinner	red	red	red	red	green	green	green	green
Amy's spinner	1	2	3	4	1	2	3	4

   b   Yes
   c   i $\frac{1}{8}$ ii $\frac{1}{4}$ iii $\frac{1}{4}$ iv $\frac{1}{4}$ v $\frac{1}{2}$ vi $\frac{1}{2}$

Index

adding 4, 31, 35, 58, 63, 88
  decimals 18, 67, 88
  fractions 126, 127, 134
angles 270, 285, 291–2, 302, 303
  acute 270
  adjacent 291, 292, 297
  alternate 295, 297, 298, 303
  base 289
  bisecting 289
  corresponding 295, 298, 303
  obtuse 270
  made with parallel lines 294–5, 303
  at a point 270, 291, 292, 303
  reflex 270
  right 270, 287, 315
  on a straight line 292, 295, 303
  supplementary 292
  of a triangle 202, 203, 270, 297–8, 303
  vertically opposite 291, 292, 295, 303
answers 466–72
approximations 368 see also estimates
Archimedes 387
area 275, 358, 375, 376, 380, 382
arithmagons 64, 204
arrowheads (deltas) 271, 309, 310, 311, 326
average 411, 414
axes (lines) of symmetry 273, 309, 332, 333, 342

bar charts 117, 419, 421, 424, 433, 434
  compound 421, 430, 434, 436
benchmarks 371
"Bermuda Triangle" 377
biased dice 458, 459
BIDMAS 76, 78, 85, 185
boxes, making 389 see also nets
brackets 77, 78, 108–9, 182, 184, 208

calculators 6, 55, 106, 107, 108–9, 144
  graphical 188, 215, 250, 251, 331, 417
capacity 274, 356, 358, 363, 364, 390
census 443
centicubes 320, 322, 383, 386
centre of enlargement 347, 351
centre of rotation 336, 350
certainty 449
checking answers 101–2, 103, 105, 111
class intervals 407, 410
clocks 274, 370
common factors 58, 116, 157, 166
commutative rule 63, 177, 187
compensation 63, 67
complements 4, 63, 67
concave shapes 311
cones 272
congruent shapes 271, 332, 336, 339, 346, 350
consecutive numbers 45
conversions 193, 367, 368
convex shapes 311
coordinates 246, 247–8, 267, 273, 330, 350
counting on and back 170, 214, 241
counting up 63, 67
cube roots 52, 59

cubed numbers 52, 59, 176
cubes (shapes) 272, 322
cuboids 272, 383–4, 387, 389, 391
cylinders 272

data
  collecting 403, 404–5, 406, 407, 409–10
  comparing 440–1, 445
  continuous 406, 407, 410, 412, 424, 430
  discrete 406, 407, 409
  primary and secondary 404, 409
decimals 1, 6, 16
  adding and subtracting 67, 88
  dividing 72, 99, 111
  to fractions 120, 133
  multiplying 72
  to percentages 138–9, 149
  recurring and terminating 123, 133
degrees of accuracy 368, 370, 390, 405
denominator 5, 77
diagonals 187, 194, 271, 309, 311, 326
difference table 228, 229, 235, 241
dividend 5
dividing 5, 40, 70, 96, 97, 110–11
  decimals 72, 99
  fractions 131
  by 10, 100 etc 1, 19, 20, 21
divisibility 3, 43, 45, 58
divisor 5, 99
dodecahedrons 272
dotty paper 308, 310, 316, 322, 382

Egyptian number system 15
emirp numbers 47
enlargement 345–6, 347, 351
equal (even) chance 449, 463
equations 175, 195–6, 205, 207
  of a straight line 248, 252, 253, 267
equivalent calculations 105, 111
equivalent fractions 120, 122, 127, 134
eqivalent ratios 156–7, 166
estimating 4–5, 80, 82, 371–2, 374, 390
events 449, 454, 463
experiments to collect data 404
expressions 169, 180, 185, 190, 207, 208
  algebraic 175, 176, 180

factors 3, 48, 49, 50, 58, 70, 72
  common 58, 116, 157, 166
formulae 169, 175, 188, 192, 194–5, 210
fractions 5, 50, 115, 116–17, 128–9, 180
  adding and subtracting 126, 127, 134
  comparing 124–5, 134
  to decimals 6, 122–3, 133
  dividing 131
  equivalent 120, 122, 127, 134
  to percentages 138–9, 149
  multiplying 128–9
  unit 116, 133
frequency diagrams 424, 430
frequency tables 24, 405, 407, 410, 412, 429
function machines 229–30, 232, 234–5, 238, 242

functions 229–30, 232, 234–5

games for a group 25, 39, 79, 184, 234, 371, 462
games for two 305, 329, 448, 452
Goldbach's conjectures 47
gradient (slope) of graphs 251, 252, 267
graghs
  conversion 255, 257, 258
  interpreting 262–3, 434. 437, 438–9, 445
  line (for data) 421, 430
  real-life 254, 258–9, 262–3, 266, 268
  straight line 247–8, 254, 267, 268

happy numbers 57
heptagons 271
hexagons 271, 315, 316
highest common factor (HCF) 50, 59, 116, 133, 157

images 332, 333, 350
imperial/metric equivalents 364, 390
impossibility 449
indices 48, 52, 78, 176
infinity 213, 241, 285, 302
input 229–30, 234, 235, 237–8
integers 1, 2, 35, 39, 40, 58
inverse operations 4, 52, 59, 103, 197–8, 199

Kaprekar's number 87
kites 271, 309, 310, 311, 326

last digits 57
length 274, 356, 363, 364, 390
like terms 177–8, 187, 208
line segments 285, 302
line symmetry 273, 342, 351
lines 285, 288, 302
  (axes) of symmetry 273, 309, 332, 333, 342
  parallel to x- and y-axes 267, 353
LOGO 315, 344, 382
lowest common multiple (LCM) 50, 59

magic squares 36, 39, 66–7, 69, 237
mapping on to 232, 237, 242
mapping diagrams 230, 242
mass 274, 356, 363, 364, 390
mean 189, 413–14, 416, 420, 429, 441
  assumed 414, 429
measurements 356, 358
  estimating 371–2, 374, 390
measures, metric and imperial 274–5
median 418, 419, 420, 430, 441
metric/imperial equivalents 364, 390
metric units 356, 358, 390
mirror line 273, 309, 332, 333, 342
mixed numbers 5, 120, 126, 137
modal class (modal group) 412, 429
mode 412, 420, 429, 441
multiples 3, 18, 19–20, 21, 31, 54
multiplying 4, 31, 40, 69–70, 91, 93
  decimals 72
  fractions 128–9
  by 10, 100 etc 1, 19–20, 21

National Lottery 461
nearly numbers 63, 67
negative numbers 2–3, 35, 40, 58, 182
nets 272, 322–3, 327, 383, 389
nth term 221, 228–9, 241
number chains 80, 172
number lines 2–3, 35, 63
number patterns 34, 35, 170, 173, 212
numerators 5, 77

objects (reflected) 332, 333
octagons 271, 315
octahedrons 272
order of operations 76, 77, 85, 185
order of rotation symmetry 309, 342, 351
ordering numbers and decimals 2, 23
orientation of a shape 339
origin of a graph 330
outcomes 452, 454, 458, 463
output 229–30, 234, 235

parallel lines 271, 285, 287, 288, 302, 303
parallelograms 271, 311, 326, 375, 376, 391
partitioning numbers 4, 63, 67, 70, 72
Pascal's triangle 217
pentagons 271, 273, 314, 315
pentominoes 345
percentages 6, 137, 142, 144, 146–7, 149
perimeter 178, 275, 375, 380, 382
perpendicular bisector 289, 303
perpendicular lines 271, 287, 289, 302–3
pie charts 140, 426–7, 430, 434
place holders 16, 31
place value 1–2, 16, 20, 31, 54, 72
planes 285, 302
polygons 187, 194, 271, 273, 308, 315
polyiamonds 316, 317
populations 433, 438–9, 440
positive numbers 2–3, 35, 40, 58
prime factors 48, 49, 50, 58
prime numbers 3, 46, 47, 48, 58
prisms 272, 322–3
probability 449, 454, 458–9, 460–1, 463, 464

proportion 6, 152, 153, 159, 161, 166
protractors 270, 318
pyramids 272, 321

quadrants of a graph 330, 350
quadrilaterals 271, 302, 318, 327
    properties of 308, 309, 310, 311, 326
    see also rectangles, rhombus, squares, trapezia
questionnaires 405
quotient 5

random 213, 454, 461
range 372, 390, 412, 429, 441
ratio 6, 155, 156–7, 159, 161, 164, 167
rectangles 271, 275, 311, 326, 375, 380
reflection 273, 332, 335, 338, 350
    symmetry 273, 342, 351
remainder 107, 111
rhombus 271, 309, 310, 311, 326
rotation 273, 336, 338, 350
    symmetry 309, 342, 273, 351
rounding 2, 26, 28–9, 82, 123, 141

samples 403, 409
satisfying a rule 246, 248, 267
scale factor 346, 347, 351
scales, reading 274, 280, 365, 390
semaphore 174
sequences 213–14, 216–17, 218, 220, 221, 224
set squares 288, 289, 303, 327
shapes
    2-D 308, 311, 326 see also polygons
    3-D 272, 320, 327, 383–4, 386
    congruent 271, 332, 336, 339, 346, 350
    concave and convex 311
    orientation 339
    made from rectangles 391
spheres 272
spiral of numbers 34
square numbers 3, 52, 54, 55, 57, 216
    algebraic 176
square roots 52, 54, 55, 57, 78
squares (shapes) 271, 309, 310, 311, 326
substituting 185, 187, 205, 208
subtracting 4, 18, 35, 58, 63, 88

decimals 67, 88
    fractions 126, 127, 134
surface area of a 3-D shape 383–4, 386, 391
surveys 141–2, 402, 403–4, 409, 443–5
symbols 15, 197, 258, 287
symmetry 273, 309, 310, 342, 351
    lines of 273, 309, 332, 333, 342

tally charts 405, 407, 410, 412
temperature 2, 367
term-to-term rule 218, 241
terms 213, 241
    like 177–8, 187, 208
tessellations 316–17, 327, 338
tetrahedrons 272, 321, 327
time 274
tonnes 356
transformations 273, 340
    see also enlargement, reflection, rotation, translation
translation 273, 339, 350, 351
trapezia 193, 271, 309, 310, 311, 326
    area 375, 376, 391
triangles 271, 306, 311, 317, 325, 326
    angles of 202, 203, 270, 297–8, 303
    area 375, 376, 391
    constructing 318, 327
    equilateral 271, 309, 311, 316, 317, 326
    isosceles 271, 298, 309, 310, 311, 326
    properties of 309, 310, 311, 326
    right-angled 271, 311, 326
triangular numbers 3, 216–17
twin primes 47

variables 175, 263
verifying an answer 197, 199, 202
vertices 272, 285
volume 387, 389, 391

websites 445
whole numbers 1, 2, 35, 39, 40, 58

x-axes 330, 350, 353

y-axes 330, 350, 353
y-intercept 252, 267